Methods in Enzymology

Volume 385
IMAGING IN BIOLOGICAL RESEARCH
Part A

METHODS IN ENZYMOLOGY

EDITORS-IN-CHIEF

John N. Abelson   Melvin I. Simon

DIVISION OF BIOLOGY
CALIFORNIA INSTITUTE OF TECHNOLOGY
PASADENA, CALIFORNIA

FOUNDING EDITORS

Sidney P. Colowick and Nathan O. Kaplan

Methods in Enzymology

Volume 385

Imaging In Biological Research

Part A

EDITED BY

P. Michael Conn

OREGON HEALTH AND SCIENCE UNIVERSITY
BEAVERTON, OREGON

ELSEVIER
ACADEMIC
PRESS

AMSTERDAM • BOSTON • HEIDELBERG • LONDON
NEW YORK • OXFORD • PARIS • SAN DIEGO
SAN FRANCISCO • SINGAPORE • SYDNEY • TOKYO
Academic Press is an imprint of Elsevier

Elsevier Academic Press
525 B Street, Suite 1900, San Diego, California 92101-4495, USA
84 Theobald's Road, London WC1X 8RR, UK

This book is printed on acid-free paper. ∞

For all information on all Academic Press publications
visit our Web site at www.academicpress.com

ISBN: 0-12-182790-9

PRINTED IN THE UNITED STATES OF AMERICA
04 05 06 07 08 9 8 7 6 5 4 3 2 1

Table of Contents

Section I. Imaging in Animal and Human Models

Section II. Imaging of Receptors, Small Molecules, and Protein–Protein Interactions

Contributors to Volume 385

Article numbers are in parentheses and following the names of contributors.
Affiliations listed are current.

DAVID L. ALEXOFF (12), *Department of Chemistry, Brookhaven National Laboratory, Upton, New York 11973*

ROBERT S. BALABAN (15), *Laboratory of Cardiac Energetics, National Heart, Lung, and Blood Institute, National Institutes of Health, Bethesda, Maryland 20892*

NICOLAU BECKMANN (14), *Novartis Institute for Biomedical Research, Analytical and Imaging Sciences Unit, CH-4002 Basel, Switzerland*

MARKUS BEU (13), *Clinic of Nuclear Medicine, University Hospital Düsseldorf, Düsseldorf 40225, Germany*

KEVIN J. BLACK (6), *Departments of Psychiatry, Neurology, and Radiology, Washington University School of Medicine, St. Louis, Missouri 63110-1093*

BRITTON CHANCE (20), *Eldridge Reeves Johnson University, Philadelphia, Pennsylvania 19104*

DELPHINE L. CHEN (17), *Washington University School of Medicine, St. Louis, Missouri 63110*

PHILIPPE CHOQUET (9), *Laboratoire de Biomecanique, Centre Hospitalier Universitaire Hautepierre, Strasbourg, France*

BRADLEY T. CHRISTIAN (11), *Department of Nuclear Medicine and Positron Emission Topography, Kettering Medical Center, Kettering, Ohio 454229*

STUART CLARE (8), *Centre for Functional Magnetic Resonance, Department of Clinical Neurology, John Radcliffe Hospital, University of Oxford, Headington, Oxford OX39DU, United Kingdom*

CHRISTIAN A. COMBS (15), *Light Microscopy Facility, National Heart, Lung, and Blood Institute, National Institutes of Health, Bethesda, Maryland 20892*

ANDRÉ CONSTANTIENSCO (9), *Laboratoire de Biomecanique, Centre Hospitalier Universitaire Hautepierre, Strasbourg, France*

BRUCE M. DAMON (2), *Department of Radiology and Radiological Sciences, Vanderbilt University Institute of Imaging Science, Nashville, Tennessee 37232*

CARMEN S. DENCE (16), *Department of Radiology, Washington University School of Medicine, St. Louis, Missouri, 63110*

DORIS J. DOUDET (10), *Department of Medicine/Neurology, Pacific Parkinson Research Center, University of British Columbia, Vancouver, British Colombia V6T2B5, Canada*

DAVID J. DUBOWITZ (7), *Department of Radiology, Center for Functional Magnetic Resonance Imaging, University of California, San Diego, La Jolla, California 92093-0677*

GUILLAUME DUHAMEL (9), *Laboratoire Mixte, Universite Joseph Fourier, Neuroimagerie Fonctionelle et Metabolique, Grenoble, France*

ix

M. R. GERASIMOV (18), *Department of Chemistry, Brookhaven National Laboratory, Upton, New York 11973*

JOHN C. GORE (2), *Department of Radiology and Radiological Sciences, Vanderbilt University Institute of Imaging Science, Nashville, Tennessee 37232*

EMMAUNELLE GRILLON (9), *Laboratoire Mixte, Universite Joseph Fourier, Neuroimagerie Fonctionelle et Metabolique, Grenoble, France*

ROBERT J. GROPLER (16), *Department of Radiology, Washington University School of Medicine, St. Louis, Missouri, 63110*

A. HEERSCHAP (3), *Department of Radiology, Medical Faculty of the University of Nijmegen, Nijmegan 6500 HB, The Netherlands*

PILAR HERRERO (16), *Department of Radiology, Washington University School of Medicine, St. Louis, Missouri, 63110*

JAMES HOLDEN (10), *Department of Medical Physics, University of Wisconsin, Madison, Wisconsin 53706*

JEAN-NOËL HYACINTHE (9), *Laboratoire Mixte, Universite Joseph Fourier, Neuroimagerie Fonctionelle et Metabolique, Grenoble, France*

N. R. JAGANNATHAN (4), *Department Head of Nuclear Magnetic Resonance, All India Institute of Medical Sciences, Ansari Nagar, New Delhi 110029, India*

SUMAN JANA (1), *Nuclear Medicine Division, Albert Einstein College of Medicine, Bronx, New York 10461*

M. KHUBCHANDANI (4), *Department of Nuclear Magnetic Resonance, All India Institute of Medical Sciences, Ansari Nagar, New Delhi 110029, India*

D. W. J. KLOMP (3), *Department of Radiology, Medical Faculty of the University of Nijmegen. Nijmegan 6500 HB, The Netherlands*

JONATHAN M. KOLLER (6), *Department of Psychiatry, Washington University School of Medicine, St. Louis, Missouri 63110-1093*

RAKESH KUMAR (1), *Department of Nuclear Medicine, All India Institute of Medical Sciences, New Delhi 110029, India**

ROLF LARISCH (13), *Clinic of Nuclear Medicine, University Hospital Düsseldorf, Düsseldorf 40225, Germany*

JEAN-LOUIS LEVIEL (9), *Laboratoire Mixte, Universite Joseph Fourier, Neuroimagerie Fonctionelle et Metabolique, Grenoble, France*

JEAN LOGAN (12), *Department of Chemistry, Brookhaven National Laboratory, Upton, New York 11973*

KATHRYN E. LUKER (19), *Mallinckrodt Institute of Radiology, Washington University School of Medicine, St. Louis, Missouri 63110*

ROBERT H. MACH (16), *Department of Radiology, Washington University School of Medicine, St. Louis, Missouri, 63110*

EVAN D. MORRIS (11), *Indiana University School of Medicine, Department of Radiology, Indianapolis, Indiana 46202*

HANS-WILHELM MÜLLER (13), *Clinic of Nuclear Medicine, University Hospital Düsseldorf, Düsseldorf 40225, Germany*

RAYMOND F. MUZIC, JR. (11), *University Hospitals of Cleveland, Case Western Reserve University, Cleveland, Ohio 44106*

Current Affiliation: Department of Radiology, Division of Nuclear Medicine, Hospital of the University of Pennsylvania, Philadelphia, Pennsylvania 19104-4283

KIYOSHI NAKAHARA (5), *Department of Physiology, The University of Tokyo School of Medicine, Tokyo 113-0033, Japan*

SUSANNE NIKOLAUS (13), *Clinic of Nuclear Medicine, University Hospital Düsseldorf, Düsseldorf 40225, Germany*

JOEL S. PERLMUTTER (6), *Departments of Neurology, Radiology, Anatomy and Neurobiology, and the Program in Physical Therapy, Washington University School of Medicine, St. Louis, Missouri 63110-1093*

DAVID PIWNICA-WORMS (19), *Mallinckrodt Institute of Radiology, Washington University School of Medicine, St. Louis, Missouri 63110*

W. K. J. RENEMA (3), *Department of Radiology, Medical Faculty of the University of Nijmegen, Nijmegan 6500 HB, The Netherlands*

JEAN-CHRISTOPHE RICHARD (17), *Washington University School of Medicine, St. Louis, Missouri 63110*

MARKUS RUDIN (14), *Novartis Institute for Biomedical Research, Analytical and Imaging Sciences Unit, CH-4002 Basel, Switzerland*

DANIEL P. SCHUSTER (17), *Washington University School of Medicine, St. Louis, Missouri 63110*

SALLY W. SCHWARZ (16), *Department of Radiology, Washington University School of Medicine, St. Louis, Missouri, 63110*

ABRAHAM Z. SNYDER (6), *Departments of Radiology and Neurology, Washington University School of Medicine, St. Louis, Missouri 63110-1093*

M. G. SOMMERS (3), *Department of Radiology, Medical Faculty of the University of Nijmegen, Nijmegan 6500 HB, The Netherlands*

PAUL VASKA (12), *Department of Chemistry, Brookhaven National Laboratory, Upton, New York 11973*

A. A. VELTIEN (3), *Department of Radiology, Medical Faculty of the University of Nijmegen, Nijmegan 6500 HB, The Netherlands*

HENNING VOSBERG (13), *Clinic of Nuclear Medicine, University Hospital Düsseldorf, Düsseldorf 40225, Germany*

MICHAEL J. WELCH (16), *Department of Radiology, Washington University School of Medicine, St. Louis, Missouri, 63110*

KARMEN K. YODER (11), *Indiana University School of Medicine, Department of Radiology, Indianapolis, Indiana 46202*

H. J. A. IN 'T ZANDT (3), *Department of Radiology, Medical Faculty of the University of Nijmegen, Nijmegan 6500 HB, The Netherlands*

ANNE ZIEGLER (9), *Center Hospitalier Universitaire, Neuroimagerie Fonctionelle et Metabolique, Grenoble, France*

Preface

As these volumes were being completed, American Paul C. Lauterbur and Briton Sir Peter Mansfield won the 2003 Nobel Prize for medicine for discoveries leading to the development of MRI.

The *Washington Post* story on October 6, 2003 announced the accolade, noting: "Magnetic resonance imaging, or MRI, has become a routine method for medical diagnosis and treatment. It is used to examine almost all organs without need for surgery, but is especially valuable for detailed examination of the brain and spinal cord." Unfortunately, the article overlooked the growing usefulness of this technique in basic research.

MRI, along with other imaging methods, has made it possible to glance inside the living system. For patients, this may obviate the need for surgery; for researchers, it becomes a noninvasive method that enables the model systems to continue "doing what they do" without being disturbed. The value and potential of these techniques is enormous, and that is why these once clinical methods are finding their way to the laboratory.

Authors have been selected based on research contributions in the area about which they have written and based on their ability to describe their methodological contributions in a clear and reproducible way. They have been encouraged to make use of graphics and comparisons to other methods, and to provide tricks and approaches that make it possible to adapt methods to other systems.

The editor wants to express appreciation to the contributors for providing their contributions in a timely fashion, to the senior editors for guidance, and to the staff at Academic Press for helpful input.

P. MICHAEL CONN

METHODS IN ENZYMOLOGY

Section I

Imaging in Animal and Human Models

[1] Positron Emission Tomography: An Advanced Nuclear Medicine Imaging Technique from Research to Clinical Practice

By RAKESH KUMAR and SUMAN JANA

Introduction

Positron emission tomography (PET) is an advanced diagnostic imaging technique, which cannot only detect and localize, but also quantify physiological and biochemical processes in the body noninvasively. The ability of PET to study various biological processes opens up new possibilities for both fundamental research and day-to-day clinical use. PET imaging utilizes β-emitting radionuclides such as ^{11}C, ^{13}N, ^{15}O, and ^{18}F, which can replace their respective stable nuclei in biologically active molecules. These radionuclides decay by positron emission. After being emitted from the nucleus, a positron will combine with a nearby electron through a process known as annihilation. Annihilation converts the mass of both particles into energy in the form of two antiparallel 511-keV γ rays. The PET detectors are arranged in a ring in order to detect these γ rays.

At present, 2-deoxy-2-[^{18}F]fluoro-D-glucose (^{18}F-FDG) is the most commonly used positron-emitting radiopharmaceutical used for PET imaging. ^{18}F-FDG is a radioactive analog of glucose and is able to detect altered glucose metabolism in pathological processes. Like glucose, FDG is transported into cells by means of a glucose transporter protein and begins to follow the glycolytic pathway. FDG is subsequently phosphorylated by an enzyme known as hexokinase to form FDG-6-phosphate.[1,2] However, FDG-6-phosphate cannot continue through glycolysis because it is not a substrate for glucose-6-phosphate isomerase. As a result, FDG-6-phosphate is trapped biochemically within the cell. This process of metabolic trapping constitutes the basis of PET imaging of the biodistribution of FDG. Because there can be a manyfold increase or decrease in the glucose metabolism of diseased tissue as compared to normal tissue, it is easy to detect such differences in metabolism using PET. This Chapter discusses general aspects of PET, including drug evaluation, biological functions evaluation, clinical applications, and future directions of PET imaging.

[1] K. M. McGowan *et al.*, *Pharmacol. Ther.* **66,** 465 (1995).
[2] R. L. Wahl, *J. Nucl. Med.* **37,** 1038 (1996).

Tracers

Position emission tomography uses radioisotopes of naturally occurring elements, such as ^{11}C, ^{13}N, and ^{15}O, in order to perform *in vivo* imaging of biologically active molecules. Although there is no radioisotope of H that can be used for PET, many molecules can replace a hydrogen or hydroxyl group with ^{18}F without changing its biological properties. Fluorine-18 can also be used as a substitute in fluorine-containing compounds such as 5-fluorouresil (5-FU), as demonstrated by Mintun *et al.*[3]

Most PET tracers utilize a radioisotope that has a short half-life and can be produced by a cyclotron (see Table I for a list of important tracers). However, there are some radiotracers, such as copper-62, that can be manufactured in a nuclear generator. Radiopharmaceuticals are produced after the radioisotope has been generated and substituted into the compound of interest. Because of the short half-lives of most PET tracers, sequential scanning on the same day is not usually possible.

Drug Evaluation

The development of new drugs presents many questions that must be addressed: is the drug sufficiently delivered to target of interest? How is normal tissue affected? At what dose is toxicity produced? How much drug is eliminated from both target tissue and normal tissue in relation to time? Does the drug affect target tissue in the predicted way? All of these questions can be answered by labeling the drug of interest with an appropriate radionuclide for PET imaging.

In addition, mathematical kinetic modeling is necessary for all aspects of drug pharmacology and to measure physiological functions, such as

TABLE I
POSITRON-EMITTER RADIOTRACERS

Radioisotope	Half-life (min)
Carbon-11	20.4
Fluorine-18	109.8
Nitrogen-13	10
Oxygen-15	2.03
Copper-62	9.7
Rubidium-82	1.25
Yttrium-86	14.7

[3] M. A. Mintun *et al.*, *Radiology* **169**, 45 (1988).

tissue perfusion, metabolic rate, and elimination. Dynamic data can be collected in a specific biological organ or tissue by defining the region of interest and recording the radioactivity over time. Input functions can be calculated by measuring tracer concentration in arterial blood. By comparing the input functions and time activity curves over the organ or tissue of interest with theoretical models, it is possible to calculate the metabolism of the applied drugs. When these drugs are not metabolized in the tissue, the calculation of drug concentration is very simple. However, most drugs will undergo some degree of metabolism to produce metabolites. If these metabolites do not include the original radionuclide, there is no problem regarding the calculation of drug metabolism. However, if these metabolites also contain a radionuclide it can be difficult for PET to distinguish signals from the parent radiopharmaceutical and those from its metabolite. Blasberg et al.[4] and Salem et al.[5] have suggested a number of mathematical calculations in order to determine the parent contribution from total measured radiotracer activity.

Pharmacokinetic studies of new pharmaceuticals are greatly simplified by labeling these compounds with a PET tracer. It is possible to measure the time-dependent, in vivo biodistribution of a new drug labeled with PET tracer in one experiment, which can be subsequently compared to many animal experiments. Furthermore, the various effects of a drug on different biological processes, such as blood flow, tissue metabolism, and receptor activation, can also be demonstrated in vivo through PET imaging.

Biological Function Evaluation

Theoretically, any biological function can be studied in vivo using an appropriately labeled PET tracer molecule. However, at present, PET tracers, which are utilized most commonly, are small molecules, which can be labeled by well-defined methods. Another important consideration is that the concentration of any PET tracer molecule should be significantly higher at the target sites than in the background. This allows biological function to be measured by determining tracer concentration over a specified time interval and by drawing a region of interest over the specified target tissue or organ region. Physiological processes such as oxygen consumption, blood flow, and tissue metabolism can also be demonstrated in vivo using PET tracers.

In the body, binding of an activating molecule (agonist) to a biochemical structure (receptor) can activate many biological functions. The same

[4] R. G. Blasberg et al., Cancer Res. **60**, 624 (2000).
[5] A. Salem et al., Lancet **355**, 2125 (2000).

process can be blocked by an antagonist, which may have a higher affinity for receptors as compared to the agonist. Receptors can be visualized and quantified by labeling receptor-binding substances (ligands) with PET tracers. Most receptors have several biochemically similar subtypes and are composed of multiple subunits. Identification of these subtypes and subunits requires specific ligands. The details of *in vivo* research of receptors and their clinical applications in the diagnosis of neurodegenerative and heart diseases are discussed later.

Clinical Applications

The resolution of computed tomography (CT) and magnetic resonance imaging (MRI) is excellent for the visualization of both normal and diseased tissues. However, all disease processes start with molecular and cellular abnormalities. Most disease processes take a long time to progress to a stage where they can be detected by these structural imaging techniques. In fact, many diseases may already be in advanced stages by the time they are detected by MRI or CT. However, the principle of PET is to detect the altered metabolism of disease processes and not the altered anatomy, such as CT and MRI. PET, as a functional molecular imaging technique, can also provide highly accurate quantitative results and therefore can be used for various research and clinical applications.

Kumar et al.[6] discussed the role of ^{18}F-FDG PET in the management of cancer patients. For these patients, PET has become important in diagnosis, staging, monitoring the response to treatment, and detecting recurrence. However, PET has also played an important role for both diagnostic and therapeutic purposes in patients with neurological and cardiological infections and inflammations, vasculitis, and other autoimmune diseases.[7]

PET in Oncology

^{18}F-FDG is the most widely used radiotracer in oncology. Because glucose metabolism is increased manyfold in malignant tumors as compared to normal cells, PET has high sensitivity and a high negative predictive value. It has a well-established role in initial staging, monitoring response to the therapy, and management of many types of cancer, including lung cancer, colon cancer, lymphoma, melanoma, esophageal cancer, head and neck cancer, and breast cancer (Table II).

[6] R. Kumar et al., *Ind. J. Cancer.* **40,** 87 (2003).
[7] M. Schirmer et al., *Expr. Gerontol.* **38,** 463 (2003).

TABLE II
Important Indications of ^{18}F-FDG PET in Oncology

Clinical application	Tumor
Differentiation of benign from malignant	Solitary pulmonary nodules
Diagnosis and initial staging	Lung cancer
	Colorectal cancer
	Lymphoma
	Esophageal cancer
	Breast cancer
	Head and neck cancer
	Melanoma
Evaluation response to chemotherapy	Lymphoma
	Breast cancer
	Lung cancer
	Head and neck cancer
	Brain tumors
	Bone tumors
Recurrence	Lymphoma
	Breast cancer
	Head and neck cancer
	Brain tumors
	Colorectal cancer
	Melanoma

Differential Diagnosis: Benign versus Malignant Lesions. FDG PET has been used successfully as a noninvasive diagnostic test for solitary pulmonary nodules (SPN) in order to distinguish benign lesions from malignancies. A meta-analysis by Gould *et al.*[8] of 40 studies that included 1474 SPNs has reported a sensitivity of 96.9% and specificity of 77.8% for detecting malignancy by PET. Studies by Matthies *et al.*[9] and Zhuang *et al.*[10] have demonstrated the advantages of dual time point imaging in the differentiation of malignant from benign lesions. These authors concluded that malignant nodules have a greater tendency to show an increase in FDG uptake over time, whereas pulmonary nodules of benign origin have a decreasing pattern of uptake over time. FDG PET imaging makes it possible to calculate a specific uptake value that is called a "standardized uptake value" (SUV). A lesion with an SUV greater than 2.5 is considered

[8] M. K. Gould *et al.*, *JAMA* **285,** 914 (2001).
[9] A. Matthies *et al.*, *J. Nucl. Med.* **43,** 871 (2002).
[10] H. Zhuang *et al.*, *J. Nucl. Med.* **42,** 1412 (2001).

to have a high probability of malignancy.[11,12] Gambhir et al.[13] suggested that biopsy or surgery of PET-positive lesions is also very cost-effective.

Cancer of Unknown Origin. Many authors have investigated the diagnostic contribution of PET in patients with unknown primary malignancies.[14–18] However, there are very few studies to date that have analyzed the impact of PET results on therapeutic management. Rades et al.[19] detected the primary site in 18 of 42 patients who had localized cancer of unknown origin by using conventional staging procedures. PET was positive for disseminated diseases in 38% of patients. In 69% of patients, PET results influenced selection of the definitive treatment.

Diagnosis and Initial Staging. The impact of FDG PET on diagnosis and initial staging has been shown for various tumors. Many PET studies for lung cancer have included nonsmall cell lung cancer (NSLC) patients in whom the regional and distant involvement of disease can change staging and guide the therapeutic approach. Dwamena et al.[20] performed a meta-analysis of staging lung cancer by PET and CT and concluded that PET was significantly more accurate than CT. They reported a sensitivity and specificity of 79 and 91% for PET and 60 and 77% for CT, respectively. Pieterman[21] reported sensitivity and specificity for the evaluation of mediastinal nodal involvement, which were 91 and 86% for PET and 75 and 66% for CT, respectively. A whole-body PET scan is able to detect more unknown distant metastases and is more accurate than conventional imaging in staging of patients with lung cancer, as shown by Pieterman[21] and Laking and Price.[22] Gambhir et al.[23] found PET to be cost-effective by avoiding surgeries that would not benefit the patient. Up to 20% of lung cancer patients are found to have an adrenal mass by CT, without necessarily confirming metastasis. It has been confirmed, however, that FDG PET can eliminate the need for a biopsy of enlarged adrenal glands in lung cancer patients.[24,25]

[11] S. B. Knight et al., *Chest* **109**, 982 (1996).
[12] V. J. Lowe, *Radiology* **202**, 435 (1997).
[13] S. S. Gambhir et al., *J. Clin. Oncol.* **16**, 2113 (1998).
[14] A. C. Kole, *Cancer* **82**, 1160 (1998).
[15] O. S. Aassar et al., *Radiology* **210**, 177 (1999).
[16] K. M. Greven et al., *Cancer* **86**, 114 (1999).
[17] U. Lassen et al., *Eur. J. Cancer* **35**, 1076 (1999).
[18] K. H. Bohuslavizki et al., *J. Nucl. Med.* **41**, 816 (2000).
[19] D. Rades et al., *Ann. Oncol.* **12**, 1605 (2001).
[20] B. A. Dwamena et al., *Radiology* **213**, 530 (1999).
[21] R. M. Pieterman, *N. Engl. J. Med.* **343**, 254 (2000).
[22] G. Laking and P. Price, *Thorax* **56**, 38 (2001).
[23] S. S. Gambhir et al., *J. Nucl. Med.* **37**, 1428 (1996).
[24] L. M. Lamki, *AJR Am. J. Roentgenol.* **168**, 1361 (1997).
[25] T. Bury et al., *Eur. Respir. J.* **14**, 1376 (1999).

PET also has high sensitivity for the preoperative diagnosis of colorectal carcinoma. However, it has no important role in early-stage patients because they require surgical diagnosis and staging. Abdel-Nabi et al.[26] reported 100% sensitivity of PET imaging in the identification of all primary lesions in 48 patients. PET was found to be superior to CT for the identification of liver metastases, with a sensitivity and specificity of 88 and 100%, respectively, for PET and 38 and 97%, respectively, for CT.

The accuracy of PET is also better than CT–MRI and Gallium scintigraphy in the staging of patients with lymphoma, as demonstrated by Sasaki et al.[27] and Even-Sapir and Israel.[28] Moog et al.[29] demonstrated 25 additional lesions using PET in 60 consecutive patients with Hodgkin's disease (HD) and non-Hodgkin's lymphoma (NHL), whereas CT found only 6 additional lesions, of which 3 were false positive. Stumpe et al.[30] had a similar experience when the accuracy of FDG PET was compared to CT. There was no significant difference in the sensitivity of PET and CT. However, PET specificity was 96% for HD and 100% for NHL, whereas CT specificity was 41% for HD and 67% for NHL. Sasaki et al.[27] showed a specificity of 99% for combined PET–CT, but sensitivity was 65% for CT alone and 92% for PET alone. These studies show that there is a large variation in sensitivity and specificity of CT, whereas these figures are not as variable for PET. Furthermore, PET results modified therapy in 25% of all lymphoma patients. In one study, 23% of patients were assigned a different stage by FDG PET imaging when compared with conventional imaging.[29,31,32]

The sensitivity of PET for primary breast cancer varies between 68 and 100% as reported by Bruce et al.,[33] Avril et al.,[34] and Schirrmeister et al.[35] A study by Schirrmeister et al.[35] showed that a whole-body FDG PET scan is as accurate as a panel of imaging modalities currently employed in detecting disease and is significantly more accurate in detecting multifocal disease, lymph node involvement, and distant metastasis. Metastasis to axillary lymph nodes is one of the most important prognostic factors in breast cancer patients. Kumar et al.[36] demonstrated that sentinel lymph node sampling has high accuracy even in multifocal and multicentric breast cancer.

[26] H. Abdel-Nabi et al., Radiology 206, 755 (1998).
[27] M. Sasaki et al., Ann. Nucl. Med. 16, 337 (2002).
[28] E. Even-Sapir and O. Israel, Eur. J. Nucl. Med. Mol. Imag. 30, S65 (2003).
[29] F. Moog et al., Radiology 203, 795 (1997).
[30] K. D. Stumpe et al., Eur. J. Nucl. Med. 25, 721 (1998).
[31] S. Partridge et al., Ann. Oncol. 11, 1273 (2000).
[32] H. Schoder et al., J. Nucl. Med. 42, 1139 (2001).
[33] D. M. Bruce et al., Eur. J. Surg. Oncol. 21, 280 (1995).
[34] N. Avril et al., J. Nucl. Med. 42, 9 (2001).
[35] H. Schirrmeister et al., Eur. J. Nucl. Med. 28, 351 (2001).
[36] R. Kumar et al., J. Nucl. Med. 44, 7 (2003).

The sensitivity of FDG PET in the detection of axillary lymph node metastasis varies from 79 to 100%.[36a,37] However, PET can fail to detect micrometastases in lymph nodes because there are fewer cells, which may have a detectable increase in glucose metabolism.

Response to Treatment. FDG PET imaging is metabolically based and is therefore a more accurate method to differentiate tumor from scar tissue. CT and other conventional imaging use shrinkage in tumor size. It is often difficult to differentiate recurrence and posttreatment fibrotic masses using CT. Bury *et al.*[25] demonstrated that PET was more sensitive and equally specific as compared to other imaging modalities for the detection of residual disease or recurrence after surgery or radiotherapy in lung cancer patients. Vitola *et al.*[38] studied the effects of regional chemoembolization therapy in patients with colon cancer using FDG uptake as a criterion and found that decreased FDG uptake correlated with response, whereas the presence of residual uptake was used to guide further regional therapy. Findlay *et al.*[39] concluded that PET was accurate for the differentiation of responders from nonresponders, both on lesion-by-lesion or patient-by-patient analysis.

Wahl *et al.*[40] demonstrated that PET can detect metabolic changes in breast cancer as early as 8 days after the initiation of chemotherapy. Several studies were able to differentiate responders from nonresponders after the first course of therapy using FDG PET imaging.[41–43] Smith *et al.*[44] correctly identified responders with a sensitivity of 100% and a specificity of 85% after the first course of chemotherapy. Vranjesvic *et al.*[45] compared PET and conventional imaging (CT–MRI–USG) to evaluate the response to chemotherapy in breast cancer patients. PET was more accurate than combined conventional imaging modalities, with positive and negative predictive values of 93 and 84%, respectively, for PET versus 85 and 59%, respectively, for conventional imaging modalities. The accuracy was 90% for FDG PET and 75% for conventional imaging modalities.

Jerusalem *et al.*[46] compared the prognostic role of PET and CT after first-line treatment in 54 NHL–HD patients. PET showed higher diagnostic

[36a] F. Crippa *et al.*, *J. Nucl. Med.* **39**, 4 (1998).
[37] M. Greco *et al.*, *J. Natl. Cancer Inst.* **93**, 630 (2001).
[38] J. V. Vitola *et al.*, *Cancer* **78**, 2216 (1996).
[39] M. Findlay *et al.*, *J. Clin. Oncol.* **14**, 700 (1996).
[40] R. L. Wahl *et al.*, *J. Clin. Oncol.* **11**, 2101 (1993).
[41] T. Jansson *et al.*, *J. Clin. Oncol.* **13**, 1470 (1995).
[42] P. Bassa *et al.*, *J. Nucl. Med.* **37**, 931 (1996).
[43] M. Schelling *et al.*, *Clin. Oncol.* **18**, 1689 (2000).
[44] I. C. Smith *et al.*, *J. Clin. Oncol.* **18**, 1676 (2000).
[45] D. Vranjesevic *et al.*, *J. Nucl. Med.* **43**, 325 (2002).
[46] G. Jerusalem *et al.*, *Blood* **94**, 429 (1999).

and prognostic values than CT (positive predictive value 100% vs 42%). The 1-year progression-free survival (PFS) survival was 86% in PET-negative patients as compared to 0% in PET-positive patients. Spaepen et al.[47] evaluated 60 patients with HD who had an FDG PET scan at the end of first-line treatment with or without residual mass. The 2-year disease-free survival (DFS) was 4% for the PET-positive compared to 85% for the PET-negative group. Kostakoglu et al.[48] showed that FDG PET can predict a response to chemotherapy as early as after the first cycle of chemotherapy.

Recurrence/Restaging. Early surgical intervention or reintervention can cure a significant number of patients with recurrent cancer. The best example in this indication is the treatment of recurrent colorectal cancer. Usually, serial serum carcinoembryonic antigen (CEA) levels are used for recurrence monitoring, but when a high serum level of CEA is encountered, imaging will be necessary to localize the site of possible recurrence. Steele et al.[49] demonstrated that CT is usually incapable of differentiating postsurgical changes from recurrence and that CT commonly misses hepatic and extrahepatic abdominal metastases. However, PET can be used to identify the metabolic characteristics of the lesions that are equivocal or undetected by CT. Valk et al.[50] demonstrated that PET was found to be more sensitive than CT for all metastatic sites except the lung, where both modalities had equivalent sensitivities. They also reported that one-third of PET-positive lesions in the abdomen, pelvis, and retroperitoneum were negative on CT. PET can also differentiate postsurgical changes from recurrence. The accuracy of PET for detection of recurrence varies from 90 to 100%, whereas for CT it is 48–65%.

For patients with breast cancer, 35% will experience locoregional and distant metastases within 10 years of initial surgery, as was demonstrated by van Dongen et al.[51] Gallowitsch et al.[52] reported sensitivity, specificity, PPV, NPV, and accuracy of 97, 82, 87, 96, and 90%, respectively, for FDG PET and 84, 60, 73, 75, and 74%, respectively, for conventional imaging. On a lesion basis, significantly more lymph nodes (84 vs 23) and fewer bone metastases (61 vs 97) were detected using FDG as compared with conventional imaging. Kamel et al.[53] analyzed the role of FDG PET

[47] K. Spaepen et al., Br. J. Haematol. **115,** 272 (2001).
[48] L. Kostakoglu et al., J. Nucl. Med. **43,** 1018 (2002).
[49] G. Steele, Jr. et al., J. Clin. Oncol. **9,** 1105 (1991).
[50] P. E. Valk et al., Arch. Surg. **134,** 503 (1999).
[51] J. A. van Dongen et al., J. Natl. Cancer Inst. **92,** 1143 (2000).
[52] H. J. Gallowitsch et al., Invest. Radiol. **38,** 250 (2003).
[53] E. M. Kamel et al., J. Cancer Res. Clin. Oncol. **129,** 147 (2003).

for 60 patients and demonstrated that overall sensitivity, specificity, and accuracy were 89, 84, and 87%, respectively, for locoregional metastasis and 100, 97, and 98%, respectively, for distant metastasis. The authors also concluded that FDG PET was more sensitive than serum tumor marker CA 15–3 in detecting breast cancer relapse. Eubank et al.[54] compared FDG PET and CT in 73 recurrent/metastatic breast cancer patients for evaluation of mediastinal and internal mammary lymph nodes metastases. In 33 patients amenable to follow-up CT or biopsy, FDG PET revealed a superior detection rate of 85% compared to CT (54%).[55]

Approximately two-thirds of patients with HD present with a mass lesion in the location of a previous tumor manifestation, but only about 20% of patients ultimately relapse.[56,57] Similarly, in patients with high-grade NHL, 50% present with mass lesion and only 25% relapse.[58] Gallium scintigraphy has proved useful in patients with recurrent disease, but has limitations in intraabdominal and low-grade lymphoma.[59] Cremerius et al.[60] reported a sensitivity of 88% and a specificity of 83% for the detection of residual disease by PET. The corresponding values for CT were 84 and 31%, respectively. A study by Mikosch et al.[61] compared PET with CT–US in detecting recurrence and reported a sensitivity of 91%, a specificity of 81%, a PPV of 79%, a NPV of 92%, and an accuracy of 85%. For CT–US, these values were 88, 35, 48, 81, and 56%, respectively.

PET in Neurology

PET allows a noninvasive assessment of physiological, metabolic, and molecular processes of brain functions. PET may become the critical test for selecting the appropriate patients for treatment when the disease process is still at the molecular level. FDG and L-[methyle-^{11}C]methionine (MET) are the most frequently used PET tracers for the evaluation of glucose and amino acid metabolism for various brain disorders. 6-[^{18}F]fluoro-L-dopa (F-DOPA) binds to dopamine transporter sites and allows for the assessment and imaging of presynaptic dopaminergic neurons.

Epilepsy. FDG PET is accepted as a useful and highly sensitive tool for the localization of epileptogenic zones. The sites of glucose hypometabolism

[54] W. B. Eubank et al., *Radiographics* **22,** 5 (2002).
[55] W. B. Eubank et al., *J. Clin. Oncol.* **19,** 3516 (2001).
[56] G. P. Canellos, *Clin. Oncol.* **6,** 931 (1988).
[57] V. Lowe and G. A. Wiseman, *J. Nucl. Med.* **43,** 1028 (2002).
[58] P. Hoskin, *Eur. J. Nucl. Med.* **28,** 449 (2002).
[59] D. Front et al., *Radiology* **214,** 253 (2000).
[60] U. Cremerius et al., *Nuklearmedizin* **38,** 24 (1999).
[61] P. Mikosch et al., *Acta Med. Austriaca* **30,** 410 (2003).

at seizure foci as shown by FDG PET correlated strongly with epilepto-genic zones at surgery. PET imaging has an accuracy of 85–90% in de-tecting epileptic focus as shown by Newberg et al.[62] Moran et al.[63] and Kobayashi et al.[64] reported that long-standing seizure episodes eventually lead to significant atrophy, which can be detected by MRI. Therefore, accurate localization of seizure foci can be obtained using a combination of MRI and FDG PET. Kim et al.[65] reported a sensitivity of 89% and a specificity of 91% for PET in the detection of epileptic foci in patients with temporal lobe epilepsy.

Alzheimer's Disease and Related Disorders. Alzheimer's disease (AD) is the most common cause of dementia in the elderly. PET imaging can differentiate AD from other forms of dementia. In patients with AD, there is a decrease in glucose metabolism in the temporoparietal lobes that is not evident in patients with other forms of dementia. The basal ganglia, thalamus, and primary sensorimotor cortex are spared in AD. Salmon et al.[66] demonstrated a sensitivity of 96 and 87% for PET in diagnosing mod-erate to severe and mild disease, respectively, in 129 cognitively impaired patients.

A new PET tracer, 2-(1-{6-[(2-[^{18}F]fluoroethyle)(methyle)amino]-2-naphthyl}ethylidene)malononitrile (^{18}F-FDDNP), has been developed to target amyloid β saline plaques and neurofibrillary tangles in AD. This tracer shows prolonged retention in affected areas of the brain.[67,68]

According to several studies, other brain disorders, such as head injury, frontal lobe dementia, and Huntington's disease, can also be assessed with high accuracy using PET.[69–72]

Movement Disorders. Several radionuclide-labeled neuroreceptors and neurotransmitters have shown excellent results with PET.[73–75] These PET radiopharmaceuticals have great potential for the assessment

[62] A. Newberg et al., Semin. Nucl. Med. **32,** 13 (2002).
[63] N. F. Moran et al., Brain **124,** 167 (2001).
[64] E. Kobayashi et al., Neurology **60,** 405 (2003).
[65] Y. K. Kim et al., J. Nucl. Med. **44,** 1006 (2003).
[66] E. Salmon et al., J. Nucl. Med. **35,** 391 (1994).
[67] E. D. Agdeppa et al., J. Nucl. Med. **42,** 65 (2001).
[68] K. Shoghi-Jadid et al., Am. J. Geriatr. Psychiatr. **10,** 24 (2002).
[69] J. C. Mazziotta et al., N. Engl. J. Med. **316,** 357 (1987).
[70] T. Kuwert et al., Psychiatr. Res. **12,** 425 (1989).
[71] D. H. Silverman et al., JAMA **286,** 2120 (2001).
[72] A. Newberg et al., Semin. Nucl. Med. **33,** 136 (2003).
[73] F. J. Vingerhoets et al., J. Nucl. Med. **37,** 421 (1996).
[74] M. R. Davis et al., J. Nucl. Med. **44,** 855 (2002).
[75] W. S. Huang et al., J. Nucl. Med. **44,** 999 (2003).

of movement disorders. 2β-Carbomethoxy-3β-(4-chlorophenyl)-8-(2-^{18}F-fluoroethyl) nortropane (FECNT) and F-DOPA both allow assessment of the integrity of presynaptic dopaminergic neurons. These compounds are able to diagnose Parkinson's disease and other diseases effectively. Davis et al.[74] demonstrated that FECNT is an excellent candidate as a radioligand for in vivo imaging of dopamine transporter density in healthy humans and subjects with Parkinson's disease.

PET in Cardiology

The detection of myocardial viability is the most important task of predicting functional recovery after medical or surgical interventions. Myocardial perfusion SPECT scintigraphy has a very high sensitivity, but has lower specificity for detection of viability as shown by Arnese et al.,[76] Pasquet et al.,[77] and Bax et al.[78] PET is considered the "gold standard" for the detection of myocardial viability. Other indications of PET in cardiology include the evaluation of ischemic heart disease, cardiomyopathies, postcardiac transplant, and cardiac receptors for the regulation of cardiovascular functions.

Myocardial Viability. The extent of viable myocardium is an important factor for both prognosis and prediction of outcome after revascularization in patients with ischemic cardiomyopathy and chronic left ventricular dysfunction, as demonstrated by Tillisch et al.,[79] Tamaki et al.,[80] Pagano et al.,[81] and Pasquet et al.[77] PET imaging shows metabolism in viable myocardial segments. Knuesel et al.[82] concluded that most metabolically viable segments on PET imaging recover function after revascularization. ^{13}N-ammonia is also being used for PET assessments of myocardial viability, but this compound has limitations due to its short half-life. Another PET tracer, rubidium-82, has shown good results in the detection of myocardial perfusion abnormalities.[83,84]

Cardiac Neurotransmission. Many pathophysiological processes take place in the nerve terminals, synaptic clefts, and postsynaptic sites in the heart. These processes are altered in many diseases such as heart failure,

[76] M. Arnese et al., Circulation **91**, 2748 (1995).
[77] A. Pasquet et al., Circulation **100**, 141 (1999).
[78] J. J. Bax et al., Circulation **106**, 114 (2002).
[79] J. Tillisch et al., N. Engl. J. Med. **314**, 884 (1986).
[80] N. Tamaki et al., Circulation **91**, 1697 (1995).
[81] D. Pagano et al., J. Thorac. Cardiovasc. Surg. **115**, 791 (1998).
[82] P. R. Knuesel et al., Circulation **108**, 1095 (2003).
[83] K. Yoshida et al., J. Nucl. Med. **37**, 1701 (1996).
[84] R. A. deKemp et al., J. Nucl. Med. **41**, 1426 (2000).

diabetic autonomic neuropathy, idiopathic ventricular tachycardia and ar-rhythmogenic right ventricular cardiomyopathy, heart transplantation, drug-induced cardiotoxicity, and dysautonomias.[85–89] Cardiac neurotrans-mission imaging can be obtained using PET. The most commonly used PET radiopharmaceuticals for imaging presynaptic activity are ^{18}F-fluoro-dopamine, ^{11}C-hydroxyphedrine, and ^{11}C-ephidrine. Postsynaptic agents include [11]C-(4-(3-*t*-butylamino-2-hydroxypropoxy)benzimidazol-1) CGP, and ^{11}C-carazolol.

PET in Infectious and Inflammatory Diseases

FDG has been used in the management of various cancers and in neurological and cardiological diseases.[90–92] Its ability to image glucose metabolism has been key to the success of PET in various disease settings. FDG PET has been used successfully in oncological imaging. However, FDG is not tumor or disease specific. Increased FDG uptake is seen in any tissue with increased glucose metabolism. Yamada *et al.*[93] demonstrated that glucose metabolism is also increased in inflammatory tissues.

Osteomyelitis. FDG PET can be used for the diagnosis of acute or chronic osteomyelitis as studied by Pauwels *et al.*[94] De Winter *et al.*[95] dem-onstrated a sensitivity of 100%, a specificity of 85%, and an accuracy of 93% for this purpose. Zhuang *et al.*[96] reported a sensitivity, a specificity, and an accuracy of 100, 87.5, and 91%, respectively, in 22 patients of sus-pected chronic osteomyelitis. Chianelli *et al.*[97] demonstrated the advantage of FDG PET over labeled leukocyte imaging. Because glucose is smaller than antibodies and leukocytes, it can penetrate faster and more easily at the lesion site.

Prosthetic Joint Infections. To detect infection in a prosthetic joint is challenging, as there are no simple modalities for this purpose. Zhuang *et al.*[98] evaluated 74 prostheses with FDG PET in order to determine its

[85] D. C. Lefroy *et al.*, *J. Am. Coll. Cardiol.* **22**, 1653 (1993).
[86] M. W. Dae *et al.*, *Cardiovasc. Res.* **30**, 270 (1995).
[87] D. S. Goldstein *et al.*, *N. Engl. J. Med.* **336**, 696 (1997).
[88] S. B. Liggett *et al.*, *J. Clin. Invest.* **102**, 1534 (1998).
[89] T. Wichter *et al.*, *Circulation* **13**, 1552 (2000).
[90] V. Salanova *et al.*, *Epilepsia* **40**, 1417 (1999).
[91] R. Bar-Shalom *et al.*, *Semin. Nuclear Med.* **30**, 150 (2000).
[92] J. J. Bax *et al.*, *Semin. Nucl. Med.* **30**, 281 (2000).
[93] S. Yamada *et al.*, *J. Nucl. Med.* **7**, 1301 (1995).
[94] E. K. Pauwels *et al.*, *J. Cancer Res. Clin. Oncol.* **126**, 549 (2000).
[95] F. De Winter *et al.*, *J. Bone Joint Surg. A* **83A**, 651 (2001).
[96] H. Zhuang *et al.*, *Clin. Nucl. Med.* **25**, 281 (2000).
[97] M. Chianelli *et al.*, *Nucl. Med. Commun.* **18**, 437 (1997).
[98] H. Zhuang *et al.*, *J. Nucl. Med.* **42**, 44 (2001).

role in this setting (38 hip and 36 knee prostheses). They reported a sensitivity of 91% and a specificity of 72% for detecting knee prosthesis infections and a sensitivity of 90% and a specificity of 89% for detecting hip prosthesis infections. However, Love et al.[99] reported a very high sensitivity of 100% but a low specificity of 47% in 26 hip and knee prostheses.

Fever of Unknown Origin. Localization of an infective focus in patients with fever of unknown origin is a difficult task. Stumpe et al.[100] demonstrated a sensitivity, specificity, and accuracy of 98, 75, and 91%, respectively, in patients of fever of unknown origin with suspected infections. On the bases of studies published by Sugawara et al.,[101] Meller et al.,[102] Blockmans et al.,[103] and Lorenzen et al.,[104] PET has greater capabilities than conventional imaging in screening 40–70% of patients with fever of unknown origin.

Future Perspectives

PET imaging has the potential to detect almost any physiological, biochemical, and molecular process in the human body and in animals. PET can describe such processes in both normal and diseased tissues. PET can also be used to observe key steps in various disease processes, including carcinogenesis. This section discusses several of these processes, including angiogenesis, apoptosis, and hypoxia.

Assessment of Multidrug Resistance

One of the most common factors for chemotherapeutic failure is multidrug resistance (MDR) in cancer patients. Overexpression of P-glycoprotein (Pgp) is responsible for MDR in many tumors, as suggested by Gottesman and Pastan[105] and Germann et al.[106] Tc99m-labeled sestamibi, tetrofosmin, and furifosmin all act as substrates for Pgp and have been shown with SPECT imaging to predict tumors expressing MDR.[107] However, these techniques are limited by a lack of quantitative data. However, PET imaging using ^{11}C-verapamil, ^{11}C-daunorubicin, and ^{11}C-colchicine as

[99] C. Love et al., *Clin. Positron Imaging* **3**, 159 (2000).
[100] K. D. Stumpe et al., *Eur. J. Nucl. Med.* **27**, 822 (2000).
[101] Y. Sugawara et al., *Eur. J. Nucl. Med.* **25**, 1238 (1998).
[102] J. Meller et al., *Eur. J. Nucl. Med.* **27**, 1617 (2000).
[103] D. Blockmans et al., *Clin. Infect. Dis.* **32**, 191 (2001).
[104] J. Lorenzen et al., *Nucl. Med. Commun.* **22**, 779 (2001).
[105] M. M. Gottesman and I. Pastan, *Annu. Rev. Biochem.* **62**, 385 (1993).
[106] U. A. Germann et al., *Semin. Cell Biol.* **4**, 63 (1993).
[107] C. C. Chen et al., *Clin. Cancer Res.* **3**, 545 (1997).

in vivo Pgp probes has been investigated in experimental studies by Elsinga et al.,[108] Hendrikse et al.,[109] and Levchenko et al.[110] Kurdziel et al.[111] demonstrated that ^{18}F-paclitaxel (FPAC) uptake is an indicator of Pgp function in tissues from rhesus monkeys. These studies may have a potential role in the selection of patients in whom the modulation of Pgp may be beneficial before and during chemotherapy.

Quantitation of Angiogenesis

Angiogenesis is one of the most important steps in tumor development. PET imaging is transforming our understanding of angiogenesis and the evolution of drugs that stimulate or inhibit angiogenesis. PET tracers such as [11]C- or ^{18}F-labeled thymidine can be used to determine cellular proliferation rates, as demonstrated by Vander Borght et al.,[112] van Eijkeren et al.,[113] Barthel et al.,[114] and Wagner et al.[115] Shields et al.[116] have shown very promising results using 3-deoxy-3-[^{18}F]fluorothymidine (FLT) to detect cell proliferation. Accordingly, FLT may be more specific for the evaluation of response to chemotherapy, as cytotoxic agents affect cell division directly.

Tumor Hypoxia

Hypoxia in tumor cells is an important prognostic indicator of chemotherapy or radiation therapy outcome. Hypoxic cells are more resistant to treatment. Therefore, patients with hypoxia in particular tumors may be pretreated with drugs to enhance oxygenation in order to improve therapeutic response. Nitroimidazole-based compounds labeled with technetium-99m, iodine-123, and iodine-131 have been evaluated for this purpose with variable success rates.[117–120] Valk et al.[121] and

[108] P. H. Elsinga et al., J. Nucl. Med. 34, 1571 (1996).
[109] N. H. Hendrikse et al., Eur. J. Nucl. Med. 26, 283 (1999).
[110] A. Levchenko et al., J. Nucl. Med. 41, 493 (2000).
[111] K. A. Kurdziel et al., J. Nucl. Med. 44, 1330 (2003).
[112] T. Vander Borght et al., Int. J. Rad. Appl. Instrum. A 42, 103 (1991).
[113] M. E. van Eijkeren et al., Acta Oncol. 35, 737 (1996).
[114] H. Barthel et al., Cancer Res. 63, 3791 (2002).
[115] M. Wagner et al., Cancer Res. 63, 2681 (2003).
[116] A. F. Shields et al., Nat. Med. 4, 1334 (1998).
[117] K. E. Linder, J. Med. Chem. 37, 9 (1994).
[118] A. Cherif et al., J. Drug Target 4, 31 (1996).
[119] J. R. Ballinger et al., J. Nucl. Med. 37, 1023 (1996).
[120] G. J. Cook et al., J. Nucl. Med. 39, 99 (1998).
[121] P. E. Valk et al., J. Nucl. Med. 33, 2133 (1992).

Rasey[122,123] demonstrated encouraging results in detecting tumor hypoxia using PET imaging with ^{18}F-fluoromisonidazole, although this compound has a relatively low uptake in hypoxic cells. Two newer ^{18}F-labeled compounds, ^{18}F-fluoroerythroimidazole and ^{18}F-fluoroetanidazole, have overcome some of the non-oxygen-dependent metabolism seen with ^{18}F-fluoromisonidazole.[124] Another ^{18}F-labeled compound, ^{18}F-EF 5, has shown good results in animal studies and may prove to be effective for non-invasive tumor hypoxia imaging.[125] ^{64}Cu-ATSM has been investigated to detect tumor hypoxia.[126] This compound is reduced in metabolically active mitochondria with oxygen-deficient electron transport chains.

Apoptosis

Detection of the process of cell death in both malignant and benign disorders by noninvasive imaging is another interesting area of study. All chemotherapeutic agents and radiation therapies induce programmed cell death in patients with malignant disease. Annexin V, an endogenous protein labeled with technetium-99m, has led the way to detect apoptosis as demonstrated by Blankenberg et al.,[127] Yang et al.,[128] and Blankenberg and Trauss.[129] The mechanism of Tc99m-annexin uptake is through its binding with phosphatidylserine, which is externalized in the cell membrane following apoptosis, as studied by Blankenberg et al.[130] Labeling annexin V with ^{18}F may further increase the utility of this promising method and may provide greater insight into the therapeutic response in patients with cancer.

Gene Expression

Advances in molecular genetic imaging allow the visualization of cellular process in normal and abnormal cells. Current PET molecular imaging studies with radiolabeled probes use HSV1-tk as the reporter gene.[131–133]

[122] J. S. Rasey, *Int. J. Radiat. Oncol. Biol. Phys.* **36,** 417 (1996).
[123] J. S. Rasey, *J. Nucl. Med.* **40,** 1072 (1999).
[124] T. Gronroos, *J. Nucl. Med.* **42,** 1397 (2001).
[125] L. S. Ziemer et al., *Eur. J. Nucl. Med. Mol. Imag.* **30,** 259 (2003).
[126] J. S. Lewis, *J. Nucl. Med.* **40,** 177 (1999).
[127] F. G. Blankenberg et al., *J. Nucl. Med.* **40,** 184 (1999).
[128] D. J. Yang et al., *Cancer Biother. Radiopharm.* **16,** 73 (2001).
[129] F. G. Blankenberg and H. W. Trauss, *Apoptosis* **6,** 117 (2001).
[130] F. G. Blankenberg et al., *Proc. Natl. Acad. Sci. USA* **95,** 6349 (1998).
[131] D. C. Blakey et al., *Biochem. Soc. Trans.* **23,** 1047 (1995).
[132] R. Weissleder, *Radiology* **212,** 609 (1999).
[133] P. Wunderbaldinger et al., *Eur. J. Radiol.* **34,** 156 (2000).

Several PET tracers, which are analogs of uracil and thymidine, are being developed by Tjuvajev *et al.*[134] and Gambhir *et al.*[135] The details of the reporter gene technique are given elsewhere in this volume.

Conclusions

Positron emission tomography is a powerful technique that provides noninvasive, quantitative, *in vivo* assessment of physiological and biological processes. Molecular imaging based on PET tracer kinetics has become a main source of information for both research purposes and patient management. In the near future, PET may become the critical modality both for diagnosing a variety of diseases and for selecting appropriate treatments when disease processes are still at the molecular level.

[134] J. G. Tjuvajev *et al.*, *Cancer Res.* **58,** 4333 (1998).
[135] S. S. Gambhir *et al.*, *Proc. Natl. Acad. Sci. USA* **96,** 2333 (1999).

[2]   Biophysical Basis of Magnetic Resonance Imaging of Small Animals

By Bruce M. Damon and John C. Gore

Introduction

Magnetic resonance (MR) images of small animals can be acquired routinely in reasonable times to portray anatomic features and functional characteristics with spatial resolution nearing true microscopic imaging. The methods available for mapping information about tissues spatially, and the various trade-offs that may be made for specific purposes in terms of image quality, resolution, and imaging strategies, are described elsewhere in this volume. This Chapter provides an overview of the nature of the underlying information that may be obtained from the most common types of images and explains the physical origins of the contrast obtainable. It should be emphasized that there are many qualitatively different types of MR images, and the signals acquired and mapped into images may be manipulated to portray or accentuate different tissue properties. The specific method used for image acquisition (i.e., the pattern of radio frequency and gradient waveforms used) determines the manner in which the image depends on different underlying characteristics, which may be regarded as analogous to the stains used in conventional microscopy. This Chapter describes only those mechanisms commonly available to modulate images based on so-called proton (^{1}H) nuclear magnetic resonance (NMR). While NMR signals from other nuclei (notably sodium) may, in principle, be used to create images, the high natural abundance and favorable nuclear properties of protons (hydrogen nuclei) make them by far the most suitable for producing high-quality images. The large majority of imaging studies of small animals rely on acquiring images of mobile hydrogen nuclei that are contained mainly in water molecules. Water comprises approximately 80% of most soft tissues, which therefore contain roughly 90 molar water protons. Most images portray an NMR property of the water within tissues, which indirectly reveals information on tissue structure or function. As we will show, many physiological or pathological processes of interest modify the properties of the water in an "MRI-visible" manner, albeit often in a nonspecific fashion. A second important type of contrast is also available from mobile nonwater protons contained mainly within lipids, which forms the basis of separate fat and water imaging.

The primary properties of tissues (or any other medium) that modify the NMR signal are the density of nuclei, the relaxation times of the nuclear magnetization (which, as we shall see, describe the times for recovery of the nuclear magnetization back to equilibrium after disturbance by applied radiofrequency pulses), the magnetic homogeneity of the environment, and the rate of transport of molecules (whether due to flow or via Brownian motion) within the medium. Each of these may in turn be affected by subtle changes in the properties of the tissue, such as the confinement of water within compartments, restrictions to water molecular diffusion, or physicochemical effects such as tissue pH and the presence of certain trace elements. Although in principle the effects of many of these influences are understood, it is often not straightforward to interpret changes in NMR signals in terms of any single or simple underlying causes. Tissue is an extremely heterogeneous medium, and the NMR signal in an image voxel represents the summed behavior of a large number of molecules undergoing a wide variety of interactions, so there is often no direct or specific interpretation of the variations within tissues depicted in MR images.

Spin Relaxation

It may be recalled that each hydrogen nucleus (proton) may be considered to possess spin, which in turn gives rise to a magnetic dipole moment. When a large number of such magnetic nuclei are placed in an external and static magnetic field, the majority tend to align themselves in the direction of the field, giving rise to a measurable macroscopic magnetization. MRI involves disturbing this magnetization using oscillating magnetic fields that alternate at a specific resonant radiofrequency (RF), the Larmor frequency, that is proportional to the strength of the applied static magnetic field (=42.6 MHz per Tesla applied field). The disturbed magnetization induces small electrical signals in coils of wire tuned to this same frequency as it recovers back to equilibrium. At any instant during this recovery there may exist components of the magnetization along the applied field direction (by convention the z direction) and orthogonal to this direction (the xy plane). The latter are responsible for signals measured during any acquisition. The relaxation times T_1 and T_2 and their corresponding rates R_1 (=$1/T_1$) and R_2 (=$1/T_2$) denote the characteristic time constants of the recovery back toward equilibrium of the z (longitudinal) and xy (transverse) components, respectively. When there are variations in the applied magnetic field within the sample or intrinsic magnetic inhomogeneities (such as occur within some tissues, as described later), the transverse component decays at an accelerated rate denoted as R_2^* (=$1/T_2^*$).

After being disturbed, the nuclear magnetization does not spontaneously recover very fast, but relies almost entirely on interactions of hydrogen nuclei with the surrounding material to reequilibrate. R_1 is called the spin–lattice relaxation rate. The "lattice" denotes the molecular environment surrounding a hydrogen nucleus and includes the remainder of the host molecule, as well as other solute and solvent molecules. Spin–lattice relaxation occurs because of magnetic interactions between nuclear spin dipoles and the local, randomly fluctuating, magnetic fields that exist on an atomic scale inside any medium. These originate mainly from neighboring magnetic nuclei, such as other hydrogen protons (e.g., within a water molecule, each hydrogen affects the neighbor) and are modulated by the motion of other surrounding dipoles in the lattice, which have components fluctuating with the same frequency as the resonance frequency. The recovery is very efficient when there is a local fluctuating field that can provide a magnetic perturbation at the Larmor frequency. For example, each proton in a water molecule has a neighboring proton that is also a magnetic dipole, which generates a magnetic field at the neighboring proton of about 5 g (0.5 mT). This field is constantly changing in amplitude and direction as the water molecule rotates rapidly and moves about in the liquid. It also changes as a result of intermolecular collision, translation, or chemical dissociation and exchange. The magnetic field experienced by any nucleus will fluctuate with a frequency spectrum that is dependent on the molecular tumbling due to the random thermal motion of the host and surrounding molecules. The mean strength of the local field is determined by the strength of the magnetic dipoles in the medium and how close they approach to the hydrogen nuclei. The component of the frequency spectrum, which is equal to the resonance frequency (or, for reasons beyond our discussion, twice as high), is effective in stimulating an energy exchange to induce recovery back toward equilibrium (i.e., spin–lattice relaxation). In liquids, the characteristic frequencies of thermal motion are of the order of 10^{11} Hz or higher, much greater than NMR frequencies of 10^7–10^8 Hz. Consequently, the component of the frequency spectrum from molecular motion that can induce spin–lattice relaxation is small and the process is slow. As the molecular motion becomes slower, either due to a lower temperature or an increased molecular size, the intensity of the fluctuations of the magnetic field at the resonance frequency increases, reaches a maximum, and then decreases again as the energy of the motion becomes increasingly concentrated in frequencies lower than the NMR range. Thus, R_1 passes through a maximum value when the molecular tumbling rate matches the Larmor frequency as the molecular motion becomes slower. It may be noted from this that relaxation rates depend on the frequency of the NMR measurement, and thus are usually shorter at low field strengths. The effect of the molecular motion is usually expressed by a correlation time, τ, characteristic of the time of rotation of a molecule or of the time of its translation into a neighboring position. Relaxation rates in simple liquids are affected, for example, by viscosity, temperature, and the presence of dissolved ions and molecules, which alter the correlation times of molecular motion or the amplitudes of the dipolar interactions.

Whereas T_1 is sensitive to radiofrequency components of the local field, T_2 is also sensitive to low-frequency components. R_2 (=$1/T_2$) is called the spin–spin relaxation rate. When an ensemble of nuclei is excited with RF, a transverse component of magnetization, orthogonal to the applied field direction, may develop, and it is this component that then rotates and induces the MRI signal in a receiver coil. T_2 reflects the time it takes for the ensemble to become disorganized and for the transverse component to decay. Since any growth of magnetization back toward equilibrium must correspond to a loss of transverse magnetization, all contributions to T_1 relaxation affect T_2 at least as much. In addition, components of the local dipolar fields that oscillate slowly, at low frequency, may be directed along the main field direction and thus can modulate the precessional frequency of a neighboring nucleus. Such frequency perturbations within an ensemble of nuclei result in rapid loss of the transverse magnetization and accelerated spin–spin relaxation. Because the low-frequency content of the local dipolar field increases monotonically as molecular motion progressively slows, although T_1 passes through a minimum value, T_2

continues to decrease and then levels off so that T_1 and T_2 then take on quite different values.

In the picture just developed, relaxation results from the action of fluctuating local magnetic fields experienced by protons, which accelerate the return to equilibrium. In pure water the dominant source of such effects is the dipole–dipole interaction between neighboring hydrogen nuclei in the same water molecule. The tumbling of each water molecule then causes the weak magnetic field produced by each proton to fluctuate randomly, and at the site of a neighboring proton these random alterations in the net field produce relaxation. The timescale characteristic of the dipolar interaction reflects molecular motion and clearly is expected to influence the efficacy of relaxation. Qualitatively, when there is a concentration of kinetic motion in the appropriate frequency range, relaxation will be efficient. We can envisage other types of motion that will be too rapid or too slow to be effective. The key important descriptor is the correlation time, τ, which measures the time over which the local fluctuating field appears continuous and deterministic. It represents the time it takes on average for the field to change significantly.

In simple liquids such as water the molecular motion is rapid and, on average, is isotropic. The motions are so fast that relaxation is not very efficient—the dipolar fields fluctuate too rapidly to be very efficient and the motion averages out any net effects of the local fields (so-called motional averaging). In pure water, T_1 is several seconds long, whereas in tissues, water relaxation times are often substantially under 1 s. In tissue, water relaxation times are shorter than in pure water because of the presence of large macromolecules and chemical species that promote relaxation. Proteins and other macromolecular structures produce, via protons on their surfaces, dipolar fields that fluctuate relatively slowly and do not average out on the NMR timescale, and these are efficient sites for relaxation. Water molecules or single protons may exchange between these sites and the rest of tissue water, thereby spreading the effect. Proteins are often considered to contain one or more layers of hydration, which is often loosely termed the "bound water" fraction, and this plays an important role in mediating interactions between the bulk aqueous medium and the macromolecular surface. Although some specific chemical side groups, notably hydroxyl and amide moieties, are known to be efficient conduits for relaxation, the large array of environments and chemical species within tissue usually make it impossible to identify specific interactions that dominate.

Solutions of macromolecules and biological tissues are chemically heterogeneous, and thus water in such media may experience a wide variety of different environments and chemical species with which to interact. Even in simple protein solutions, there may be different ranges and distributions of correlation time, coupling strengths and molecular dynamics, that affect the local dipolar fields experienced by water protons. An even greater variety of different scales and types of constituents occur within cells and whole tissues. Tissues contain diverse, freely tumbling solute ions and molecules, such as small proteins and lipids, as well as relatively immobilized or even rigid macromolecular assemblies, such as membranes and mitochondria. Tissues are also spatially inhomogeneous, containing many different types of cells or structures, and there may exist multiple compartments that are not connected or in which water transport is restricted. Nonetheless, although tissues are markedly heterogeneous at the cellular level, NMR relaxation will still reflect the average character of local dipolar fields experienced by water protons.

Effects of Exchange and Compartmentation

At any time only a small proportion of water protons may be in close juxtaposition to efficient relaxation sites. Then the average water proton relaxation rate (which is what is measured) will depend on how effectively, and at what rate, these effects are spread through the rest of the water population. Such exchange processes have profound effects on the observable NMR relaxation phenomena. In a time of 50 ms, water molecules diffuse distances of the order of 20 μm so that they sample many different environments on the cellular level within the timescale of relaxation. In many (although by no means all) situations, very rapid exchange may occur between bulk water and bound and interfacial water in biological systems. Suppose, for example, there are two types of environment in exchange; fraction f of water at any instant occupies sites with relaxation rate R_a, while the remaining fraction $1 - f$ is bulk solvent with relaxation rate R_b. If the exchange is very rapid, then the average relaxation rate is R

$$R = fR_a + (1 - f)R_b \tag{1}$$

so that $R = R_b + f(R_a - R_b)$.

If $R_a \gg R_b$, then R is very sensitive to small changes in f, which in turn increases with protein content in tissues. This is believed to be the origin for many increases in T_1 or T_2 in various pathologies, such as edematous changes following insults to tissue, or in rapidly dividing cells that have higher water fractions. Changes in tissue water and protein content in general will affect relaxation.

Note that the existence of water in separate compartments that are only slowly exchanging gives rise to more complex behavior that is not described adequately by a single relaxation rate. For example, the transverse decay may be measured as a sum of exponential terms, but by appropriate analysis of the decay curve some inferences may be made about the sizes of the contributing compartments and the rate of water exchange between them.

An additional contribution to R_2 can arise when there are protons present in different chemical species for whom the resonance frequencies are slightly different. Several surface groups (e.g., amides on proteins), may exchange protons with water, and they start with slightly different resonance frequencies. This mixing causes a further increase in the transverse relaxation rate, and so R_2 is sensitive to pH and other factors that affect the rate of such mixing.

We have suggested that relaxation in tissues is affected by interactions that occur between water protons and protons at or near the surface of macromolecules. Longitudinal proton magnetization can be exchanged between water and neighboring nuclei (whether interfacial water that is hydrogen bonded to the surface or protons within other chemical groups that are part of the macromolecule) by direct through-space dipolar couplings, as well as by the so-called chemical exchange of protons. Chemical dissociation of protons occurs at rates that are pH dependent, providing possible interchanges between water and surface sites (hydroxyls, amides, and so forth) as intact water molecules move constantly in and out of the hydration layers of macromolecular surfaces and exchange protons. These mechanisms give rise to so-called *magnetization transfer* between pools of water in different environments, and images may be produced that are sensitized to these processes.

Paramagnetic Relaxation

Paramagnetic agents may be administered to animals or they may arise naturally in some conditions, and they reduce the relaxation times of tissue water. Paramagnetic agents such as manganese, gadolinium, and several other transition and rare-earth metal ions are materials that, on the atomic scale, generate extremely strong local magnetic fields. The origin of their strong local fields lies in the fact that they contain unpaired electrons that have not been "matched" (paired off) in a chemical bond with spins of opposite character so that there is a net residual magnetic dipole moment from the electrons. The electron magnetic dipole is 658 times greater than the proton essentially because it has a smaller radius but the same charge so any water molecules that approach close to an unpaired electron will experience an intense interaction that can promote relaxation. For example, a 1 mM solution of gadolinium, even when chelated with diethylenetriamine penta-acetic acid (DTPA), reduces the T_1 of water to under 250 ms.

Susceptibility Contrast and BOLD Effects

Another important factor that may modulate the MRI signal intensity is the magnetic homogeneity of the local environment in which water molecules reside. In particular, if there are variations in the magnetic susceptibility within the sample, then the transverse magnetization, and therefore the measured signal, decays more rapidly. Variations in the susceptibility arise, for example, at interfaces between air and tissue. Within tissues there may also be small-scale variations in susceptibility due to the presence of metals within tissues, such as iron-containing proteins such as hemosiderin and hemoglobin. MRI may be usefully employed for the noninvasive detection of iron deposition in animal models.

A specific example of susceptibility variations arises from the vasculature within tissues because of the presence of hemoglobin in the circulation. This particular variation leads to the so-called blood oxygenation level-dependent (BOLD) effect, which has been found very useful for studies of brain activation and tissue oxygenation. In the brain, the physical origins of BOLD signals are reasonably well understood, although their precise connections to the underlying metabolic and electrophysiological activity need to be clarified further. It is well established that increasing neural activity in a region of the cortex stimulates an increase in the local blood flow in order to meet the increased demand for oxygen and other substrates. At the capillary level, there is a net increase in the balance of oxygenated arterial blood to deoxygenated venous blood. Essentially, the change in tissue perfusion exceeds the additional metabolic demand so the concentration of deoxyhemoglobin within tissues decreases. This decrease has a direct effect on the signals used to produce MR images. While blood containing oxyhemoglobin is not very different in terms of its magnetic susceptibility to the rest of tissues or water, it transpires that deoxyhemoglobin is significantly paramagnetic, and thus deoxygenated blood differs substantially in its magnetic properties from surrounding tissues. When oxygen is not bound to hemoglobin, the difference between the magnetic field applied by the MRI machine and that experienced close to a molecule of the blood protein is much greater than when the oxygen is bound. On a microscopic scale, replacing deoxygenated blood by oxygenated blood makes the local magnetic environment more uniform. The longevity of the signals used to produce MR images depends directly on the uniformity of the magnetic field experienced by water molecules—the less uniform the field, the greater the mixture of different signal frequencies that arise from the sample, and therefore the quicker the overall signal decays. The result of having lower levels of deoxyhemoglobin present in blood in a region of brain tissue is therefore that the MRI signal from that region decays less rapidly and so is stronger when it is recorded in a typical MR image acquisition. This small signal increase is the BOLD signal recorded in functional MRI and is typically around 1% or less, although this varies depending on the applied field strength (one of the reasons why higher field MRI systems are being developed). As can be predicted from the aforementioned explanation, the magnitude of the signal depends on the changes in blood flow and volume within tissue, as well as the change in local oxygen tension, so there is no simple relation between the signal change and any single physiological parameter. Furthermore, as neurons become more active, there is a time delay before the necessary vasodilation can occur to increase flow and for the washout of deoxyhemoglobin from the region. Thus the so-called hemodynamic response detected by BOLD has a delay and a duration of several seconds following a stimulating event. BOLD effects also occur in other tissues and may be manipulated, for example, by alterations in the oxygen content of the inspired gas of the animal.

Effects of Spin Motion

Water molecules are in constant motion, and there are various ways in which their movements may affect the signals used to produce MR images. For example, the Brownian motion of molecules is random and gives rise to the self-diffusion of water within an aqueous compartment. In the presence of applied magnetic field gradients, Brownian motion causes the net signal from water to attenuate to a degree that depends on the rate of diffusion. Experimental measurements typically involve a specific timescale over which the effects of diffusion affect the signal: in small animal imaging, this is typically a few milliseconds. MR measurements of diffusion rely on the fact that the net distance that a molecule moves away from a starting position increases with time—in free diffusion, the mean squared distance moved is proportional to time and the self-diffusion coefficient D. If, during the time of the measurement, water that is initially freely diffusing encounters an obstructing boundary (such as a membrane that is not permeable), then its ability to move away from the original position is reduced and thus the *apparent diffusion coefficient* (ADC) is reduced. The manner in which ADC falls below the intrinsic value of D is a measure of the sizes and permeabilities of the cellular compartments in which water resides. Theory and experiments have shown that ADC in the brain is affected by the sizes of the intra- and extracellular water spaces, and it alters when fluid shifts between them, as in ischemia, seizure, and other conditions. Images can be acquired that depict areas of higher ADC as darker than areas of lower ADC, and such images are very sensitive to physiological and pathological perturbations. Moreover, some tissues are markedly anisotropic, and diffusion is faster in some directions than in others. For example, muscle fibers and neuronal tracts in white matter contain structures that permit water to move more freely along their length than orthogonally. The distances between restricting boundaries are much greater along a fiber or tract than across them, so the ADC is anisotropic and is properly described by a tensor. By measuring diffusion in different directions and noting how the tensor in adjacent pixels changes from point to point, it is possible to trace the principal direction of diffusion and thereby assess the direction of the corresponding fibers or tracts.

There are additional methods for sensitizing the MRI signal to other types of spin motion. In particular, a variety of methods have been devised in which the bulk motion of blood within vessels can modulate the signal within an MR image in such a way that flow velocity can be quantified. Others are sensitive to the variation of flow directions and speeds within a small volume element, which is more typical of microvascular perfusion than large vessel laminar flow. An additional and important dynamic effect in images may come from the transient passage of a bolus of paramagnetic contrast material, such as occurs following an intravenous injection. Images taken at frequent intervals can be used to record the transient signal changes caused by the temporary changes in relaxation times that occur as the material passes through the vasculature and tissue. This time course reflects the kinetic behavior of the agent, and appropriate analysis of the concentration–time curve can be used to derive measurements of flow and transit time. In some circumstances, such as occur when capillary epithelium is damaged, the rate of extravasation of the agent from the blood and into the interstitial space can be measured and quantities such as the permeability–surface area product of the vasculature can be derived.

An Illustrative Example: MRI of Exercising Skeletal Muscle

T_2-weighted imaging of exercising skeletal muscle is an excellent example of how biophysical influences on NMR signal can introduce physiological significance into an image. Isolated skeletal muscle has long been a preferred tissue for studying fundamental biophysical NMR properties because it is accessed easily by dissection, has a well-understood physiology, and, in the case of muscles from poikilotherms, can be maintained easily in good physiological condition *ex vivo.* Isolated muscle preparations allow well-controlled manipulations of basic physiological properties (such as temperature, pH, and water content), creation of exercise conditions that validly model the intramuscular changes that occur during exercise *in vivo,* and access to NMR methods that allow the detailed evaluation of the effects of such manipulations on water compartmentation, relaxation times, and diffusive and chemical exchange of magnetization. Within limits, the results of isolated tissue studies can be extended theoretically and experimentally to the more complex *in vivo* condition.

Effects of Exercise on Muscle T_2: Introduction

For almost 40 years, it has been recognized that the T_2 of muscle water increases during and following exercise.[1] Some of the earliest imaging studies reported similar changes in T_1,[2] and in 1988 Fleckenstein *et al.*[3] reported signal intensity increases from exercised muscles in T_2-weighted images (those designed to provide contrast on the basis of T_2 differences between structures). Since that time, investigation of the mechanism of exercise-induced increases in T_2 has led to considerable insight into how physiological variables influence NMR relaxation. In addition, measuring T_2 changes during exercise has been proposed as an indicator of the extent and spatial pattern of neural activation in human studies because the image intensity changes are localized spatially and relate directly to both exercise intensity[4–7] and electromyographic measures of neural activation.[4,8] In order to understand more specifically what these changes may indicate about the physiology of exercising muscle, this Chapter explores four critical concepts: water compartmentation and its evaluation using transverse relaxation and the effects of physiological changes in the intracellular, interstitial, and vascular spaces on NMR relaxation.

Transverse Relaxation Studies of Water Compartmentation

Since the early 1970s, it has become well established that transverse relaxation measurements of muscle water protons can be used to study the compartmentation of muscle water. The basis of such measurements

[1] C. B. Bratton, A. L. Hopkins, and J. W. Weinberg, *Science* **147**, 738 (1945).
[2] J. M. S. Hutchison and F. W. Smith, *in* "Nuclear Magnetic Resonance (NMR) Imaging" (C. L. Partain, A. E. James, F. D. Rollo, and R. R. Price, eds.), p. 231. Saunders, Philadelphia, 1983.
[3] J. L. Fleckenstein, R. C. Canby, R. W. Parkey, and R. M. Peshock, *AJR Am. J. Roentgenol.* **151**, 231 (1988).
[4] M. J. Fisher, R. A. Meyer, G. R. Adams, J. M. Foley, and E. J. Potchen, *Invest. Radiol.* **25**, 480 (1990).
[5] G. R. Adams, M. R. Duvoisin, and G. A. Dudley, *J. Appl. Physiol.* **73**, 1578 (1992).
[6] G. Jenner, J. M. Foley, T. G. Cooper, E. J. Potchen, and R. A. Meyer, *J. Appl. Physiol.* **76**, 2119 (1994).
[7] T. B. Price, R. P. Kennan, and J. C. Gore, *Med. Sci. Sports Exerc.* **30**, 1374 (1998).
[8] T. B. Price, G. Kamen, B. M. Damon, C. A. Knight, B. Applegate, J. C. Gore, K. Eward, and J. F. Signorile, *Magn. Reson. Imaging* **21**, 853–861 (2003).

is that the transverse relaxation of muscle water protons is multiexponential.[9] In studies of isolated frog muscle performed at 25°, Belton et al.[9] identified three components to the proton NMR signal of water: 18% of the signal had a very short T_2 (10 ms), 67% of the signal had an intermediate T_2 ($\approx$40 ms), and the remaining 15% of the signal had a long T_2 ($\approx$170 ms). Many other studies of isolated muscle have since demonstrated similar relaxation times and relative volume fractions.[9–15] Substantial effort has been placed into identifying the origins of these signals.

It is now generally accepted that the source of signals in the short T_2 component is the water bound to macromolecules (so-called hydration water), discussed earlier.[9–12,14,16–18] To demonstrate this, Belton et al.[9] froze the tissue; this caused 80% of the signal to disappear. This signal loss probably corresponded to the freezing of free water; as a solid, ice has a extremely short T_2, rendering it unobservable in typical data acquisitions. The remaining 20% did not freeze, even at temperatures as low as −30°. Because water does not freeze when it is absorbed to surfaces (such as proteins or phospholipids), Belton et al.[9] speculated that the short T_2 component is composed of macromolecular hydration water. This water has a short T_2 because the macromolecules in skeletal muscle undergo slow, anisotropic motion; as discussed earlier, low-frequency molecular motion results in static magnetic field inhomogeneities and causes transverse relaxation very effectively. Water bound to the macromolecule adopts this motion and so relaxes efficiently as well.

Likewise, a consensus has emerged concerning the other T_2 components, with most authors accepting an anatomical compartmentation theory in which the intermediate and long components represent intracellular and interstitial water, respectively. This was originally proposed by Belton et al.,[9] who based this conclusion on the similarity between the relative volume fractions of the intermediate and long T_2 components and the

[9] P. S. Belton, R. R. Jackson, and K. J. Packer, *Biochim. Biophys. Acta* **286**, 16 (1972).

[10] P. S. Belton and K. J. Packer, *Biochim. Biophys. Acta* **354**, 305 (1974).

[11] C. F. Hazlewood, D. C. Chang, B. L. Nichols, and D. E. Woessner, *Biophys. J.* **14**, 583 (1974).

[12] B. M. Fung and P. S. Puon, *Biophys. J.* **33**, 27 (1981).

[13] J. F. Polak, F. A. Jolesz, and D. F. Adams, *Invest. Radiol.* **23**, 107 (1988).

[14] W. C. Cole, A. D. LeBlanc, and S. G. Jhingran, *Magn. Reson. Med.* **29**, 19 (1993).

[15] B. M. Damon, A. S. Freyer, and J. C. Gore, "Proceedings of the International Society for Magnetic Resonance in Medicine 10th Scientific Meeting and Exhibition," p. 619, 2002.

[16] G. Saab, R. T. Thompson, and G. D. Marsh, *Magn. Reson. Med.* **42**, 150 (1999).

[17] G. Saab, R. T. Thompson, and G. D. Marsh, *J. Appl. Physiol.* **88**, 226 (2000).

[18] G. Saab, R. T. Thompson, G. D. Marsh, P. A. Picot, and G. R. Moran, *Magn. Reson. Med.* **46**, 1093 (2001).

space determinations made using inulin and other sugars.[19] Direct experimental support for this proposition was provided in subsequent studies. Belton and Packer[10] showed that dessicating the muscle progressively caused first the loss of signal from the putative interstitial T_2 component and next from the putative intracellular component. Finally, the bound water component shrank in absolute size over time, but did not completely dehydrate. Overall, these data are consistent with a unidirectional flow of water from the "innermost" to "outermost" of three compartments connected in series to the atmosphere (intracellular-bound water → free intracellular water → interstitial water → atmosphere). The volume fractions in multicomponent T_2 analysis also correspond to the relative compartment sizes implied by the analysis of sodium and chloride ion efflux curves.[20] In addition, Cole et al.[14] showed that tissue maceration eliminates T_2 compartmentation. The reasons for the different T_2 values for the intracellular and interstitial spaces (~40 vs ~150 ms) are considered later.

It is noteworthy that other models of anatomical compartmentation have also been proposed. First, Le Rumeur et al.,[21] Noseworthy et al.,[22] and Stainsby and Wright[23] each proposed that intermediate T_2 component represents all of the tissue parenchyma (i.e., intracellular and interstitial water together), whereas the long T_2 component represents blood. The evidence for these conclusions is that changes in venous volume affect the apparent volume fraction of the long component and that changes in blood oxygen content affect its T_2. Conversely, in vivo transverse relaxation data reported by Saab and colleagues[16–18] obtained using high signal-to-noise ratio acquisition methods suggest the existence of two water compartments with T_2 values greater than 100 ms, which they proposed to be interstitial and vascular water; these were resolvable from the intracellular water components. In attempting to reconcile these observations with each other and the compartmentation model presented earlier, we note that the volume fraction associated with the long T_2 component was as much as 10% in the studies by Le Rumeur et al.,[21] Noseworthy et al.,[22] and Stainsby and Wright.[23] This is much larger than the capillary volume fraction (~1.5%) calculated from optical measurements[24,25] or the total vascular volume

[19] R. O. Law and C. F. Phelps, J. Physiol. **186**, 547 (1966).

[20] M. C. Neville and S. White, J. Physiol. **288**, 71 (1979).

[21] E. Le Rumeur, J. De Certaines, P. Toulouse, and P. Rochcongar, Magn. Reson. Imaging **5**, 267 (1987).

[22] M. D. Noseworthy, J. K. Kim, J. A. Stainsby, G. J. Stanisz, and G. A. Wright, Magn. Reson. Imaging **9**, 814 (1999).

[23] J. A. Stainsby and G. A. Wright, Magn. Reson. Med. **45**, 662 (2001).

[24] C. A. Kindig, W. L. Sexton, M. R. Fedde, and D. C. Poole, Respir. Physiol. **111**, 163 (1998).

[25] M. M. Porter, S. Stuart, M. Boij, and J. Lexell, J. Appl. Physiol. **92**, 1451 (2002).

fraction (2 to 4%) implied by application of the intravoxel incoherent motion model of Le Bihan *et al.*[26] to skeletal muscle by Morvan.[27] The inclusion of all of the tissue parenchyma into a single relaxation component is also not consistent with work by Landis *et al.*,[28] indicating an *in vivo* intracellular residence time for skeletal muscle water of 1.1 s. This result indicates that the trans-sarcolemmal water exchange is sufficiently slow that distinct intracellular and interstitial T_2 components *must* exist. It is therefore possible that in attempting the difficult task of making multiexponential transverse relaxation measurements in skeletal muscle *in vivo,* the methods used by Wright and colleagues provided too few data points (48) and/or too low of a signal-to-noise ratio (670) to distinguish the interstitial and vascular components, and the long T_2 component that they measured in fact represented the entire extracellular space (interstitial + vascular water). We also note that Saab *et al.*[16–18] have suggested that there may in fact be an additional intracellular T_2 component, but a definitive identification of such a water compartment has not been demonstrated experimentally.

We will therefore continue with a model for transverse relaxation of muscle water that includes four components: water bound to macromolecules (which is not typically observed in most transverse relaxation measurements), free intracellular water, free interstitial water, and vascular water. In order to discuss them unambiguously, we will refer to them as $T_{2,Bound}$, $T_{2,Intra}$, $T_{2,Inter}$, and $T_{2,Blood}$. As shown later, the latter three are subject to considerable variation as a result of normal physiological changes in the muscle. This has taught us much about the biological influences on NMR relaxation and how in the future we might use NMR relaxation measurements to make inferences about the underlying tissue biology.

Physiological Influences on NMR Relaxation in the Intracellular Space

The two variables that have the most important effects on NMR relaxation in the intracellular space of skeletal muscle under physiological conditions are free intracellular water content and intracellular pH. The most likely mechanism through which free intracellular water content effects T_1 and T_2 is through a dilution of the intracellular protein content. As discussed previously, the protein concentration is among the most important influences on relaxation times in all cells. This is particularly true in muscle

[26] D. Le Bihan, E. Breton, D. Lallemand, M. L. Aubin, J. Vignaud, and M. Laval-Jeantet, *Radiology* **168,** 497 (1988).

[27] D. Morvan, *Magn. Reson. Imaging* **13,** 193 (1995).

[28] C. S. Landis, X. Li, F. W. Telang, P. E. Molina, I. Palyka, G. Vetek, and C. S. Springer, *Magn. Reson. Med.* **42,** 467 (1999).

because it has a very high protein content ($\sim$20% by weight[29]) and in order to function appropriately, the contractile proteins are oriented parallel to the long axis of the cell at all times. As discussed earlier, the slow anisotropic motion of a large amount of cellular protein leads to very effective transverse relaxation of both the protein itself and the water bound to it. Because rapid exchange exists between free intracellular water and hydration water and/or between free intracellular water and titratable protons on the protein,[11,30] relaxation of the free intracellular water is also affected. This shortens $T_{2,Intra}$ considerably relative to $T_{1,Intra}$ ($T_{2,Intra} \approx$ 40 ms; $T_{1,Intra} \approx 1000$ ms; aforementioned discussion and Landis et al.[28]). Consistent with this high protein content, there is also a significant magnetization transfer effect in muscle[31-33] that originates primarily from the intracellular space.[34]

A number of studies have demonstrated outright the effect of free intracellular water content on $T_{2,Intra}$. First, although not the principal purpose of their study, the tissue dessication studies of Belton and Packer[10] demonstrated the importance of water content in determining the $T_{2,Intra}$. As the time of dessication progressed and the absolute volume of water in the intracellular space decreased, the $T_{2,Intra}$ did as well. The results are plotted in Fig. 1. Note that data have been plotted as the transverse relaxation rate (R_2) because concentration effects are expected to be linear with R_2 but not T_2. For all conditions in which dessication did not affect the amount of macromolecular hydration water (intracellular volume $\geq$8% of initial volume), the relationship between $R_{2,Intra}$ and intracellular volume is highly linear. Damon et al.[15] tested this hypothesis explicitly as well by manipulating the osmolarity of the bathing solution. By modifying the Ringer's solution around frog sartorius muscles to be hypotonic, we caused water to enter cells and $T_{2,Intra}$ to increase; by adding sucrose to the solution to make it hypertonic, we caused water to leave the cell and decrease $T_{2,Intra}$. Figure 1 includes data from this study. The slope of the relationship is greater in our study than in the Belton and Packer study, probably because of the different magnetic field strengths employed; this point is expanded upon later.

The importance of osmotically induced water shifts in determining muscle T_2 has been shown in in vivo imaging studies as well. Among the

[29] D. R. Wilkie, "Muscle." Edward Arnold, London, 1976.

[30] M. M. Civan, A. M. Achlama, and M. Shporer, Biophys. J. 21, 127 (1978).

[31] R. S. Balaban and T. L. Ceckler, Magn. Reson. Q. 8, 116 (1992).

[32] X. P. Zhu, S. Zhao, and I. Isherwood, Br. J. Radiol. 65, 39 (1992).

[33] H. Yoshioka, H. Takahashi, H. Onaya, I. Anno, M. Niitsu, and Y. Itai, Magn. Reson. Imaging 12, 991 (1994).

[34] R. Harrison, M. J. Bronskill, and R. M. Henkelman, Magn. Reson. Med. 33, 490 (1995).

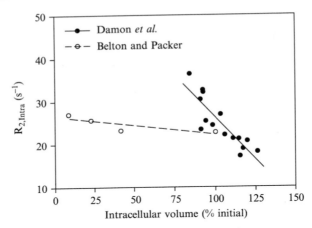

FIG. 1. Relationship between intracellular volume changes and $R_{2,Intra}$ in skeletal muscle. Data shown have been replotted from studies by Damon et al.[37] (filled circles and solid line) and from Belton and Packer[10] (open circles and dashed line). The more robust relationship reported by Damon et al.[37] may result from a greater magnetic field strength used in this study.

most creative demonstrations of this was a study performed by Meyer et al.[35] in which muscle contractions were stimulated electrically in the tail muscles of lobsters (which are osmoconformers with an intracellular osmolarity of ~1 Osm) and freshwater crayfish (which are osmoregulators with an intracellular osmolarity of 340 mOsm). Thus an accumulation of osmolytes during exercise would be expected to have an approximately threefold smaller effect in osmoconformers than in osmoregulators, and the T_2 change should be smaller as a result. This is exactly what was observed.

Several studies have also shown that intracellular pH may be an important influence on $T_{2,Intra}$ as well. In isolated strips of rabbit psoas muscle permeabilized by incubation with glycerin, Fung and Puon[12] showed that decreases in intracellular pH cause $T_{2,Intra}$ to decrease. This has also been shown in intact isolated frog sartorius muscles made acidic by exposure to NH_4^+ and subsequent washout of NH_3.[15] The results of both studies are shown in Fig. 2, again plotting the relaxation data as $R_{2,Intra}$. Again, a more robust relationship is observed at the higher magnetic field strength used in the Damon et al.[15] study (7.05 Tesla) than in the Fung and Puon[12] study (~0.5 Tesla). Conversely, other studies[35,36] have provided evidence that

[35] R. A. Meyer, B. M. Prior, R. I. Siles, and R. W. Wiseman, *NMR Biomed.* **14**, 199 (2001).
[36] B. M. Prior, L. L. Ploutz-Snyder, T. G. Cooper, and R. A. Meyer, *J. Appl. Physiol.* **90**, 615 (2001).

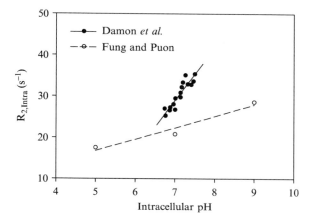

FIG. 2. Relationship between intracellular pH changes and $R_{2,Intra}$ in skeletal muscle. Data shown have been replotted from studies by Damon et al.[37] (filled circles and solid line) and from Fung and Puon[12] (open circles and dashed line). The more robust relationship reported by Damon et al.[15] may result from a greater magnetic field strength used in this study.

intracellular pH changes do not influence the whole muscle T_2, as measured in images.

The field strength dependences of $T_{2,Intra}$ on intracellular pH and volume are important clues to the specific relaxation mechanism that is active, as well as to why these studies have reported different conclusion concerning the role of intracellular pH in helping determine the $T_{2,Intra}$. A field strength dependence of T_2 could, in principle, result from either chemical exchange between sites that differ in Larmor frequency or diffusion through magnetic field gradients. Diffusion though intracellular magnetic field gradients is unlikely to depend on pH changes and has also been ruled experimentally.[15] In addition, Damon et al.[37] and others[38] have provided direct experimental evidence that chemical exchange between sites differing in Larmor frequency is important to at least the pH effect. Because the quantitative importance of exchange processes on $T_{2,Intra}$ lessens when the transverse relaxation decay is sampled at low frequencies, such as those used during imaging,[39] this may explain the absent or very small effects of pH on $T_{2,Intra}$ reported by some investigators[35,36] It is also likely that the effect of intracellular volume changes also acts through the

[37] B. M. Damon, C. D. Gregory, K. L. Hall, H. J. Stark, V. Gulani, and M. J. Dawson, *Magn. Reson. Med.* **47,** 14 (2002).

[38] E. Moser, E. Winklmayr, P. Holzmuller, and M. Krssak, *Magn. Reson. Imaging* **13,** 429 (1995).

[39] Z. Luz and S. Meiboom, *J. Chem. Phys.* **39,** 366 (1963).

chemical exchange mechanism, as changing the relative amount of free and macromolecular hydration water or exchangeable protons would affect the kinetics of the exchange. This mechanism would therefore also explain the field strength dependence of the effect of intracellular volume on $T_{2,Intra}$ depicted in Fig. 1.

A working interpretation of the aforementioned results is that both intracellular free water content and intracellular pH affect the $T_{2,Intra}$, but in each case the actual quantitative importance depends on magnetic field strength and data acquisition conditions. Given this, one would anticipate that glycolysis would be a major contributor to T_2 changes during exercise because it results in both a substantial accumulation of osmolytes (lactate and sugar phosphates) and a severe acidosis (up to one pH unit). The importance of glycolysis to the T_2 change was first demonstrated by Fleckenstein et al.[40] in their study of myophosphorylase deficiency patients, who do not undergo a T_2 change following exercise. In principle, the lack of T_2 change in these patients could also relate to an absence of glycogen metabolism (glycogen binds ~4 g of water for each 1 g of glucose, and so the release of this water with glycogenolysis might be expected to increase $T_{2,Intra}$). However, Price and Gore[41] showed that glycogen depletion affects neither the resting T_2 nor the T_2 change of exercise. In subsequent studies of isolated frog muscle, the principal reason for the smaller change in $T_{2,Intra}$ in muscles poisoned with iodoacetic acid (a glycolytic inhibitor) than in those poisoned with NaCN (an oxidative phosphorylation inhibitor) was the greater lactate and hydrogen ion accumulation in the latter metabolic condition. Thus, at exercise durations long enough for glycolysis to be activated significantly, the primary cause of the increase in the $T_{2,Intra}$ appears to be the end products of anaerobic metabolism.

Physiological Influences on NMR Relaxation in the Interstitial Space

As discussed earlier, $T_{2,Inter}$ is lengthened relative to $T_{2,Intra}$; this and a relatively slow exchange of water across the sarcolemma allow the resolution of these two T_2 components using appropriate data acquisition and analysis methods. Both the relatively longer value for T_2 and the finding that there is less magnetization transfer in this T_2 component than in the intracellular space are consistent with the lower protein concentration in the interstitium.[42] It is known that $T_{2,Inter}$ increases during exercise

[40] J. L. Fleckenstein, R. G. Haller, S. F. Lewis, B. T. Archer, B. R. Barker, J. Payne, R. W. Parkey, and R. M. Peshock, *J. Appl. Physiol.* **71**, 961 (1991).
[41] T. B. Price and J. C. Gore, *J. Appl. Physiol.* **84**, 1178 (1998).
[42] C. N. Karatzas and C. G. Zarkadas, *Poult. Sci.* **68**, 811 (1989).

ex vivo[15]; this may be due to osmotically induced water entry into the interstitium secondary to lactate accumulation.[43] Apart from this, there has been little investigation of the physiological influences on $T_{2,Inter}$.

During exercise conditions, the positive contribution of changes in the interstitial space to T_2-weighted signal intensity is likely to be quite small because the its low volume fraction. In fact, the application of negative pressure to the leg increases the appearance of the interstitial T_2 component in multiexponential T_2 analyses dramatically, but causes only a modest increase in signal intensity in T_2-weighted images.[44] However, after severe exercise in which muscles lengthen under load (so-called "eccentric" contractions), T_2 can remain elevated for almost 2 months.[45,46] These latter changes appear to reflect increases in interstitial volume brought about by muscle damage.

Physiological Influences on NMR Relaxation in the Vascular Space

As revealed by near-infrared spectroscopy studies and a long history of arterial and venous oxygen content measurements, blood oxygen extraction increases during exercise. There are two possible ways through which blood oxygenation changes could affect image signal intensity. The first is through an effect on the transverse relaxation of the blood itself (the intravascular BOLD effect). A number of authors[47-49] have shown that the $T_{2,Blood}$ decreases with decreasing values of oxyhemoglobin saturation. This results from a change in the magnetic susceptibility of the hemoglobin molecule when oxygen is released[50]; the difference in magnetic susceptibility between water in the red cell and in the plasma, which exchange rapidly, accelerates transverse relaxation. As the transcapillary exchange is slow in skeletal muscle capillaries, a decrease in $T_{2,Blood}$, such as that caused by increased oxygen extraction, would attenuate the signal in a T_2-weighted image by a factor directly proportional to the blood volume fraction.

[43] G. Sjogaard and B. Saltin, *Am. J. Physiol.* **243**, R271 (1982).

[44] L. L. Ploutz-Snyder, S. Nyren, T. G. Cooper, E. J. Potchen, and R. A. Meyer, *Magn. Reson. Med.* **37**, 676 (1997).

[45] K. Nosaka and P. M. Clarkson, *Med. Sci. Sports Exerc.* **28**, 953 (1996).

[46] J. M. Foley, R. C. Jayaraman, B. M. Prior, J. M. Pivarnik, and R. A. Meyer, *J. Appl. Physiol.* **87**, 2311 (1999).

[47] K. R. Thulborn, J. C. Waterton, P. M. Matthews, and G. K. Radda, *Biochim. Biophys. Acta* **714**, 265 (1982).

[48] M. E. Meyer, O. Yu, B. Eclancher, D. Grucker, and J. Chambron, *Magn. Reson. Med.* **34**, 234 (1995).

[49] W. M. Spees, D. A. Yablonskiy, M. C. Oswood, and J. J. Ackerman, *Magn. Reson. Med.* **45**, 533 (2001).

[50] L. Pauling and C. Coryell, *Proc. Natl. Acad. Sci. USA* **19**, 349 (1936).

Second, the mismatch in magnetic susceptibility between blood and tissue parenchyma caused by increased oxygen extraction would promote the dephasing of spins in the tissue parenchyma (the extravascular BOLD effect). As discussed previously, a number of authors have investigated this phenomenon theoretically, including our own group.[51,52] The magnitude of the magnetic susceptibility difference between vasculature and tissue parenchyma depends on a number of tissue physiological and architectural parameters, including capillary size, spacing, and geometry; the hematocrit of the blood; oxyhemoglobin saturation; and the diffusion coefficient for water.

Using reasonable values for each of these parameters (listed in Table I), we can estimate the extravascular BOLD effect in three muscles: the biceps brachii, a fusiform muscle in which the fibers and capillaries would be oriented essentially parallel to the magnetic field in typical imaging experiments; the anterior tibialis, in which the fiber and capillary orientations would be slightly oblique to the field; and the soleus, in which the fibers and capillary orientation would be slightly oblique to the field; and the soleus, in which the fibers and capillary orientation would be significantly different from the magnetic field. In each case, the relative blood volumes

TABLE I

MODEL PARAMETERS USED IN PREDICTING EXTRAVASCULAR BOLD EFFECTS IN SKELETAL MUSCLE (FOOTNOTES GIVE THE REFERENCES USED IN ASSUMING THESE VALUES)

	Muscle		
Parameter	Soleus	Anterior tibialis	Biceps brachii
Blood volume[a]	1.5%	1.5%	1.5%
Capillary diameter[b]	5.4 μm	5.4 μm	5.4 μm
Capillary orientation[c]	40°	15°	0°
Transverse diffusion coefficient[d]	1.0×10^{-5} cm$^2 \cdot$ s^{-1}	1.0×10^{-5} cm$^2 \cdot$ s^{-1}	1.0×10^{-5} cm$^2 \cdot$ s^{-1}
Capillary hematocrit[e]	0.22	0.22	0.22

[a] Kindig et al.[24] and Porter et al.[25]
[b] Kindig et al.[24] and Porter et al.[25]
[c] Y. Kawakami, Y. Ichinose, and T. Fukunaga, J. Appl. Physiol. **85**, 398 (1998) and C. N. Maganaris and V. Baltzopoulos, Eur. J. Appl. Physiol. Occup. Physiol. **79**, 294 (1999).
[d] B. M. Damon, Z. Ding, A. W. Anderson, A. S. Freyer, and J. C. Gore, Magn. Reson. Med. **48**, 97 (2002).
[e] Kindig et al.[24]

[51] R. P. Kennan, J. Zhong, and J. C. Gore, Magn. Reson. Med. **31**, 9 (1994).
[52] L. A. Stables, R. P. Kennan, and J. C. Gore, Magn. Reson. Med. **40**, 432 (1998).

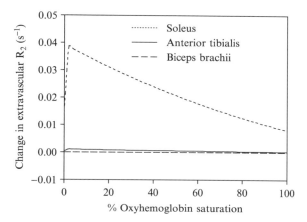

FIG. 3. Predicted contribution to extravascular R_2 due to changes in blood oxygenation. Data are based on theoretical models of Kennan *et al.*[51] and Stables *et al.*[52] and are shown for the soleus, anterior tibialis, and biceps brachii muscles. Model parameters are given in Table I.

can be assumed to be similar because each muscle is composed of predominantly slow, oxidative fibers. The results of these calculations are shown in Fig. 3. For the biceps brachii, the predominant orientation of the capillaries parallel to the field causes us to predict that there would be no extravascular effect of blood oxygenation changes. Conversely, for the soleus muscle, in which the fibers and capillaries are oriented ~40° to the field, there is a strong extravascular effect that is predicted. The effect diminishes in magnitude at low oxygen tensions because myoglobin, an oxygen-binding protein in muscle cells, abruptly deoxygenates at ~4 mmHg, causing a better magnetic susceptibility match between vasculature and tissue parenchyma. For the anterior tibialis, an intermediate effect is predicted. Clearly, full interpretation of these effects *in vivo* requires an understanding of tissue architecture.

Conclusion: How Can We Use Transverse Relaxation Measurements to Learn about Exercising Muscle?

An overall conclusion to be drawn from the aforementioned discussion is that changes in T_2-weighted signal intensity in MR images of exercising skeletal muscle primarily reflect the metabolic and hemodynamic responses to exercise. Moreover, it is likely that the relative contributions of these events change as functions of exercise during and intensity. Because a sufficient theoretical understanding of the exact contributions of different physiological and biochemical responses to exercise does not yet exist, specific conclusions about the intensity of the metabolic or hemodynamic responses (or the neural output to the muscle that is ultimately responsible for them) cannot be made. However, end-exercise T_2 measurements can certainly be used in human studies to detect muscles that have been activated during a given exercise task.

For small animal MRI experiments, there is not yet an overly compelling reason to measure T_2 changes during stimulated exercise tasks. While these changes could also be used to detect the spatial pattern of muscle activation during stimulated exercise protocols, it is likely that the stimulation pattern would already be known on the basis of the placement of stimulating electrodes over a muscle (in cases of direct muscle stimulation) or by the anatomical distribution of the motor nerve (in cases of indirect muscle stimulation via the nerve). In addition, many of the fundamental contributions to T_2-weighted signal intensity, such as metabolite accumulation, pH, and blood flow, can be measured more directly and with similar time resolution through other means.

Why, then, might we measure T_2 in skeletal muscles during small animal experiments? The answer to this question lies in the prediction that by developing a comprehensive understanding of the biophysical influences on NMR relaxation in muscle, we will be able to image parameters not otherwise accessible through noninvasive methods. For example, changes in capillary geometry, distribution, and size that occur in diabetes[24] would be expected to influence the magnitude of the extravascular BOLD effect (and therefore the ratio of R_2 to R_2^*). Such measurements might therefore serve as a marker of capillary morphologic changes in animal models of this disease. Additionally, measurements of sarcolemma water permeability, as well as an understanding of how this parameter affects the overall T_2 of the muscle, would be useful in experimental models of muscular dystrophy and muscle damage. Such measurements will require continued efforts to develop a comprehensive understanding of the biological influences on NMR relaxation in skeletal muscle.

[3] Nuclear Magnetic Resonance in Laboratory Animals

By A. Heerschap, M. G. Sommers, H. J. A. in 't Zandt,
W. K. J. Renema, A. A. Veltien, and D. W. J. Klomp

Introduction

Nuclear magnetic resonance (NMR) is a very versatile scientific and diagnostic tool. After the discovery of the NMR phenomenon in 1946 by Bloch and Purcell,[1,2] it has proven useful in physics, chemistry, biochemistry, and biomedicine. Nowadays its most widespread application is in medical diagnosis under the name magnetic resonance (MR) imaging, based on conceptions for which Lauterbur and Mansfield received the 2003 Nobel prize in Physiology or Medicine.

After some early attempts, NMR of live intact animals started to be explored more widely in the 1970s.[3–5] Since then the potentials of NMR to study anatomy, physiology, and biochemistry *in vivo* in animals have been utilized in numerous studies. NMR machines designed for animal investigations are found in major biomedical research institutions and are becoming a standard research instrument at most advanced medical faculties. Although NMR has been applied to a wide range of animals, including pinnipeds,[6] birds,[7] sheep,[8] and monkeys,[9] among others, by far the majority of animals undergoing NMR examinations are rats and mice as these serve as the main model systems in biomedical research, mostly for practical reasons. Interest in NMR studies of mice has increased substantially due

---

[1] F. Bloch, *in* "Encyclopedia of Nuclear Magnetic Resonance" (D. M. Grant and R. K. Harris, eds.), Vol. 1, p. 215. Wiley, New York, 1996.

[2] E. M. Purcell, *in* "Encyclopedia of Nuclear Magnetic Resonance" (D. M. Grant and R. K. Harris, eds.), Vol. 1, p. 551. Wiley, New York, 1996.

[3] P. C. Lauterbur, *Pure Appl. Chem.* **40,** 149 (1974).

[4] W. S. Hinshaw, E. R. Andrew, P. A. Bottomley, G. N. Holland, and W. S. Moore, *Br. J. Radiol.* **51,** 273 (1978).

[5] J. J. H. Ackerman, T. H. Grove, G. G. Wong, D. G. Gadian, and G. K. Radda, *Nature* **283,** 167 (1980).

[6] P. W. Hochachka, *Comp. Biochem. Physiol. A Mol. Integr. Physiol.* **126,** 435 (2000).

[7] A. Van der Linden, M. Verhoye, V. van Meir, I. Tindemans, M. Eens, P. Absil, and J. Balthazart, *Neuroscience* **112,** 467 (2002).

[8] A. M. Van Cappellen, G. P. van Walsum, M. Rijpkema, A. Heerschap, B. Oeseburg, J. G. Nijhuis, and H. W. Jongsma, *Pediatr. Res.* **54,** 747 (2003).

[9] N. K. Logothetis, J. Pauls, M. Augath, T. Trinath, and A. Oeltermann, *Nature* **412,** 150 (2001).

to the progress in the rapid generation of transgenic mice and mice modified by random mutation, which provide powerful models in studies of *in vivo* protein functions and as models of human disease. By these new approaches, large numbers of specifically modified mice have become available and efficient phenotyping has become a real research bottleneck. Because noninvasive imaging is attractive for mice phenotyping, dedicated mouse imaging centers are currently being established. NMR is a main modality in these centers because of excellent noninvasive imaging possibilities[10] and the potential for parallel mice examinations with relatively high throughput.[11] These developments, together with continuing progress in diversification and innovation of MR methods and increased sensitivity due to better hardware and higher-field magnets, have given a strong impetus to NMR of small laboratory animals. Although no specific textbooks on animal NMR have been published, some excellent introductions to *in vivo* NMR for scientific purposes, including animal studies, have become available.[12,13]

It is beyond the scope of this Chapter to give a comprehensive overview of all NMR studies in laboratory animals, but rather we focus on more practical aspects of *in vivo* NMR of animals (i.e., practicalities and optimization of *in vivo* NMR spectroscopy of the mouse with some illustrative examples and anesthesiology and physiological monitoring of animals during NMR experiments). This is preceded by a brief description of the basics and available methods of NMR as far as relevant for animal applications.

Nuclear Magnetic Resonance Basics

Some atomic nuclei possess a magnetic property that can be detected in an NMR experiment. The most important of these nuclei for biological applications are ^1H (hydrogen or protons), which occur in almost all biological compounds; ^{31}P (phosphorus); present in compounds such as ATP; and ^{13}C (carbon), which is also present in most body compounds. Because of their magnetic property, these "NMR-sensitive" nuclei can be regarded as tiny bar magnets. Outside a magnetic field these "bar magnets," called spins in NMR terminology, have a random orientation,

[10] G. A. Johnson, G. P. Cofer, S. L. Gewalt, and L. W. Hedlund, *Radiology* **222,** 789 (2002).
[11] N. A. Bock, N. B. Konyer, and R. M. Henkelman, *Magn. Reson. Med.* **49,** 158 (2003).
[12] D. G. Gadian, "Nuclear Magnetic Resonance and Its Applications to Living Systems." Oxford Univ. Press, New York, 1995.
[13] R. J. Gillies (ed.), "NMR in Physiology and Biomedicine." Academic Press, San Diego, 1994.

but when placed in a strong magnetic field they align with the direction of that field or are opposed to it. Applying an oscillating magnetic field can bring about transitions between these two orientations. For this purpose, a radiofrequency (RF) coil is used fitting the whole animal or wrapped around a specific part of the animal. After application of the oscillating field, which is only turned on for a short moment (RF pulse), the "tiny bar magnets" return to their original orientations. During this period, energy is emitted that induces a small current in the radiofrequency coil, which is detected by the NMR spectrometer. The frequency, ω, of this current is proportional to the strength of the main magnetic field B_0 and the type of nucleus (represented by its gyromagnetic ratio γ) according to the Larmor equation: $\omega = \gamma \times B_0$. At the field strength of 7 Tesla, which at present is used commonly for small animal studies, the resonance frequency for ^1H is about 300 MHz. The exact resonance frequency is also sensitive to the particular chemical environment of the nucleus. ^1H nuclei in water have an equivalent environment and produce a single signal. Other compounds with NMR nuclei in different environments, such as lactate with methyl and methylene protons and ATP with three chemically different phosphate groups, generate signals at distinctly different frequencies, the so-called chemical shift. Together with other specific NMR spectral characteristics, this provides a fingerprint for a compound that is used for its spectral identification. The intensity of an NMR signal of a compound is proportional to its amount present in the volume seen by the NMR coil.

The main hardware components of an *in vivo* NMR system are the magnet with adjustable field gradients, radiofrequency coils, and spectrometer (Fig. 1).

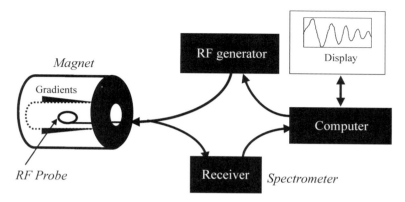

FIG. 1. Schematic overview of the main elements of an NMR system.

In Vivo NMR Imaging and Spectroscopy of Animals

While it may be useful to apply volume RF coils to image whole animals, it is more common to use local RF coils or multiple coil arrays to select particular body parts or tissues, as this allows for better sensitivity and spatial resolution. The variable distribution of NMR signals over these body parts can be scanned to create an NMR image by imposing one or more orthogonal linear field gradients on the animal while exciting nuclear spins with RF pulses. After collection of data with a variety of gradient fields, deconvolution yields a one-, two-, or three-dimensional image of that part of the animal located within the sensitive area of the RF coil. Typically, the image is formed on the basis of the NMR signal from protons of water. Thanks to the abundant presence of water in living soft tissue, an image with high spatial resolution can be constructed. The signal intensity in a given volume is a function of the water concentration and spin relaxation times (T1 and T2). Local variations in these three parameters provide the vivid anatomical contrast observed in NMR images. In small living animals, a spatial resolution of about 100 μM can be routinely achieved in brain studies within acceptable measurement times.

Next to detailed morphological information, NMR may also uncover more functional aspects of the animal body. These applications of NMR are rapidly growing and are often presented under umbrella names such as functional, molecular, or cellular NMR. An important characteristic of NMR is its sensitivity to water movement, which can be exploited in various ways. For instance, MR angiograms can be obtained.[14,15] An example of our own work on mouse brain is shown in Fig. 2. Diffusion and perfusion processes can also be monitored.[13,16] With so-called diffusion tensor imaging it is possible to follow axonal tracts.[17] The transient binding of water to macromolecules is exploited in magnetization transfer experiments.[18] Furthermore, the water signal may be affected by "magnetic active" compounds, of which the clinically used lanthanide-based NMR contrast agent Gd-DTPA is best known. When this compound is administered intravenously to the animal it only has an effect on the water signal at locations where it passes by and thus acts as a contrast agent. This property can be

---

[14] W. R. Bauer, K. Hiller, F. Roder, S. Neubauer, A. Fuchs, C. Grosse Boes, R. Lutz, P. Gaudron, K. Hu, A. Haase, and G. Ertl, *Circulation* **92**, 968 (1995).

[15] N. Beckmann, *Magn. Reson. Med.* **44**, 252 (2000).

[16] F. A. Van Dorsten, R. Hata, K. Maeda, C. Franke, M. Eis, K. A. Hossmann, and M. Hoehn, *NMR Biomed.* **12**, 525 (1999).

[17] P. Van Zijl and D. le Bihan, *Special Issue NMR Biomed.* **15**, 431 (2002).

[18] R. M. Henkelman, G. J. Stanisz, and S. J. Graham, *NMR Biomed.* **14**, 57 (2001).

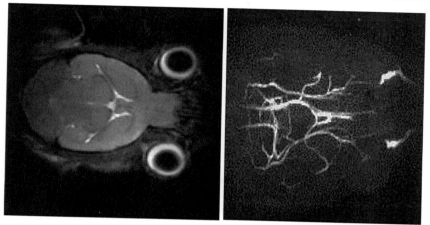

FIG. 2. NMR images of the mouse brain. (Left) Transversal T2-weighted NMR image. (Right) NMR angiogram of the same mouse showing major blood vessels. This image was acquired with a three-dimensional time-of-flight method using a Tr of 22 ms.

employed to study vascularity and vessel wall permeability such as in tumors.[19] Other particles can also be used to enhance specific contrasts such as manganese for neuronal tracts[7,20] or iron in ultrasmall particles of iron oxide for vasculature.[21,22] An example of the latter application is shown in Fig. 3. In fact, the production of new smart NMR contrast agents is a very active field of research with great promises (e.g., to image gene expression) that is explored in many animal studies.[23,24] Another exciting development is the possibility of following cell migration after labeling of the cells with contrast agents.[25]

An endogenous contrast agent is provided by deoxyhemoglobin, which affects the intensity of the water signal, using particular MR image recordings. When it is replaced by oxyhemoglobin, the intensity of the water signal changes in these MR images, which can be visualized in difference

[19] B. P. van der Sanden, T. H. Rozijn, P. F. Rijken, H. P. Peters, A. Heerschap, A. J. van der Kogel, and W. M. Bovee, *J. Cereb. Blood Flow Metab.* **20,** 861 (2000).

[20] Y. J. Lin and A. P. Koretsky, *Magn. Reson. Med.* **38,** 378 (1997).

[21] W. Leenders, B. Kusters, J. Pikkemaat, P. Wesseling, D. Ruiter, A. Heerschap, J. Barentsz, and R. M. de Wall, *Int. J. Cancer* **105,** 437 (2003).

[22] S. P. Robinson, P. F. Rijken, F. A. Howe, P. M. McSheehy, B. P. van der Sanden, A. Heerschap, M. Stubbs, A. J. van der Kogel, and J. R. Griffiths, *J. Magn. Reson. Imag.* **17,** 445 (2003).

[23] T. J. Meade, A. K. Taylor, and S. R. Bull, *Curr. Opin. Neurobiol.* **13,** 1 (2003).

[24] R. Weissleder, A. Moore, U. Mahmood, R. Bhorade, H. Benveniste, E. A. Chiocca, and J. P. Basilion, *Nat. Med.* **6,** 351 (2000).

[25] J. W. Bulte, I. D. Duncan, and J. A. Frank, *J. Cereb. Blood Flow Metab.* **22,** 899 (2002).

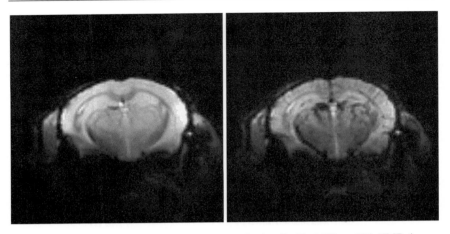

FIG. 3. Coronal NMR images of the mouse brain. (Left) A256 × 256 NMR image; Te = 15 ms, Tr = 3500 ms, 0.5-mm slice and FOV 40 mm. (Right) After the application of ultrasmall particles of iron oxide. Note the specific decrease in signal intensity, reflecting well-vascularized regions of the brain.

images. In this way, changes in local blood level oxygen-dependent (BOLD) contrast can be viewed (e.g., at activated brain locations),[26] which has become known as functional MRI (Fig. 4).

NMR or MR imaging has to do with the spatial distribution of the strong ^{1}H signals of body water and fat. When NMR is performed on other compounds by ^{1}H NMR, ^{31}P NMR, or ^{13}C NMR, one usually speaks of NMR spectroscopy. As these compounds are much less abundant than water and fat, the associated signals are much less intense. The sensitivity of the NMR experiment *in vivo* requires that compounds are present at concentrations of about 0.1 mM or more. Furthermore, they usually also have to be small (molecular weight 500 or less) to be detectable. With these restrictions, about 30–50 metabolites remain, which may be observed by *in vivo* NMR spectroscopy. Thus it is possible to monitor aspects of biochemistry and physiology related to these compounds *in vivo* in animals.[13] For the acquisition of localized NMR spectra, special methods have been developed, but employing essentially the same hardware (i.e., RF coils and magnetic field gradients).[12,27]

[26] F. Hyder, K. L. Behar, M. A. Martin, A. M. Blamire, and R. G. Shulman, *J. Cereb. Blood Flow Metab.* **14,** 649 (1994).
[27] R. A. de Graaf, "*In Vivo* NMR Spectroscopy: Principles and Techniques." Wiley, Chichester, 1988.

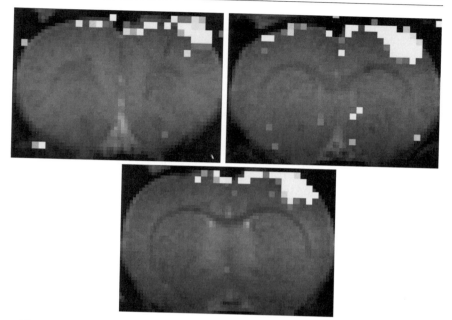

FIG. 4. Functional activation of rat brain after left forepaw stimulation. Gradient echo NMR images are obtained before and during electrical stimulation. The images were subtracted; the subtraction image was color coded and overlaid on a control image. Areas of brain activation are visible in the right cortex. (See color insert.)

Because NMR is a nondestructive technique, it can be used in longitudinal experiments. In this way, NMR imaging or spectroscopy may be applied to the same animal several times during a certain period such as postnatal development[28,29] or tumor growth,[30] or in one experimental session NMR images or spectra may be obtained sequentially to follow dynamic changes (e.g., changes in brain oxygenation[26] or in energy metabolism in skeletal muscle during electrical stimulation[31]). This nondestructive character also allows NMR experiments to be combined with other

[28] A. Heerschap, A. H. Bergman, J. J. Van Vaals, P. Wirtz, H. M. Th, Loermans, and J. H. Veerkamp, NMR Biomed. 1, 27 (1988).
[29] H. J. Zandt, A. J. Groof, W. K. Renema, F. T. Oerlemans, D. W. Klomp, B. Wieringa, and A. Heerschap, J. Physiol. 548, 847 (2003).
[30] J. J. Van Vaals, A. H. Bergman, H. J. van den Boogert, A. Heerschap, A. J. van der Kogel, A. C. Ruifrok, and H. J. J. A. Bernsen, NMR Biomed. 4, 125 (1991).
[31] J. Van Deursen, A. Heerschap, F. Oerlemans, W. Ruitenbeek, P. Jap, H. ter Laak, and B. Wieringa, Cell 74, 621 (1993).

imaging modalities (optical or nuclear methods) or with invasive analysis such as histopathology (e.g., in the case of tumors[19]).

Magnet and Gradient System

Both vertical and horizontal magnets with a clear bore diameter of about 9 cm or more can be used for small animal studies, but for practical and physiological reasons, magnets allowing a horizontal position of the animals are often preferred. Available field strengths of these magnets, commonly used for small animal NMR, range from 4.7 up to 9.4 T. Horizontal magnets up to 11.7 T are also offered, whereas suitable vertical bore magnets can be obtained up to 17.6 T. Although higher field strengths have some obvious advantages, such as increased intrinsic signal-to-noise ratio (SNR) and chemical shift dispersion, the overall performance of a system is determined by many factors (such as RF quality and field-dependent changes in relaxation times), and the choice of a magnet with a particular field strength should be the balanced outcome of these factors in relation to the anticipated usage and costs of the magnet, which is proportional to bore size and field strength.

Likewise, shielded gradient systems have better potential performance with high field amplitude and rapid rise times, and maximal field gradient amplitudes are commonly in the order of 100–1000 mT/m with rise times between about 100 and 500 μs, very much depending on the size of the clear bore.

Also of importance are the shim coils in the magnet to optimize field homogeneity, which should include second-order terms at sufficient power levels.[32] To work with an unstable system in terms of drifting resonance frequency is a frustrating experience. Proper room temperature stability is crucial in this respect.

Practicalities and Optimized Radio Frequency Hardware for *In Vivo* NMR Spectroscopy of the Mouse

This section gives an overview of the different setups developed in our laboratory in order to perform proton, carbon, and phosphorous MR spectroscopy focusing on brain and skeletal muscle tissue of the mouse in a horizontal magnet at a field strength of 7.0 T (Magnex Scientific, Abingdon, England) interfaced to a spectrometer (Surrey Medical Systems, Surrey, England).

---

[32] R. Gruetter, S. A. Weisdorf, V. Rajanayagan, M. Terpstra, H. Merkle, C. L. Truwit, M. Garwood, S. L. Nyberg, and K. Ugurbil, *J. Magn. Reson.* **135,** 260 (1998).

General Aspects

Scaling down from human to mouse size corresponds to a decrease by a factor of about 15 in linear dimension. If one wishes to have NMR resolution in proportion to this scaling for spatially related metabolic or morphological information with a similar signal-to-noise ratio, this means that sensitivity has to be increased by 15^3. Thus the measurement of NMR images or spectra from small volumes in the mice requires specific optimization of hardware. Higher sensitivity is mainly achieved by employing local RF detection coils at the dimensions of the animals. For example, in ^{31}P NMR studies of skeletal muscle of transgenic mice we used a 0.8-cm-diameter coil that fits the hind limb of the animals.[31] Sometimes for high sensitivity and spectral resolution it is necessary to perform semi-invasive experiments; for example, to study the salivary glands of the rat by ^{31}P NMR we had to build a special coil in which the glands could be placed after surgical exposure, leaving the neurovascular system intact.[33] In addition to coil optimization, it is important to minimize interference from RF sources located outside the magnet. Furthermore, in general, sensitivity can also be increased by performing NMR of mice at the highest possible field strength.

A potential problem when measuring on the millimeter scale is motion, which can be minimized by optimizing the mechanical setup of the probe, including a fixation device, and to anesthetize the mouse. As a consequence, the physiological condition of the mouse has to be monitored.

Standard Setup

All of our coils are mounted in a standard frame of Perspex with copper disks mounted at both ends to close the Faraday cage of the magnet bore. Tune and match sticks and cables are guided through tubes mounted in these disks. In the middle of the frames, a construction can be installed containing the RF coil for NMR. In this manner RF coils may be changed rapidly without needing separate frames for every coil. The frame is kept half open in the middle to allow easy access to the coil and the animal.

Many NMR spectroscopy experiments last more than 2 h and occasionally can take up to 6 h. Up until now we have mostly used inhalation anesthetics, applied by a nose cone, as an easy way to keep the mouse under anesthesia for such a long time period. In our setup, a gas mixture of isoflurane, oxygen, and nitrogen is used. The composition of this mixture is monitored continuously using a gas analyzer. The animal is kept warm using a warm water circuit, and the rectal temperature of the animal is

[33] A. Heerschap, J. J. Van Vaals, A. H. Bergman, J. den Boef, P. van Gerwen, and E. J. s-Gravenmade, *Magn. Reson. Med.* **8,** 129 (1988).

monitored using a fluoroptic thermometer. This thermometer is connected to a computer using a RS232 port, which allows display of the temperature of the animal as a function of time graphically. Breathing of the animal is guarded using an optical respiratory gating apparatus (Sirecust 401, Siemens). Parameters available are the amplitude and the frequency of the respiration. With this information, tiny adjustments in the percentage of isoflurane in the gas mixture can be performed, which proved to be necessary when the animal is under anesthesia for a long time (>2 h).

Radiofrequency Shielding

For RF shielding, an aluminium cage is connected to the rear side of the aluminium frame of the magnet cryostat. All shim and gradient cables enter this cage through low-pass II-filters[34] with cutoff frequencies of about 1 MHz. The gradient coil (Magnex) enters the magnet from the rear side and its conductive extension is fixed to the aluminium mechanics of the cryostat. At the front side of the magnet a similar cage is mounted. These extensions couple capacitively with the 6-cm-long copper disks at both ends of the probe tube that close the Faraday cage (Fig. 5). The RF signal from and to the RF coils enters through standard SMA connectors (Suhner). All nonconductive materials, such as tuning sticks, warm water and gas hoses, and glass fibers, are fed through relatively long conductive tubes. If the length and the diameter of such a tube are chosen carefully, the tube acts as a waveguide with a typical high-pass filter behavior.[34] These tubes are positioned at both sides of the probe to enable active removal of gases, thus preventing a local increase in the concentration of anesthetics or CO_2.

The transmit/receive switches and small band preamplifiers, placed in an aluminum cabinet, are situated at the left side of the passively shielded magnet at a distance of approximately 1 m from the probe connector. A double-shielded coaxial cable of this length still does not suppress environmental interference enough. Therefore, at the front side of the magnet and the cabinet, an extension with copper plate boxes is placed (Fig. 5). A copper tube where the coaxial cable is guided through interconnects these boxes. After tuning, matching, and selecting the desired RF path, the boxes are closed to create a Faraday cage.

General Coil Design

NMR spectroscopy experiments can be carried out with different nuclei, which makes it a powerful and attractive tool to study metabolism.

---

[34] E. C. Young, "Dictionary of Electronics." Penguin, New York, 1988.

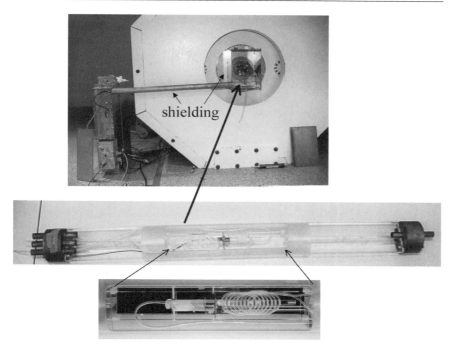

Fig. 5. NMR probe and shielding of the magnet. (Bottom) Measurement probe with RF coil, tube for anesthetics, and plastic tubes containing warm water to keep the mouse at the right temperature. (Center) Probe holder made from a large Perspex tube, which is closed at both ends by copper disks with smaller copper tubes. The measurement probe, containing the RF coil, is placed in the open middle part of the holder. (Top) Front of the 7-T magnet. After mounting of the animal, the probe holder is inserted into the magnet bore. Iron shielding around the cryostat and copper shielding at the end of the bore and around the preamplifier (at the left of the magnet) assure minimal magnetic and electric disturbances.

If, however, nuclei other than protons are considered, the RF probe should have the possibility to operate both at the frequency of protons and at that of the nucleus of interest. By using the strong signal of water protons from the tissue itself, the local homogeneity of the magnetic field can be optimized (shimming) and NMR images can be made as guides for localization. In general, additional losses due to the noise coupling between both frequencies should always be minimized because it degrades the performance for the nucleus of interest. Therefore, for every nucleus dedicated, RF coils are designed next to a ^1H coil. A high signal-to-noise ratio was obtained by positioning the main RF coil closely to the object to be measured. The coil was made resonant and balanced at the desired frequency by using fixed capacitors (American Technical Ceramics) inside the coil loop. The fine adjustment of the resonance frequency and the matching to 50 Ω is done by small, variable capacitors (Voltronics) positioned to the coil as close as possible to minimize feed losses. Extended tuning sticks are used to tune and match the coil when it is inside the magnet. The ^{1}H coil is designed to have an orthogonal RF field with respect to the first coil and the main magnetic field. This coil can therefore be used for shimming, localizing, and decoupling without any practical interference with the main coil.

Radiation losses are minimal because the designed coils are relatively small.[35] Therefore, it is not necessary to use a closely positioned RF screen. If, however, such a RF screen is needed to suppress signals arising from other tissue, the screen is slit and closed by a relatively large capacitance for RF to minimize eddy current effects.

^{31}P NMR Spectroscopy of Mouse Hind Limb

In most experiments, the mouse leg is placed in a vertical position as shown in Fig. 6, with the mouse body in a natural horizontal position. To create the highest sensitivity combined with a homogeneous B_1 field, a three-turn solenoid coil with a diameter of 8 mm is used. Outside this solenoid coil, an Alderman–Grant type of coil[36] is placed for shimming

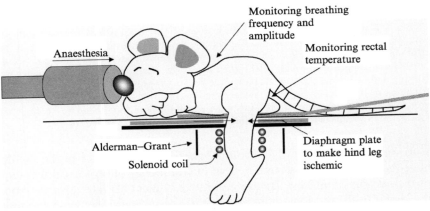

FIG. 6. Schematic overview of the probe for ^{31}P NMR of hind limb skeletal muscle of the mouse. The hind leg is positioned in the solenoid coil, and the foot is fixed. The mouse is kept under anesthesia while temperature and respiration are monitored. A copper plate prevents the acquisition of signals arising from the whole mouse instead of the skeletal muscle of the hind leg. A diaphragm plate can be used to apply ischemia on the hind leg.

[35] J. J. Jackson, "Classical Electrodynamics." Wiley, New York, 1962.
[36] D. W. Alderman and D. M. Grant, *J. Magn. Reson.* **36,** 447 (1979).

only (Fig. 6). A minimum distance of 5 mm is maintained to prevent coupling between the proton and the phosphorous coil. In addition, a copper plate is used to prevent undesirable signals arising from the body of the mouse instead of the muscles of its hind leg. For metabolic challenging, we often applied ischemia to the hind leg using a diaphragm plate. The position of the plate can be manipulated from outside the magnet. The force by which the plates push is calibrated at 5–10 Newton, which was found to make the hind leg ischemic without damaging muscle tissue. To minimize the displacement of the hind leg, the force is applied from two sites simultaneously. Otherwise, the quality of the shimming and the quality of the tuning of the coil may decrease dramatically. An example of a ^{31}P NMR experiment performed with this coil is shown in Fig. 7. It demonstrates the use of a magnetization transfer method to identify the absence of enzyme activity in creatine kinase knockout mice.

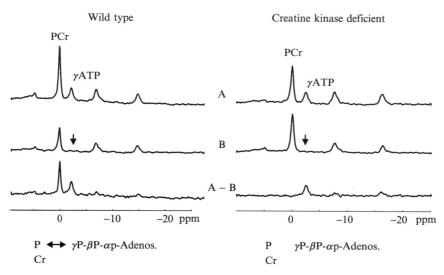

FIG. 7. Phosphorus-31 NMR spectra of mouse skeletal muscle obtained with the setup shown in Fig. 6. (A) Spectra of a wild-type mouse (left) and a creatine kinase-deficient mouse (right). Three ATP peaks and a peak for phosphocreatine (PCr) are clearly seen. (B) Spectra obtained with irradiation on the γATP peak for 1 s. This magnetic labeling can be transferred via the creatine kinase reaction to the phosphate of PCr (see bottom left). (A and B) Difference spectra clearly showing this transfer in the wild-type muscle (left) but not in the creatine kinase-deficient muscle (right). Exchange between γATP and PCr is not possible anymore (bottom right).

^{1}H-Decoupled ^{13}C NMR Spectroscopy on Mouse Hind Limb

For ^{13}C NMR experiments, the same coil setup is used as described earlier for ^{31}P NMR spectroscopy. Because the NMR frequency, at 7 T, of ^{13}C (76 MHz), is lower than that of ^{31}P (122 MHz), an additional turn on the solenoid is possible, resulting in a slightly better quality (Q) factor of the resonator. An issue in ^{13}C NMR spectroscopy is the interaction between carbon and proton spins. The so-called J coupling may complicate ^{13}C NMR spectra and decreases the SNR as the signal integral of one carbon spin is divided over several peaks. To decouple this interaction, RF power is transmitted at the proton frequency (300 MHz) during reception of the ^{13}C NMR signal. During this period, the amplifier is not blanked and therefore the noise level at the ^{13}C NMR frequency may increase enormously. This additional noise has to be minimized to allow proper acquisition. The first step is to minimize the electromagnetic coupling between the two coils. An Alderman–Grant type of coil was used,[36] which has an orthogonal RF field with respect to the solenoid coil and leads to a coupling of −40 dB. The coaxial cable to the ^{1}H coil is not matched to 50 Ω at the ^{13}C NMR frequency. Therefore, the cable also acts as a transmitter for ^{13}C RF noise. This ^{13}C RF noise is filtered by a homemade seven-pole Chebyshev (high pass) filter through the RF screen with a cutoff frequency of 235 MHz. The homemade, small-band preamplifier is designed for low power levels only. During reception of the ^{13}C NMR signal, too much RF power at the ^{1}H frequency will enter into the preamplifier and degrade the performance of the amplifier. A high-order, low-pass filter (Trilithic) with a cutoff frequency of 88 MHz was used to prevent this problem. Altogether this only results in a (additional) noise level of less than 5% due to the ^{1}H decoupling. We have used this setup to measure the uptake of ^{13}C4-labeled creatine in mouse skeletal muscle[29] and are currently in progress to realize a similar setup for ^{13}C NMR of mouse brain.[37]

^{1}H NMR of Mouse Skeletal Muscle

The coil construction depicted in Fig. 6 is not suitable to achieve the desired quality for localized proton NMR spectroscopy of the mouse muscle. The distance between the Alderman–Grant coil and the mouse leg is too large. For that reason, a smaller, anatomically shaped Alderman–Grant type of coil was constructed (Fig. 8). This coil has the advantage that the angle of the coil and the mouse leg with respect to the main magnetic field

[37] W. K. J. Renema, A. A. Veltien, F. Oerlemans, B. Wieringa, and A. Heerschap, *Proc. Int. Soc. Magn. Res. Med.* **11** (2003).

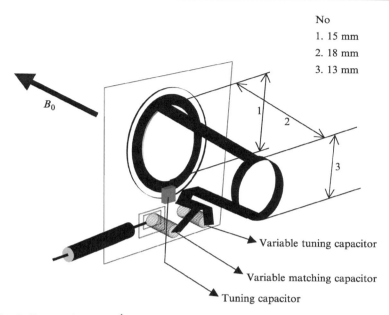

No
1. 15 mm
2. 18 mm
3. 13 mm

Variable tuning capacitor

Variable matching capacitor

Tuning capacitor

FIG. 8. Construction of the ^{1}H coil for the hind limb skeletal muscle of the mouse. The coil is balanced by a fixed tuning capacitor. Fine tuning and matching are done by variable capacitors.

can be changed without paying too much of a penalty with respect to the signal-to-noise ratio (Fig. 9). This proved to be very useful in studies where it is desirable to change the angle of the leg with respect to B_0.[38] Because the coil is mainly polarized in the x direction, a rotation of the coil around this axis only results in a small signal loss. Finite element analyses based on Maxwell equations indicate that the signal loss is less than 3% (Maxwell software by Ansoft). Localized ^{1}H MR spectroscopy requires large switching gradient fields, which can cause eddy-current artifacts. The end rings of the coil are therefore opened by relative large capacitances (470 pF), which are closed for RF. When the angle of the coil is fixed at larger angles ($>40°$), the height of the whole insert is adjusted in order to put the center of the coil in the center of the magnetic field for an optimal homogeneity of the magnetic field.

[38] H. J. A. in 't Zandt, D. W. Klomp, F. Oerlemans, B. Wieringa, C. W. Hilbers, and A. Heerschap, *Magn. Reson. Med.* **43**, 517 (2000).

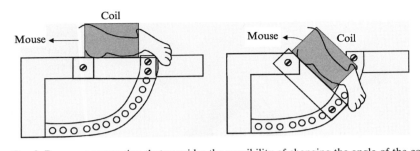

FIG. 9. Perspex construction that provides the possibility of changing the angle of the coil and the leg with respect to the main magnetic field direction. The desired angle of the coil is fixed with plastic screws. If the angle of the leg changes much ($>40°$), the height of the whole setting is changed to position the center of the coil in the center of the magnetic field for optimal homogeneity.

^{31}P NMR Spectroscopy of Mouse Brain

In order to measure ^{31}P NMR-detectable metabolites in the mouse brain, a surface coil is used most of the time.[39,40] The advantage of this concept is a simple coil design combined with high sensitivity close to the coil. Localization is achieved due to the limited penetration depth of the RF field of a surface coil. To avoid unwanted signal contributions of nonbrain tissue containing significant amounts of ^{31}P NMR-detectable metabolites, the skin and underlying muscle tissue are often removed. This invasive operation makes longitudinal studies problematic. In studies on genetically altered mice, it might be unwanted to sacrifice the animal if further studies on the same animal have to be performed or, if due to certain phenotypic consequences of a genetic lesion, only a small number of animals are available. This was one of the main reasons to develop a technique to perform ^{31}P NMR in a completely noninvasive manner. As a consequence, proper localization with field gradients is needed. For the RF setup, this demands a ^{1}H coil for shimming and imaging and a ^{31}P coil with enough RF penetration depth to cover the whole brain. The probe housing should also contain the setup to keep the mouse under anesthesia, to monitor its condition, and have a fixation device to prevent motion artifacts. The challenge is to realize this setup on the small scale of the mouse brain.

Preliminary experiments with a single surface coil revealed that the signal-to-noise ratio of ^{31}P NMR spectra was not satisfactory. A way to

[39] D. Holtzman, E. McFarland, T. Moerland, J. Koutcher, M. J. Kushmerick, and L. J. Neuringer, *Brain Res.* **483**, 68 (1989).
[40] D. Holtzman, R. Meyers, E. O'Gorman, I. Khait, T. Wallimann, E. Allred, and F. Jensen, *Am. J. Physiol.* **272**, C1567 (1997).

improve the signal-to-noise ratio is the concept of quadrature transmission and reception of the RF signal. It will make the transmission of RF power more efficient during transmission and increases the signal-to-noise ratio during reception.

When an oscillating RF field is applied at, for example, the x axis, the magnitude of the oscillating B_1 field varies in amplitude. It is straightforward to show that the efficiency is then

$$\int_0^{\pi/4} \cos\theta \; d\theta = \frac{1}{2}\sqrt{2} \approx (70\%)$$

The efficiency can be increased using two perpendicular (or quadrature) coils, each fed with alternating currents 90° out of phase. If the precession direction of the B_1 field is anticlockwise, like the movement of the effective B field, the applied B_1 field is always parallel to B_{eff} and the efficiency will be 100%. During reception, the addition of two signals after phase matching will result in signal enhancement by a factor of 2. The inductive coupling between the two coils is zero as the fields are orthogonal. In this situation, the noise from these coils is not correlated. This results in a noise enhancement of only a factor of $\sqrt{2}$. Effectively, the signal-to-noise ratio will therefore increase by a factor of $2/\sqrt{2} = \sqrt{2}$. The 90° phase shifting during both transmission and reception is realized by a narrow-band quadrature hybrid.[41]

To realize a quadrature mode configuration, we have selected the concept of two slightly overlapping surface coils.[42] Both surface coils with 16-mm diameters are shaped anatomically to fit as closely as possible to the brain.[43] This complete configuration is then shifted inside a cylindrical ^{1}H MR volume coil, based on the Alderman–Grant concept.[36] To maintain sufficient image quality for good localization, this coil has a diameter of only 33 mm. Additional stereotactic fixation is realized by a teeth bar and ear plugs of Perspex (Fig. 10).

Mouse Brain Proton NMR

For high ^{1}H NMR sensitivity, a surface coil is used with an elliptical shape (15 × 11 mm) optimized for the geometry of the mouse brain. The mechanical setup, including stereotactic fixation, is essentially the same

[41] C. H. Chen and D. I. Hoult, "Biomedical Magnetic Resonance Technology." IOP Publishing, New York, 1989.
[42] G. Adriany and R. Gruetter, *J. Magn. Reson.* **125**, 178 (1997).
[43] D. W. J. Klomp, H. J. A. in 't Zandt, H. J. van den Boogert, F. Oerlemans, B. Wieringa, and A. Heerschap, *Proc. Intl. Soc. Magn. Res. Med.* **7**, 2069 (1999).

Alderman–Grant coil

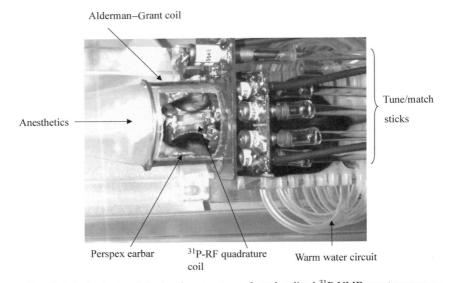

Anesthetics

Tune/match

sticks

Perspex earbar ^{31}P-RF quadrature   Warm water circuit
coil

FIG. 10. Mechanical and electronic setup to perform localized ^{31}P NMR spectroscopy on the mouse brain. The mouse with the two-loop ^{31}P quadrature coil around the ears is shifted into the Alderman–Grant ^{1}H coil and is kept under anesthesia using isoflorane. The two ^{31}P quadrature coils and the proton coil yield six tuning/matching sticks.

as described for the localized ^{31}P NMR experiments. Examples of ^{1}H NMR spectra of the mouse brain obtained with this configuration are shown in Figs. 11 and 12.

Anesthesia and Physiological Monitoring of Animals in NMR Experiments

The actual time to record a routine NMR image of water in the body varies between about 1 s and a few minutes, and longer if high-resolution images are needed. To obtain an NMR spectrum of metabolites, up to 60 min may be needed. Unpredictable movements of the animal if awake may spoil these recordings. Therefore, it is necessary to keep the animals under (mild) anesthesia during the measurements. In the majority of cases, anesthesia is used to achieve hypnosis in the presence of autonomic stabilization. Occasionally, analgesia is also used or muscle relaxation in more demanding experiments.

In the case of plain MRI, standard types of simple anesthesia are sufficient. However, if metabolic or physiological processes are studied, accurate maintenance of homeostasis becomes important and the anesthetic approach has to be designed very carefully because it may interfere with

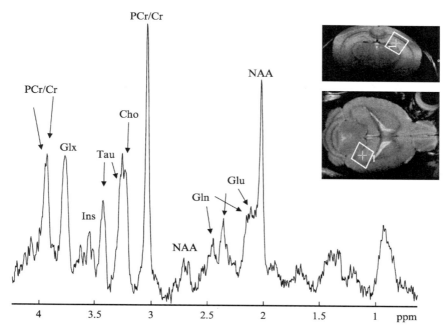

FIG. 11. Proton NMR spectra obtained from a 4-μl voxel in the hippocampus/cortex area of the living mouse brain (the voxel is indicated in the NMR images). NMR localization was performed with a STEAM sequence (Te = 10 ms, Tm = 15 ms, Tr = 5 s). Signals are observed for various compounds, such as N-acetylaspartate (NAA), glutamate (Glu), glutamine (Gln), phosphocreatine and creatine (PCr/Cr), choline (Cho), taurine (Tau), inositol (Ins), and several other compounds not indicated in the spectrum.

metabolic or physiological processes under investigation.[44–47] First, a choice has to be made in the way the anesthetics will be applied. One may select between injection (intraperitoneal, intravenous, or continuous infusion) and inhalation. In the latter case, it is possible to let the animal breathe freely an anesthetic gas mixture via a nose cone or to ventilate the animal (oral intubation or tracheotomy). The first way is the easiest, but control over the adequacy of respiration is hardly possible; the last

[44] P. A. Flecknell, "Laboratory Animal Anaesthesia: A Practical Introduction for Research Workers and Technicians." Academic Press, New York, 1996.
[45] D. F. Kohn, S. K. Wixson, W. J. White, and G. J. Benson, "Anesthesia and Analgesia in Laboratory Animals." Academic Press, New York, 1997.
[46] R. D. Miller, E. D. Miller, Jr., J. G. Reves, M. F. Roizen, J. J. Savarese, R. R. F. Cucchiara, and A. Ross (eds.) "Anesthesia," 5th Ed. Churchill Livingstone, New York, 2000.
[47] M. Zhao, L. G. Fortan, and J. L. Evelhoch, *Magn. Reson. Med.* **33,** 610 (1995).

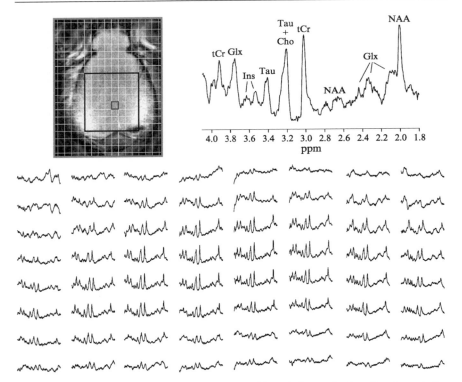

FIG. 12. Proton spectroscopic NMR image (multivoxel data set) of a slice through the mouse brain. (Top left) Transversal localizer image showing the grid from which the spectroscopic imaging data set was obtained. Data were selected by the STEAM sequence from the large box. (Bottom) Individual NMR spectra from voxels in this box. (Top right) An enlarged spectrum of the indicated voxel shows the same resonance as in Fig. 11. tCr is PCr and Cr; Glx is glutamine and glutamate.

approach gives much better modes of anesthesia control, but requires more expertise, especially in smaller animals such as rats and mice. Furthermore, the choice of the anesthesia itself is important (e.g., in brain studies concerning metabolism, function, or pathological processes such as ischemia). Common anesthetics such as isoflurane, halothane, ketamine, propofol, and pentobarbital may have very different effects on cerebral blood pressure and blood flow (CBF) and on cerebral oxygen consumption. Cerebral oxygen consumption is reduced variably by these anesthetics, leading to less neuronal damage during ischemia; it may also reduce the extent of the BOLD signal in fMRI experiments. More dramatic are the differences in CBF changes with these anesthetics, and because this is part of the BOLD phenomenon, they require careful attention in fMRI studies. Blood pressure in the brain may be shifted to critical autoregulation limits with increasing concentration of the anesthetic. In studies of tumor models, the location of the tumor in the animal, in combination with the selected anesthetics, is important when vascularization, hypoxic conditions, or effects of therapy are of interest. In the common subcutaneous xenograft models, attention should be given to the usually occurring reduction in systemic vascular resistance, which may cause peripheral vasodilatation. Accumulation of carbon dioxide often will occur during inhalation anesthesia with a nose cone because of respiratory depression. This will also cause peripheral blood vessel dilatation and may blunt the effect of vasodilator approaches, such as treatment with carbogen (95% O_2 and 5% CO_2) breathing. Thus ventilation may be a better choice to start at normocapnic conditions. Equally important as fine control over parameters such as oxygen and carbon dioxide levels is control over the temperature of the animal, because this is likely to drop due to anesthesia. Temperature may affect various NMR parameters, such as diffusion and chemical shift. The control of temperature is usually achieved with a bed or blanket under the animal with circulating warm water, and some cases, such as subcutaneous xenografts extruding from the body, may need special attention.

From the aforementioned overview of anesthesia, it becomes clear that physiological monitoring of the animals during an NMR experiment is needed under all circumstances. Monitoring devices have to be MR compatible. Furthermore, in the case of small animals, devices have to deal with high-frequency heart and respiratory rates and low air volume and flow during ventilation. Very basic parameters that can be monitored (e.g., for plain NMRI) are body temperature (e.g., using a fiber-optic rectal probe) and regularity of respiration (e.g., via a strain gauge). In more advanced physiological experiments, it is also advisable to monitor blood oxygen (e.g., via a pulse oximeter attached to the hind paw), heart rate (ECG with filtered leads to be MR compatible), and carbon dioxide levels (e.g., via a capnograph or mass spectrometer). The latter monitoring only makes sense in combination with ventilation of the animal (Fig. 13). Advanced invasive monitoring may include blood pressure measurements or arterial blood gas analysis. In the course of an NMR experiment, which may last for several hours, physiological changes can be counteracted via the regulation of temperature, ventilation, or anesthesia level before these become detrimental.

If anesthesia is not desired, some investigators mount the animal in the NMR probe in such a way that it cannot move using a three-point fixation with a teeth bar and earplugs. To prevent high stress levels, extensive training of the animal is usually required to adapt to the fixation and high noise levels inside the scanner.

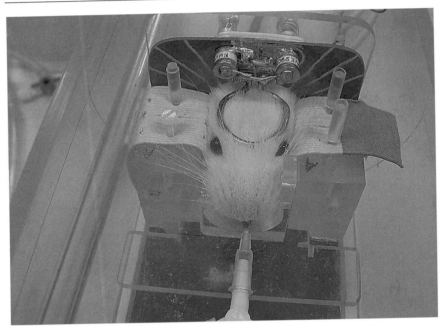

FIG. 13. Setup for NMR of rat brain. The ^{1}H coil for the brain is clearly seen, as well as the oral tube for ventilation of the animal. (See color insert.)

Because the anesthesiology and monitoring field of animals has expanded tremendously during the last few decades and an NMR researcher usually needs to focus his attention fully on the course of the NMR experiment during more complicated examinations, it is generally a good idea to employ a skilled person who is only concerned with animal anesthesia and monitoring. This may be a biotechnician or veterinary anesthetist, who may also assist in small operations or in inserting arterial infusion lines for the administration of contrast media.

NMR and Animal Use

Because of the noninvasive nature of the NMR method, its use has several particular advantages. Most importantly, animals can be investigated repeatedly over periods of time in longitudinal studies. Therefore, use of a large number of animals is not needed. However, it is necessary to keep the animals under anesthesia during NMR experiments to avoid movement artifacts, which may pose considerable discomfort to the animals, especially in longitudinal experiments.

As the animal remains intact, NMR can be combined with other (invasive) analyzing methods. This is important in optimizing the scientific outcome of a study with a limited number of animals. The impact of NMR on the reduction or increase in the number of animals used in experimentation, on the refinement of experiments, and on replacement by other assays[48] depends on the particular employment of the method.[49]

Acknowledgments

We thank Drs. J. Pikkemaat, G. Gambarota, H. van Laarhoven, M. Philippens, W. Leenders, J. van Egmond, and J. van Asten and the biotechnicians and staff members of the Central Animal Laboratory in Nijmegen (director Dr. J. Koopman) for collaboration and support in animal studies. We also thank the Dutch National Science Organization (NWO) for contributions to the animal NMR facility.

[48] M. Balls, J. H. Goldberg, C. L. Fentem, Burch, R. L. Broadhead, M. F. W. Festing, J. M. Frazier, C. F. M. Hendriksen, M. Jennings, M. D. O. van der Kamp, D. B. Morton, A. N. Rowan, C. Russell, W. M. S. Russell, H. Spielmann, M. L. Stephens, W. S. Stokes, D. W. Straughan, J. D. Yager, J. Zurlo, and B. F. M. van Zutphen, *ATLA* **23,** 838 (1995).
[49] A. Heerschap, *in* "Animal Alternatives, Welfare and Ethics" (L. F. M. van Zutphen and M. Balls, eds.), p. 303. Elsevier Science, New York, 1997.

[4] Simultaneous Electrophysiology and Functional Magnetic Resonance Imaging Studies in Conscious Rats

By M. Khubchandani and N. R. Jagannathan

Introduction

The use of functional MRI (fMRI) in clinical and physiological research has been phenomenal in recent years.[1-4] It is a noninvasive procedure for studying the localization of brain activity and measures the local changes in cerebral hemodynamics.[5,6] These studies have been applied to map the

[1] S. G. Kim and K. Ugurbil, *J. Neurosci. Methods* **74,** 229 (1997).
[2] P. M. Matthews, S. Clare, and J. Adcock, *J. Inherit. Metab. Dis.* **22,** 337 (1999).
[3] K. Ugurbil, G. Adriany, P. Andersen, W. Chen, R. Gruetter, X. Hu, H. Merkle, D. S. Kim, S. G. Kim, J. Strupp, X. H. Zhu, and S. Ogawa, *Annu. Rev. Biomed. Eng.* **2,** 633 (2000).
[4] J. A. Detre and T. F. Floyd, *Neuroscientist* **7,** 64 (2001).

human cerebral cortex in several cognitive functions.[7–11] Being noninvasive, fMRI offers several advantages in comparison to other functional neuroimaging techniques, such as positron emission tomography (PET) or single photon emission-computed tomography (SPECT), as it has superior soft tissue contrast and good spatial resolution. However, temporal resolution is limited by the hemodynamic response time. Hence, to map brain functions with good temporal resolution requires techniques that are not dependent on secondary events such as hemodynamic changes, but rather are directly sensitive to electromagnetic fields induced by the nervous system. This can be accomplished by measuring electric potentials (e.g., EEG) or by measuring magnetic induction (e.g., EMG). However, neither of these techniques is strictly tomographic. In view of this, there has been growing interest for combining the two techniques (i.e., electrophysiology and fMRI). With the integration of these techniques, it will be possible to obtain both spatial and temporal resolution unattainable from either one alone.[12] Such integration of techniques finds application in several experimental paradigms, which include studies on epilepsy[13] or those involving transient brain states such as attention or sleep.[14]

In these studies, electrophysiological tracings are used to characterize the brain state (i.e., the presence or absence of seizure activity, sleep stage, etc.) during fMRI data acquisition. fMRI data can then be analyzed as a function of brain state. In other words, fMRI acquisitions are correlated with spontaneous electrophysiological events (i.e., with epileptiform discharges and physiological events such as evoked potentials or monitoring the state of arousal or sleep) and thus help identify the generators of these events.[15–18]

[5] J. W. Belliveau, B. R. Rosen, H. L. Kantor, R. R. Rzedzian, D. N. Kennedy, R. C. McKinstry, J. J. Vevea, and B. R. Rosen, *Magn. Reson. Med.* **14,** 538 (1990).

[6] S. Ogawa, T. M. Lee, A. S. Nayak, and P. Glynn, *Magn. Reson. Med.* **14,** 68 (1990).

[7] P. A. Bandettini, E. C. Wong, R. S. Hinks, R. S. Tikofsky, and J. S. Hyde, *Magn. Reson. Med.* **25,** 390 (1992).

[8] J. W. Belliveau, D. N. Kennedy, R. C. McKinstry, B. R. Buchbinder, R. M. Weisskoff, M. S. Cohen, J. M. Vevea, T. J. Brady, and B. R. Rosen, *Science* **254,** 716 (1991).

[9] J. Frahm, H. Bruhn, K.-D. Merbolt, and W. Hanicke, *J. Magn. Reson. Imaging* **2,** 501 (1992).

[10] S. G. Kim, J. Ashe, K. Hendrich, J. M. Ellermann, H. Merkle, K. Ugurbil, and A. P. Georgopoulos, *Science* **261,** 615 (1993).

[11] K. K. Kwong, J. W. Belliveau, D. A. Chesler, I. E. Goldberg, R. M. Weisskoff, B. P. Poncelet, D. N. Kennedy, B. E. Hoppel, M. S. Cohen, and R. Turner, *Proc. Natl. Acad. Sci. USA* **89,** 5675 (1992).

[12] L. Lemieux, K. Krakow, and D. R. Fish, *Neuroimage* **14,** 1097 (2001).

[13] A. S. Haddadi, M. Merschhemke, L. Lemieux, and D. R. Fish, *Neuroimage* **16,** 32 (2002).

[14] C. Portas, K. Krakow, P. Allen, O. Josephs, and J. L. Armony, *Neuron* **28,** 991 (2000).

However, studies involving the integration of electrophysiology and MRI are associated with significant technical problems, leading to artifacts in the MR images and electrophysiological recordings. Artifacts in electrophysiological recordings are mainly due to the gradient magnetic field, the static magnetic field, and the radiofrequency (RF) pulses used during MR acquisition. Another important source of artifacts is any motion in the magnetic field that generates voltage interfering with the electrophysiological signal. This is an important issue for studies involving animal models due to difficulties in animal handling during MR data acquisition. It is a serious limitation because animal studies are essential not only for fundamental research, but also for investigating many disease processes that affect humans. Artifacts in MR images due to the movement of the animal can at times be mistaken for stimulus-induced signal changes in brain activity.[19] Hence, anesthetics are normally used to immobilize animals during MRI procedures. It is known that anesthetics depress central nervous system (CNS) metabolic activity and reduce basal cerebral blood flow (CBF)[20] and, consequently, reduce the blood oxygenation level dependent (BOLD) signal in fMRI. α-Chloralose is used routinely as an anesthetic in fMRI studies because it does not depress cortical activity and provides a stable CBF. Nonetheless, changes in CBF are more robust in awake animals.

Thus it is desirable to develop methods and procedures to obtain fMR images from conscious animals.[21] For this purpose, a stereotaxic apparatus with a head and body restrainer is essential in immobilizing a conscious animal. In addition, suitable nonmagnetic electrodes and anchoring screws have to be implanted on the head of the rat to facilitate the recording of electrophysiological recordings in a magnetic environment. EEG, EOG, and EMG are used for the assessment of sleep–wakefulness in the

[15] F. R. Huang-Hellinger, H. C. Breiter, G. McCormack, M. S. Cohen, K. K. Kwong, J. P. Sutton, R. L. Savoy, R. M. Weisskoff, T. L. Davis, J. R. Baker, J. W. Belliveau, and B. R. Rosen, *Hum. Brain Mapp.* **3**, 13 (1995).

[16] K. Krakow, F. G. Woermann, M. R. Symms, P. J. Allen, L. Lemieux, G. J. Barker, J. S. Duncan, and D. R. Fish, *Brain* **122**, 1679 (1999).

[17] M. Seeck, F. Lazeyras, C. M. Michel, O. Blanke, C. A. Gericke, J. Ives, J. Delavelle, X. Golay, C. A. Haenggeli, N. de Tribolte, and T. Landis, *EEG Clin. Neurophys.* **106**, 508 (1998).

[18] S. Warach, J. R. Ives, G. Schlaug, M. R. Patel, D. G. Darby, V. Thangaraj, R. R. Edelman, and D. L. Schomer, *Neurology* **47**, 89 (1996).

[19] J. V. Hajnal, R. Myers, A. Oatridge, J. E. Schwieso, I. R. Young, and G. M. Bydder, *Magn. Reson. Med.* **31**, 283 (1994).

[20] M. Ueki, G. Mies, and K. A. Hossmann, *Acta Anaesthesiol. Scand.* **36**, 318 (1992).

[21] M. Khubchandani, H. N. Mallick, N. R. Jagannathan, and V. Mohan Kumar, *Magn. Reson. Med.* **49**, 962 (2003).

conscious rat.[22] However, the recording of electrophysiological parameters during MR requires several procedures to be standardized. This Chapter presents technical details and various methodological issues involved in the design, fabrication, and procedures developed and standardized for the simultaneous acquisition of electrophysiology and fMRI from conscious animals.

Materials and Methods

fMRI and electrophysiology are performed on adult male Wistar rats (150–175 g). Rats are given food and water *ad libitum*. The study is approved by the Institutional Animal Ethics Committee. Details of the various standardization procedures carried out on 40 rats, which could be trained successfully for simultaneous electrophysiology recordings and fMRI, are presented.

Fabrication of Stereotaxic Apparatus

Figure 1 shows the schematic line diagram and the dimensions of a non-magnetic stereotaxic apparatus with a head holder and a body restrainer. The design consists of a base plate or a chassis (where the rat rests), a head restrainer (for restraining the head of a conscious animal), and a body restrainer (to restrain the body movements). This apparatus fits to an existing RF volume resonator of 72 mm in diameter available with a 4.7-T animal MR scanner. Figure 2 shows the cross-sectional view of the RF resonator with the stereotaxic apparatus.

The head restrainer has four lateral bars (two on each side of the head; each 1.5 cm in length) that fit exactly into the dummy receptacle/mold (described later) of the head of the rat. For fixing and removing these lateral bars while restraining and releasing the rat, additional screws are used as shown in Fig. 1. The height of the head restrainer is fixed at 4.5 cm by taking into consideration the size of the dental mold to be prepared (later) on the head of the rat. This indirectly limits the size (and weight) of the rat to be used in this study. Using this stereotaxic apparatus, the head of the conscious rat can be restrained from vertical, horizontal, and lateral movements during fMRI.

A Teflon plate surrounds the belly of the rat and is situated at the rear of the base plate, which restricts the movements of the body, paws, and limbs of the conscious animal. The design allows proper unrestricted

---

[22] C. Timo-Iaria, N. Negrao, W. R. Schmidek, K. Hoshino, C. E. Loboto de Menezes, and T. Leme da Rocha, *Physiol. Behav.* **5,** 1057 (1970).

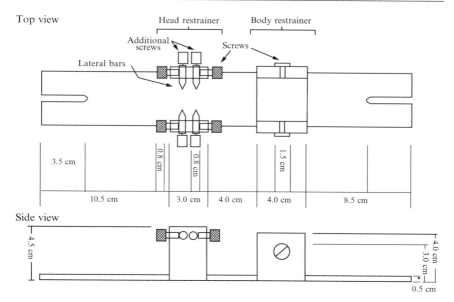

Fig. 1. Schematic line diagram of the fabricated stereotaxic apparatus. Reproduced with permission from Khubchandani et al.[21]

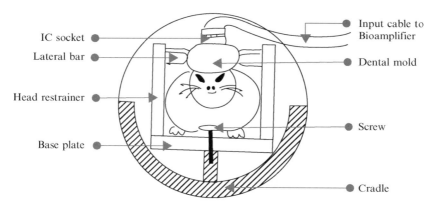

Fig. 2. Cross-sectional view of the RF resonator with stereotaxic apparatus in it.

respiration. There should be no indication of discomfort or stress, and the rat should not struggle after successful training. Figure 3 shows the side view of the stereotaxic apparatus. Using this assembly, a rat in the weight range of 150–175 g can be restrained easily.

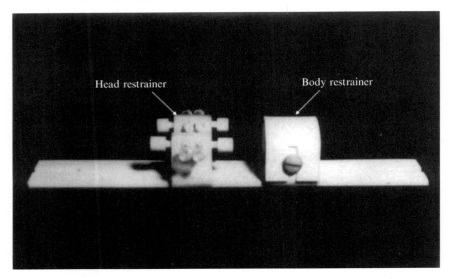

FIG. 3. The fabricated nonmagnetic stereotaxic apparatus. Reproduced with permission from Khubchandani *et al.*[21]

Selection of Suitable Materials for Implantation

In simultaneous studies using fMRI and electrophysiology, several issues pertaining to the quality of MR images obtained with the use of electrode assembly need to be evaluated. Materials used as electrodes should be completely nonmagnetic, and those tested include stainless steel wire, copper wire, and silver wire. These materials were investigated for their nonmagnetic property, and the quality of images obtained using spin echo (SE) and gradient echo (GE) RF sequences were evaluated. Silver wire gives artifact-free MR images and thus was chosen for use as electrodes for electrophysiological recordings.

The leads used for electrophysiological recordings may interfere with the RF field. Hence, RF-shielded annealed tinned copper (ATC) wires insulated with PVC were chosen because they are both nonmagnetic and biocompatible. Some amount of detuning was observed in implanted rat/phantom during MRI. However, the effect was found to be negligible and does not affect the quality of MR images obtained.

Anchoring screws are essential to anchor the acrylic dental cement mold on the skull of the rat along with the electrodes implanted. Screws made of stainless steel, brass, and brass coated with nickel, Teflon, and plastic/polycarbonate were tested for this purpose. Of these, plastic/polycarbonate

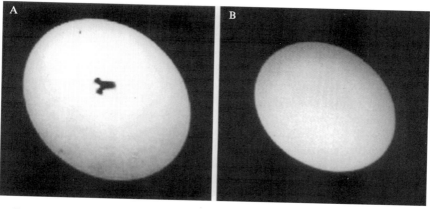

FIG. 4. (A) A transverse MR image showing a nonmagnetic polycarbonate screw embedded in a 1% agarose gel phantom. (B) Transverse MR image obtained from a 1% agarose gel phantom with a superficially placed IC socket showing no artifacts.

screws are nonmagnetic, biocompatible, and also relatively easier to fix on the skull of the rat. Figure 4A shows a transverse MR image obtained using a SE sequence of the polycarbonate screw embedded in 1% agarose gel phantom, showing no artifacts.

An IC socket (8 pin) is implanted on head of the rat to serve as a junction box between the electrodes implanted on the rat skull and the input cable of the bioamplifier used for electrophysiological recordings. For this purpose, several IC sockets were tested for their nonmagnetic properties. The socket obtained from Bruker (Germany) gives artifact-free MR images in 1% agarose gel. Locally available IC sockets give artifacts when placed inside the phantom. However, if these are placed superficially above the phantom, no artifacts in MR images were observed, as shown in Fig. 4B. In the actual study on rats, the IC socket is placed on the skull of the animal (well above the region of interest, i.e., brain); hence, these locally available IC sockets (BEL, Bangalore, India) were used for the final study.

Implantation of Electrodes

The implantation of electrodes to record EEG, EMG, and EOG in the skull of the rat is carried out under aseptic conditions, which are acclimatized to the gradient magnetic fields (see later). For this purpose, rats are anesthetized with sodium pentabarbitone (40 mg/kg, ip). In order to minimize the possibility of infection and to facilitate surgery, the head of the animal is shaved and then fixed to a surgical stereotaxic instrument (INCO, Ambala, India). The angle of the head fixed in the surgical stereotaxic instrument is in accordance with De Groot.[23] Respiration is monitored visually throughout the implantation procedure. In case of respiratory distress, the animal is removed and necessary resuscitation is followed until respiration comes back to normal.

A dorsal midline incision is made (1.5 cm) using a sterile surgical blade, and connective tissues are scraped off with a blunt spatula to expose a small portion of frontal bone, bregma, coronal sutures, sagittal suture, lambda, and a small portion of the occipital bone. Five holes are drilled on the skull for fixing the polycarbonate-anchoring screws. These anchoring screws help hold the electrodes and the dental mold in place on the skull of the rat. For this purpose, one hole is drilled in the frontal region and two holes each (bilaterally) are made caudal to the bregma and rostral to the lambda. Care is taken such that the screws are not inserted farther than the thickness of the skull, which is around 2 mm.

The nonmagnetic electrodes are then implanted,[24,25] and two types of electrodes are prepared for implantation, namely, the loop and screw design. Because EMG and EOG electrodes have to be sewn to the muscle, loop design electrodes are used, which are prepared by soldering loop of a silver wire (31 gauge) to the ATC wire. Screw design electrodes used as EEG and ground are prepared by soldering silver wire (21 gauge) to the ATC wire.

EOG electrodes (loop design) for recording the eye movement are sewn to the external eye canthus muscle bilaterally. Similarly, EMG electrodes (loop design electrode) are sewn to the neck muscle of the rat bilaterally for recording neck muscle activity. EEG electrodes are fixed to the skull 2 mm anterior to the bregma and 4 mm lateral to the midsagittal suture. A unilateral screw electrode is fixed similarly, approximately 5 mm away from the EEG electrodes, rostrally at the nasofrontal portion of the skull that serves as the ground.

A thin film of acrylic cement is then applied on the skull to fix the recording electrodes, as shown in Fig. 5. Distal ends of the electrodes are trimmed and soldered to an IC socket. The ground electrode is connected to the first pin of the socket. Two bilateral EOG electrodes are soldered to the second pair of pins of the IC socket (left and right, respectively), whereas the EEG and EMG electrodes are connected to the third and fourth pair of pins of the socket.

[23] J. De Groot, *Trans. R. Neth. Acad. Sci.* **52,** 1 (1959).
[24] J. John, V. Govindaraju, P. Raghunathan, and V. M. Kumar, *Brain Res. Bull.* **48,** 273 (1996).
[25] V. M. Kumar, J. John, V. Govindaraju, N. A. Khan, and P. Raghunathan, *Neurosci. Res.* **24,** 207 (1994).

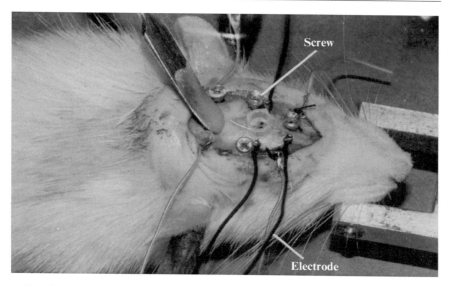

Fig. 5. An anesthetized rat fixed to the stereotaxic frame with implanted screws and electrodes.

Later, a mold, as shown in Fig. 6, is prepared using acrylic dental cement for painless restraining of the head of the rat in the stereotaxic apparatus as per the procedure described by Nakamura and Ono.[26] For this, liquid cement is poured over the skull of the rat in such a way that it spreads around the anchoring screws. A brass head holder is used for preparing the dummy mold on the skull. For preparation of the dummy mold on the skull of the rat, petroleum jelly is applied on the surface of the (fake) bars of the brass head holder and is screwed to the fixed ear bar of the surgical stereotaxic frame. Thereafter, thin layers of dental cement are applied over these bars. After the dental cement hardens, the brass head holder is removed to give two conical depressions (holes) on both sides to fix the head of the animal.

During subsequent restraining, nonmagnetic lateral bars are inserted into these hollow conical impressions of the dental mold for painless fixing of the head of the rat in the fabricated stereotaxic restrainer apparatus. Gentamycin (50 mg/kg body weight) is given intramuscularly for 3 days postoperatively. The implanted rats housed in separate cages are monitored periodically for 4 to 7 days during the period of recovery. Food and water are provided *ad libitum*. Recovery from postoperative trauma is

[26] K. Nakamura and T. Ono, *J. Neuro. Physiol.* **55**, 163 (1986).

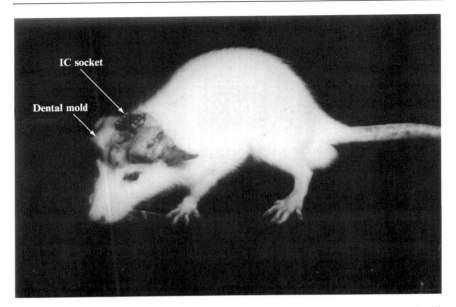

FIG. 6. An implanted rat showing the dental mold and the IC socket. Reproduced with permission from Khubchandani *et al.*[21]

assessed by observing their food and water intake and by their general behavior. The rats are handled at least once a day during the recovery period so that they become docile and easy to handle subsequently. This is important in this type of study because it involves handling of conscious animals during subsequent restraining and training. The rats are familiarized to the MR environment once again (i.e., to the sound of gradient magnetic fields).

Training of Animals

This is an essential aspect of the study, and the procedure of training is carried out in different stages. Before the implantation of electrodes, normal rats in cages are placed inside the MR suite for 3–4 days to acclimatize to the gradient magnetic field noise. Electrodes for the electrophysiological recordings are implanted as described earlier on those rats acclimatized to the noise. During and after postoperative recovery, the rats (in cages) are again kept in the MR suite for 2 to 3 days.

Initially, rats are positioned in the stereotaxic restrainer apparatus with the body flap only for an hour for 2 to 3 days; later the amount of time for which they are kept in the restrainer is increased to 3 to 4 h/day. Thereafter, the head of the animal is also restrained with the help of lateral bars

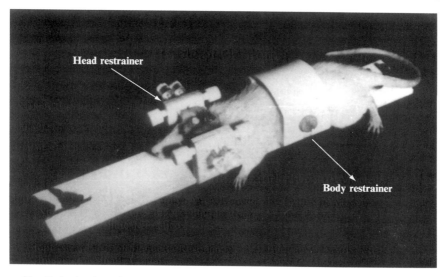

Fɪɢ. 7. An implanted rat restrained in the fabricated nonmagnetic stereotaxic apparatus. Reproduced with permission from Khubchandani *et al.*[21]

that are inserted gently into the two conical depressions of the dental mold. Figure 7 shows an implanted rat restrained in the fabricated stereotaxic restrainer apparatus.

No anesthesia or sedation is used to position the rats in the restrainer. No food and water are given during the period of restraining. However, they are rewarded with glucose water after every successful restraining session. The time during which the rats are kept in the restrainer is increased slowly from a few minutes initially to a few hours per day. At the end of the training session the restrained conscious rats can remain immobilized inside the magnet bore for 4 to 5 h. Well-trained rats cooperate, and there is no difficulty positioning them in the restrainer. However, some rats remove their dental mold during the training period.

Sleep Deprivation

Rats are sleep deprived by placing them in a sleep deprivation chamber that consists of a circular cage (wire mash) connected electrically to a motor, and the cage rotates at a rate of one revolution per minute. This prevents the rat from staying in one position and from falling asleep. Water and food are provided to the rats inside the cage. While the diet is kept inside the chamber, water is provided to the rat through a hole in the wire mesh at frequent intervals.

Initially, the rats are familiarized to the sleep deprivation chamber by placing them in it for 2 to 3 h. Later, the time of sleep deprivation is increased to 18 to 24 h slowly in several steps. Then they are trained to remain motionless in the stereotaxic restrainer apparatus after sleep deprivation inside the MR environment. This training is important, as most of the rats become restless after sleep deprivation and also do not cooperate further for restraining. In the next step, sleep-deprived rats are exposed to magnetic field gradient noises by keeping them inside the magnet room for extended periods of time (6 to 7 h/day). With this training, they get acclimatized to sleep with gradient magnetic-field noise after sleep deprivation. This training period takes another 2 to 3 weeks. In total, about 2 months are required to train a rat. Once the rats are trained successfully, simultaneous electrophysiology and fMRI experiments can be carried out during sleep–wakefulness. An outline of the various steps undertaken in order to carry out simultaneous electrophysiological and MRI recordings in the rat brain is given as a flowchart in Table I.

Electrophysiological Recordings

Electrophysiological recordings are performed for the assessment of sleep–wakefulness (S–W) using a bioamplifier (S-75; Coulbourn Instruments, Allentown, PA) with low and high filters set at 0.1 and 13 Hz for EOG, 1 and 150 Hz for EEG, and 1 and 1000 Hz for EMG. Output signals from these amplifiers are recorded using a thermal array recorder (WR 7700; Graphtec Corp., Japan). The sensitivity of the different channels is kept at 75 μV/cm for EEG, 20 μV/cm for EMG, and 150 μV/cm μV/cm for EOG. The final calibration is taken keeping the calibrator switch at 100 μV. The final calibrations are 100 μV/1 cm for EEG, 100 μV/3 cm for EMG, and 100 μV/0.5 cm for EOG by adjusting the gains. Calibrations are taken before and after S–W recordings. The S–W recordings are carried out at 2 and 5 mm/s. Animals are also monitored visually during the recordings.

The electrophysiological tracings of S–W are analyzed with certain modifications[21] in comparison to the classical method.[22] Low-frequency, high-amplitude EEG and decreased EOG and EMG activity are characterized as the sleep state, whereas high-frequency, low-amplitude EEG and increased EOG and EMG activity are characterized as the wakeful state.

MRI Experiments

MRI experiments are carried out on a 4.7-T animal MR scanner (BIOSPEC, Bruker, Germany) with a gradient strength of 231 mT/m and a rise time of 450 μs. A proton RF volume resonator of 72 mm diameter

TABLE I
FLOWCHART OF THE VARIOUS STEPS USED TO CARRY OUT SIMULTANEOUS ELECTROPHYSIOLOGICAL
AND fMRI RECORDINGS IN CONSCIOUS RATS

Habituating normal rats to gradient noises and MR environment
⇓
Implantation of EOG, EEG, and EMG electrodes
⇓
Postoperative recovery from surgery for 5–7 days
⇓
S–W electrophysiological recordings (EEG, EOG, EMG) and handling the animal
⇓
Habituating the implanted rat to the restrainer
⇓
Restraining the implanted rat using a head and body restrainer
⇓
Habituating the rat to remain restrained inside magnet bore with gradient noises
⇓
Electrophysiological recordings from restrained rat placed outside the magnet
⇓
Electrophysiological recordings from the restrained rat placed inside the magnet
bore with gradients on
⇓
Acclimatizing the rat to the sleep deprivation chamber
⇓
Acclimatizing the rat to the restrainer after 24 h of sleep deprivation
⇓
Acclimatizing the rat to gradient noises after sleep deprivation
⇓
Simultaneous electrophysiological and fMRI recordings from a 24-h sleep-deprived rat
⇓
Analysis of electrophysiological recordings for pure epochs of sleep or wakeful state
⇓
Analysis of fMRI data

is used. Global shimming is performed for all the rats so as to achieve the line width of 40–60 Hz. MR images of the rat brain are obtained using the fast T_2-weighted rapid aquisition by relaxation enhancement (RARE)[27] and T_2-weighted gradient echo Fourier imaging (GEFI) sequence. The MR acquisition parameters used for RARE are Tr, 2000 ms; Te, 25 ms; slice thickness, 2 mm; matrix size, 256×256; FOV, 6 cm; number of slices, 3; RARE factor (number of echoes), 8; and total experimental time for single average, 1 min 10 s. For two averages, the experimental time is 2 min 20 s. The MR acquisition parameters for GEFI are Tr, 300 ms;

[27] J. Hennig, A. Nauerth, and H. Friedburg, *Magn. Reson. Med.* **3,** 823 (1986).

Te, 5.8 ms; pulse length, 2000 μs; slice thickness, 2 mm; matrix size, 256 × 256; FOV, 6 cm; number of slices, 3; and total experimental time for single average, 1 min 16 s. Because the ultimate goal is to study the sleep–wakefulness in a conscious rat, three transverse slices, with one of them passing through the preoptic area, are acquired using the aforementioned sequences. Adequate care is taken to maintain the body temperature of the rat by covering it with cotton padding.

Results and Observations

In Vivo MR studies require anesthesia of the animal in order to position them correctly with the region of interest (ROI) at the iso center of the magnet/RF probe, as well as to avoid any body movement during the course of the experiment. Because the objective of the present study is to study sleep–wakefulness, no anesthesia was used to position the rats. Also, no drugs were given to the rats for sleep induction during the entire course of the experiment. Wyrwicz *et al.*[28] reported motion artifact-free images in conscious rabbits during visual activation. Similarly, Lahti *et al.*[29] reported motion-free images from the conscious rat brain using a holder. However, in their study, the animals were anesthetized lightly with an ip injection of chloral hydrate for positioning them in the restrainer assembly. Tabuchi *et al.*[30] reported fMRI studies on conscious animals during drinking activity. However, no details were given as how the animals were trained to remain motionless in the restraining assembly. In general, simultaneous electrophysiological recordings and fMRI provide a unique method to study the brain state with good spatial and temporal resolution, but such studies are associated with a number of technical limitations.

Artifacts in MR Images

Random movement or motion of the animal induces artifact in images and is a source of concern in MR investigations.[29] This is especially true in fMRI, where movement can be mistaken for brain activation. Bite plates and head rests have been used successfully in human studies. However, such devices do not have any relevance in experimental research using animal models. Thus it is important to restrict movements of the animal

[28] A. M. Wyrwicz, N. Chen, L. Li, C. Weiss, and J. F. Disterhoft, *Magn. Reson. Med.* **44**, 474 (2000).

[29] K. M. Lahti, C. F. Ferris, F. Li, C. H. Sotak, and J. A. King, *Magn. Reson. Med.* **41**, 412 (1998).

[30] E. Tabuchi, T. Yokawa, H. Mallick, T. Inubushi, T. Kondoh, T. Ono, and K. Tori, *Brain Res.* **951**, 270 (2002).

by restraining it in a stereotaxic apparatus for functional studies, which was accomplished successfully in this study. With proper training it is possible to restrain (immobilize) a conscious animal in the stereotaxic restrainer continuously for 4 to 5 h as demonstrated in this work.

The evaluation of artifacts (if any) due to random motion of the animal during MR data acquisition and between different sampling/acquisitions over a period of time was carried out in the following manner. Motion artifact during a single data acquisition was assessed visually (blurring of the image) and between data samples were assessed by subtracting the images of each data set from the initial baseline MR image on a pixel-by-pixel basis. This was accomplished by acquiring the MR images at different periods of time (e.g., every 30 min) up to 5 h using both RARE and GEFI sequences. Subtraction of the images acquired at various time points showed no signals (blank image), as shown in Figs. 8 and 9, confirming that images without motion artifacts could be acquired reliably even after 5 h.

Even though the head of the rat is immobilized in the stereotaxic head holder, any slight movement would induce artifacts in the MR images acquired. Despite the training given to rats, occasional head and/or body movements are observed. Both electrophysiological tracings and MR images (Fig. 10) acquired during movements clearly showed that these artifacts could be identified and discarded easily.

Another area of concern with this new method of immobilization is the stress to the animals. Habituating the animals in a restrainer for increasing periods of time before the actual MR study decreased both the movement and the stress.[31] Restraint stress may change brain activity in limbic structures involved in the perception and emotional response to stress. Physiological effects of restraint stress are currently under investigation. Preliminary measurements of blood gases, mean arterial blood pressure, heart rate, respiration rate, and plasma cortisol (a well-established indicator of stress) demonstrated that acclimatization reduces restraint stress markedly.[32]

Susceptibility and Other Artifacts

GE sequences are more sensitive than SE sequences in detecting BOLD phenomenon and are used commonly in most simultaneous EEG–fMRI studies.[33] However, GE sequences are sensitive to susceptibility artifacts, especially when conductive materials are used to record the

[31] M. Konarska, R. E. Sterwart, and R. McCarthy, *Physiol. Behav.* **45,** 255 (1989).
[32] Q. Shen, K. Sicard, and T. Q. Duong, *11th Proc. ISMRM* abstract 352, 2003.
[33] C. R. Fisel, J. L. Ackerman, R. B. Buxton, L. Garrido, J. W. Belliveau, B. R. Rosen, and T. J. Brady, *Magn. Reson. Med.* **17,** 336 (1991).

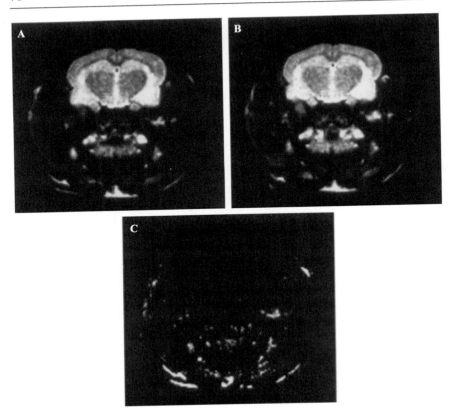

FIG. 8. Conscious rat brain images obtained using the RARE sequence at different time intervals: (A) at 0 h, (B) at 5 h, and (C) subtraction of [B–A] showing no movement artifacts.

electrophysiological tracings. Thus it is important to investigate the effect of individual components of the electrophysiological assembly on the quality of images obtained using the GE sequence on phantoms. The magnetic susceptibility differences in electrode assembly can lead to local signal dropout in MR images. However, with the use of nonmagnetic and biocompatible materials as electrodes, susceptibility artifact-free MR images from rat brain could be obtained.

Busch et al.[34] reported no susceptibility artifacts due to the use of nonmagnetic electrodes in MR images in their study on the simultaneous

[34] E. Busch, M. Hoehn-Berlage, M. Eis Manfred, M. Gyngell, and K. A. Hossmann, NMR Biomed. **8,** 59 (1995).

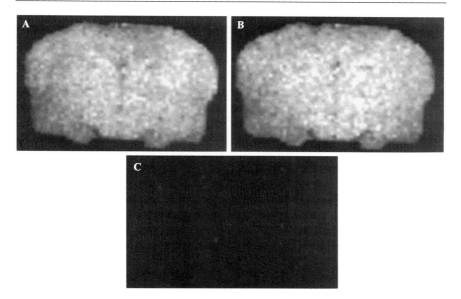

FIG. 9. ROI of rat brain images obtained using a GEFI pulse sequence at different time intervals: (A) at 0 h, (B) at 5 h, and (C) subtraction of [B–A] showing no movement artifacts. Reproduced with permission from Khubchandani *et al.*[21]

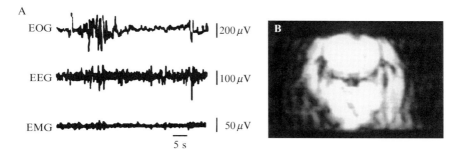

FIG. 10. Simultaneous electrophysiological and MRI acquisition: Electrophysiological tracings for an awake epoch (A) and transverse MR image (B).

recording of EEG, DC potential, and MRI in anesthetized rats. Similar observations of artifact-free images using electrodes have also been reported in humans.[18,35]

[35] R. M. Muri, J. Felblinger, K. M. Rosler, B. Jung, C. W. Hess, and C. Boesch, *Magn. Reson. Med.* **39,** 18 (1998).

A certain amount of detuning of the coil was noticed when rats with electrophysiological assembly (implanted electrodes and input leads) were placed in the magnet bore compared with rats that were not implanted with electrodes. This detuning may lead to a reduced signal-to-noise ratio (SNR) ratio. The SNR obtained using GE and SE sequences was compared (implanted versus normal rat), and a decrease of 7–10% in SNR was observed. This slight reduction in SNR did not affect fMRI data adversely.

Quality of Electrophysiological Recordings

Another area of concern is the distortion of the electrophysiological recordings during MRI. The electrical potentials generated in the brain are generally weak signals with an amplitude of only few microvolts. The recording of such signals is complicated by artifacts that arise mainly from the static and gradient magnetic fields and the RF pulses employed during MR acquisition. In order to verify whether artifact-free electrophysiological recordings could be obtained from rats placed inside the MR scanner, several pilot experiments with free-moving and sleep-deprived restrained rats were carried out. For this purpose, the electrophysiological recordings were obtained from sleep-deprived restrained rats placed outside and inside the magnetic environment with the gradient magnetic fields switched off (Fig. 11C and D) and on (Fig. 11E and F). There were no distortions in the electrophysiological tracings due to static magnetic field or gradient magnetic fields, and both the sleep and the awake states were distinguished easily in all the rats. The quality of the electrophysiological recordings obtained allowed proper analysis of different brain states.

To further check the validity of electrophysiological recordings, a simple experiment was carried out on implanted sleep-deprived rats kept inside the magnet bore. The tracings were recorded with and without gradient magnetic fields using both SE and GE sequences. As soon as sleep epoch was observed, a stimulus was given to the rat (such as twitching its whisker), in which the rat became awake, which was evident visually, as well as from analysis of EEG waveforms. From Fig. 12 it is clear that high-amplitude, low-frequency EEG waveforms (sleep state) changed to low-amplitude, high-frequency waveforms (awake state). Based on experiments carried out on 40 rats ($n = 40$), it is our experience that with the procedures developed, distortionless electrophysiological recordings could be obtained.

Other Precautionary Measures Adopted

In addition, several other precautions were followed. Care was taken to ensure that no loops were present in the leads used for electrophysiological recordings. This is because the time-varying gradient magnetic fields used

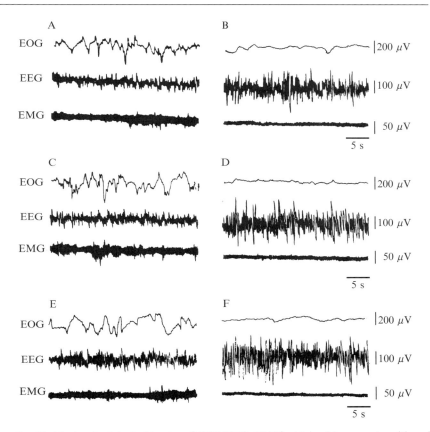

FIG. 11. Electrophysiological tracings (EOG, EEG, EMG) obtained from a rat positioned in the restrainer. (A) Outside the magnetic environment/magnet room showing the wakeful state. (B) Outside the magnetic environment/magnet room showing the sleep state. (C) Inside the magnet bore with gradients switched off showing the wakeful state. (D) Inside the magnet bore with gradients switched off showing the sleep state. (E) Inside the magnet bore with gradients switched on during the wakeful state. (F) Inside the magnet bore with gradients switched on showing the sleep state. Reproduced with permission from Khubchandani et al.[21]

during MR acquisition may induce an electromotive force (EMF) in a wire loop perpendicular to the main magnetic field (B_0) field direction. This EMF, by Lenz's law, is proportional to the cross-sectional area of the wire loop and to the rate of change of the perpendicular magnetic field (dB_0/dt). Thus, when leads used for electrophysiological recordings are placed inside the MR scanner, the rapidly changing gradient magnetic fields and the RF pulses may induce voltages that may obscure the electrophysiological

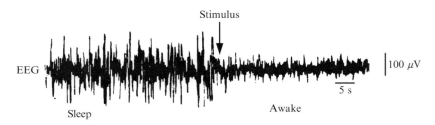

FIG. 12. EEG waveforms obtained from a rat placed inside the magnet bore showing transition from the sleep state to the awake state in response to stimulus.

signal.[35] Moreover, the induced EMF can heat the electrodes and may cause burns to the subject.

The RF-shielded cable used for electrophysiological recordings was fixed to a stationary surface inside the magnet. This was necessary because any motion in the magnetic field may generate voltages that may interfere with the electrophysiological signal, leading to artifacts in the bioelectrical signals. These arise as a consequence of the Lorentz force.[36] This effect may be physically manifest because of voluntary head movement of the subject, or cardiac-related pulsations of body, head, and blood flow.[15,35,37,38] Motion of the leads themselves within the static field of the magnet also induces EMF.

Because the electrodes in our study were implanted on the skull of the rat, no special precautions were felt necessary to avoid RF interference. The electrophysiological recordings obtained from rats placed inside the magnet were of acceptable quality, and both sleep and awake states were distinguished easily.

Safety Aspects

The safety-related aspect of subjects is another area of concern in these studies. These are due to the induction of currents in the electrophysiological electrodes and leads by the various electromagnetic fields used in fMRI. This involves the risk of skin burning due to the RF interaction with the electrodes.[39] Although general guidelines for physiological monitoring,

[36] J. D. Jackson, "Classical Electrodynamics." Wiley, New York, 1975.
[37] J. R. Ives, S. Warach, F. Schmitt, R. R. Edelman, and D. L. Schomer, *EEG Clin. Neurophysiol.* **87,** 417 (1993).
[38] R. I. Goldman, J. M. Stern, J. Engel, Jr., and M. S. Cohen, *Clin. Neurophysiol.* **111,** 1974 (2000).
[39] L. Lemieux, P. J. Allen, F. Franconi, M. R. Symms, and D. R. Fish, *Magn. Reson. Med.* **38,** 943 (1997).

mainly derived from experience with ECG monitoring inside MR scanners in human subjects, have been published,[40–42] there is a very limited amount of literature available for EEG monitoring inside MR scanners in animal systems.

Whereas currents are invariably induced in the body in the course of normal MR imaging, resulting in a certain amount of heating (which is limited to specific levels under current safety regulations), conducting loops provide a concentration of current in metallic components and therefore a high current density in the adjacent tissue. These represent a health risk and can result in ulcers, electric shock, or stimulation and heating.[43] The safety precautions followed here consisted of avoiding loops in electrode leads and putting current-limiting resistors in series with each electrode lead.[15,18]

The major concern is during conventional SE imaging, which has the highest RF transmission rate, leading to burning sensations in subjects[40] above SAR limitations. Care was taken such that the power disposition to the animal is minimal. This is especially important for the high-power disposition pulse sequences such as spin echo (RARE). Moreover, none of the trained rats showed any discomfort or any signs of physical injuries during simultaneous electrophysiological recordings and fMRI.

Conclusions

This Chapter described the various methodological procedures developed for simultaneous electrophysiology and fMRI in conscious rats. The design and fabrication of a stereotaxic restraining apparatus using nonmagnetic materials for the immobilization of conscious rats for MRI experiments were presented. Further, it was demonstrated that with the use of nonmagnetic electrodes, artifact-free fMR images could be obtained from the implanted rat. Similarly, no distortions were observed in the electrophysiological recordings obtained during MRI and the tracings were easily interpretable. Such studies provide a new approach in studying the relationship between electrophysiological activity and MR visible changes in blood flow, blood oxygenation, and their relationship with sleep and other physiological parameters.

[40] F. G. Shellock and E. Kanal, "Magnetic Resonance: Bioeffects, Safety, and Patient Management." Lippincott-Raven, Philadelphia, 1996.
[41] F. G. Shellock and G. L. Slimp, Am. J. Roentgenol. 153, 1105 (1989).
[42] T. R. Brown, B. Goldstein, and J. Little, Am. J. Phys. Med. Rehabil. 72, 166 (1993).
[43] D. J. Schaefer, in "The Physics of MRI—1992 AAPM Summer School Proceedings" (M. J. Bronskill and P. Sprawls, eds.). AAPM, Woodbury, New York, 1993.

Acknowledgments

The grant from the Indian Council of Medical Research (ICMR–No. 55/3/99-BMS II) for carrying out this research work is acknowledged. The Senior Research Fellowship to MK is provided by the Council for Scientific and Industrial Research (CSIR). We thank Professor V. Mohan Kumar and Dr. H. N. Mallick for providing the facility to carry out implantation and for fruitful discussions. We also thank Dr. Takashi Kondoh, Chief Physiologist, Institute of Life Sciences, Ajinomoto Co., Inc., Japan, for providing us with nonmagnetic plastic screws.

[5] Functional Magnetic Resonance Imaging of Macaque Monkeys

By Kiyoshi Nakahara

Introduction

Functional magnetic resonance imaging (fMRI) is a noninvasive functional brain-imaging method that utilizes blood oxygenation level-dependent (BOLD) signals as a measure of brain activation.[1,2] This method can be applied to human subjects while they perform certain cognitive tasks and can be used to visualize a map of regional brain activation correlated with the cognitive functions required to perform the task across the whole brain. Although not much time has passed since the discovery of the BOLD signal, this method has been used in thousands of researches on a wide variety of human brain functions, from perception to high-level cognitive functions, and has become a major tool in cognitive neuroscience.[3–5] In the late 1990s, some laboratories began to apply fMRI to macaque monkeys.[6–10] Why should monkey fMRI be developed in addition to human fMRI? What are the advantages?

---

[1] S. Ogawa, T. M. Lee, A. R. Kay, and D. W. Tank, *Proc. Natl. Acad. Sci. USA* **87,** 9868 (1990).

[2] S. Ogawa, D. W. Tank, R. Menon, J. M. Ellermann, S.-G. Kim, H. Merkle, and K. Ugurbil, *Proc. Natl. Acad. Sci. USA* **89,** 5951 (1992).

[3] S. M. Courtney and L. G. Ungerleider, *Curr. Opin. Neurobiol.* **7,** 554 (1997).

[4] R. S. J. Frackowiak, K. J. Friston, C. D. Frith, R. J. Dolan, and J. C. Mazziotta, "Human Brain Function." Academic Press, San Diego, 1997.

[5] M. E. Raichle, *Proc. Natl. Acad. Sci. USA* **95,** 765 (1998).

[6] D. J. Dubowitz, D.-Y. Chen, D. J. Atkinson, K. L. Grieve, B. Gillikin, W. G. Bradley, Jr., and R. A. Andersen, *Neuroreport* **9,** 2213 (1998).

[7] L. Stefanacci, P. Reber, J. Costanza, E. Wong, R. Buxton, Z. Stuart, L. Squire, and T. Albright, *Neuron* **20,** 1051 (1998).

[8] N. K. Logothetis, H. Guggenberger, S. Peled, and J. Pauls, *Nat. Neurosci.* **2,** 555 (1999).

Potential of Functional MRI in Macaque Monkeys

First, this method offers opportunities to make direct comparisons of the functional architectures of the brains of humans and monkeys. Macaque monkeys have been used over the past half century as a major experimental model for humans, and a vast amount of physiological and anatomical knowledge of brain function has been accumulated using invasive but informative methods, such as microelectrode recordings, tracer injections, microstimulation, inactivation, and experimental lesions. Obviously, comparisons and integration of results from human studies with fMRI and monkey studies with these methods are very important for obtaining a more precise view of our brain mechanisms. However, such attempts have been hampered by species and methodological differences. A straightforward solution is to apply the same physiological method under the same experimental design to both monkey and human subjects. fMRI of macaque monkeys is an ideal tool for such investigations and could provide a bridge between humans and monkeys.[11,12]

Second, this method could be used to clarify the relationships between the BOLD signals and the electrical activities of neurons. Thousands of studies on human cognitive function using fMRI have already been reported. However, the physiological origin and the nature of the BOLD signal are largely unknown. Measuring the electrical activities and the BOLD signal simultaneously in macaque monkeys will be a promising approach to this issue. One such study reported that the BOLD signal showed a greater correlation with the local field potential compared with multi- or single unit activity.[13] Further experiments in line with this kind of approach will clarify a precise interpretation of the observed BOLD signal, which seems to be essential for bringing fMRI to completion as a "physiological" tool.[14]

Third, this method can be a strong complement to traditional microelectrode recordings. Microelectrode recordings provide microscopic information about electrical activities from the level of single neuron to local neuronal circuits. However, cognitive functions are executed by the coordinated activities of distributed neuronal networks across several cortical areas, and the use of microelectrode recordings alone presents difficulties

[9] T. Hayashi, S. Konishi, I. Hasegawa, and Y. Miyashita, *Eur. J. Neurosci.* **11,** 4451 (1999).

[10] W. Vanduffel, D. Fize, J. B. Mandeville, K. Nelissen, P. Van Hecke, B. R. Rosen, R. B. H. Tootell, and G. A. Orban, *Neuron* **32,** 565 (2001).

[11] K. Nakahara, T. Hayashi, S. Konishi, and Y. Miyashita, *Science* **295,** 1532 (2002).

[12] R. B. H. Tootell, D. Tsao, and W. Vanduffel, *J. Neurosci.* **23,** 3981 (2003).

[13] N. K. Logothetis, J. Pauls, M. Augath, T. Trinath, and A. Oeltermann, *Nature* **412,** 150 (2001).

[14] S.-G. Kim, *Proc. Natl. Acad. Sci. USA* **100,** 3550 (2003).

in seeing how local neuronal activities participate in the global networks. fMRI in monkeys can depict a macroscopic activation map across the whole brain, which could complement the microscopic information obtained by microelectrode recordings.[15] Moreover, this method can serve as a guide for choosing brain areas for microelectrode recordings. When researchers want to start research using microelectrode recordings, and if the responsive brain regions are unknown, the choice of recording sites usually relies on previous information from anatomical and lesion studies, which often give only an estimation. Using fMRI, researchers can identify multiple responsive brain regions rapidly and obtain a navigation map for the appropriate choice of electrode penetration sites.

Finally, other than these inherent potential benefits, fMRI in monkeys would be of unexpected value to researchers in cognitive neuroscience.[16] Its potential seems to be even greater when it is combined with other invasive methods, such as experimentally placed lesions, microstimulation, reversible inactivation, and labeling of fiber projection using contrast agents.[17]

Brief History of fMRI in Macaque Monkeys

The earliest applications of fMRI to macaque monkeys were reported independently by Dubowitz et al.[6] and Stefanacci et al.[7] in 1998. They scanned awake macaque monkeys with conventional clinical 1.5-T MR scanners while the subjects observed visual stimuli passively and demonstrated the stimulus-related BOLD signal in the visual cortex of the monkey. Some laboratories made efforts to develop MR systems specially adapted to monkey experiments. Logothetis's group developed a customized MR system equipped with a high-field (4.7 T) vertical magnet specially designed for monkey studies and achieved much progress in several aspects of imaging techniques, including simultaneous fMRI microelectrode recordings and ultra high-resolution fMRI (voxel size = 0.0113 μl) using implantable radiofrequency (RF) coils.[8,13,17,18] Orban's group took a different approach to improving fMRI sensitivity in a normal magnetic field (1.5–3 T) rather than using a high-field magnet.[10,19] They injected an MR

[15] D. A. Leopold, Y. Murayama, and N. K. Logothetis, *Cerebr. Cortex* **13**, 422 (2003).

[16] Y. Miyashita and T. Hayashi, *Curr. Opin. Neurobiol.* **10**, 187 (2000).

[17] K. S. Saleem, J. M. Pauls, M. Augath, T. Trinath, B. A. Prause, T. Hashikawa, and N. K. Logothetis, *Neuron* **34**, 685 (2002).

[18] N. K. Logothetis, H. Merkle, M. Augath, T. Trinath, and K. Ugurbil, *Neuron* **35**, 227 (2002).

[19] F. P. Leite, D. Tsao, W. Vanduffel, D. Fize, Y. Sasaki, L. L. Wald, A. M. Dale, K. K. Kwong, G. A. Orban, B. R. Rosen, R. B. H. Tootell, and J. B. Mandeville, *Neuroimage* **16**, 283 (2002).

contrast agent (dextran-coated monocrystalline iron oxide nanoparticles; MION) intravenously into monkey subjects and achieved increments of signal sensitivity and more precise signal localization relative to the conventional BOLD technique. Our laboratory also started fMRI of macaque monkeys at an early stage. Our first monkey fMRI experiment was reported in 1999, where we mapped somatosensory activation in anesthetized monkeys using a conventional 1.5-T machine and demonstrated the ability to discriminate topographical organization in two adjacent functional areas: hand and face representations in primary and secondary somatosensory cortices.[9] In the next study, we succeeded in using fMRI of conscious monkeys performing a high-level cognitive task and made a direct comparisons of the functional organization of the prefrontal cortex of humans and monkeys.[11] The next section introduces current experimental procedures using a 1.5-T scanner with anesthetized and awake, task-performing monkeys (Fig. 1).

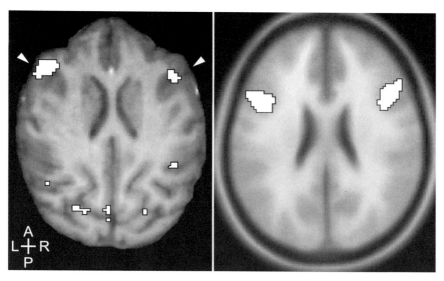

FIG. 1. Comparative fMRI of macaque monkeys and humans. During the fMRI scans, both the monkey and the human subjects performed the Wisconsin card-sorting task, which required cognitive set shifting, a characteristic function of the prefrontal cortex. Event-related BOLD activation correlated with the cognitive set shifting was analyzed with SPM99 software, and cortical areas that showed significant activation were superimposed on transverse sections of the anatomical images (white pixels). Prominent prefrontal activation is observed in the bilateral bank of the arcuate sulcus in monkeys (left) and in the bilateral inferior frontal sulcus in humans (right).[11] The statistical threshold was set at $p < 0.001$ (uncorrected) with conjunction analysis in monkeys and with random-effect analysis in humans. A, anterior; P, posterior; R, right; L, left. Arrowheads; arcuate sulcus.

Experimental Procedures

Animal Experiments

All of our experimental procedures are carried out in full compliance with the NIH guidelines for the care and use of laboratory animals and the regulations of the University of Tokyo School of Medicine.

fMRI in Anesthetized Monkeys: Anesthetic Procedures

Because some anesthetic agents can influence cerebral blood flow or oxygen consumption, special care is needed in the selection of the anesthetic procedures for monkey fMRI. As a matter of fact, in our pilot studies, we failed to acquire BOLD signals in monkeys anesthetized with pentobarbital. After a survey of several anesthetic agents, we found that the BOLD signal can be obtained reproducibly in monkeys anesthetized with droperidol or propofol. Droperidol was reported to show no significant effect on cerebral blood flow and oxygen consumption in humans.[20,21] Although propofol was reported to decrease both cerebral blood flow and oxygen consumption, it preserved vascular reactivity in humans and was feasible in fMRI of rats.[22,23] A careful use of gas anesthesia together with muscle relaxant was also reported to be successful in high-field monkey fMRI.[8] In our current protocol, monkeys are anesthetized with either intravenous droperidol (0.5–1.5 mg/kg/h; Sankyo, Tokyo, Japan) or intravenous propofol (3–7.5 mg/kg/h; Zeneca, Tokyo, Japan) during fMRI scans.

fMRI in Awake Monkeys

For fMRI in awake monkeys, a major concern is how to make the subjects behave cooperatively in the noisy and small space of the bore of the MR machine. The animals must not only keep still, but also perform a cognitive task during fMRI scans. To realize this, monkeys are trained in three steps. In the first step, monkey subjects are trained on a cognitive task of interest while sitting vertically in a standard primate chair. After they achieve good performances (>90%) on the task, they go to the second step,

[20] A. Sari, Y. Okuda, and H. Takeshita, *Br. J. Anaesth.* **44,** 330 (1972).
[21] J. E. Cottrell and D. S. Smith, "Anesthesia and Neurosurgery." Mosby, St. Louis, 1994.
[22] H. Stephan, H. Sonntag, H. D. Schenk, and S. Kohlhausen, *Anaesthesist* **36,** 60 (1987).
[23] B. E. Scanley, R. P. Kennan, S. Cannan, P. Skudlarski, R. B. Innis, and J. C. Gore, *Magn. Reson. Med.* **37,** 969 (1997).

where the subjects are habituated in performing the task in the posture and noise of MR scans. In this stage, the monkeys are laid in a custom-built horizontal monkey container, which is also used in the actual scans, and trained to perform the task in it. The sound of recorded MR noise is sometimes played back during the second step training. Use of a mock-up MR bore at this stage makes habituation training more effective. Finally, the subjects are trained to perform the task in a real MR machine. With these habituation procedures, monkeys can adapt easily to performing the cognitive tasks quietly in the MR bore without showing considerable additional body movements.

Some Apparatus

All of the apparatus used in MR bore must be made of nonferrous materials and should be preexamined for effects on image quality using a control MR phantom. We usually use acrylic, ceramic, polysulphone, or Delrin (DuPont) according to the purpose. During scans, monkeys lie in a custom-made acrylic monkey container, which is anchored to the bed of the MR machine. The monkey container can be equipped with fiber-optic-based MR-compatible buttons (Omron, Kyoto, Japan), which are used for the responses of the subjects. Prior to MR scans, custom-made head posts (polysulphone) are implanted in the skull of the monkey using ceramic screws (zirconium oxide; Kyocera, Kyoto, Japan) under general anesthesia in aseptic conditions. With these head posts, the head of the monkey is fixed to a custom-built head holder (acrylic), which is anchored to the bed and makes the head position constant relative to the magnet. Usually the observed head movements are less than 0.5 mm throughout each scan session (about 10 min). A custom-built acrylic bite bar is used to minimize licking movements. A liquid reward can be delivered through this bite bar. Visual stimuli are back projected from an LCD projector (Sony, Tokyo, Japan).

Scan Procedures

Functional imaging of monkeys is performed on a clinical 1.5-T scanner (Stratis II; Hitachi Medical Corp., Tokyo, Japan). A knee coil (quadrature, with an inner diameter of 190 mm) is used as a radio-frequency probe. Scan parameters should be optimized carefully for each MR system and each experimental design to obtain the best images. Usually, we use four-segmented gradient-echo echo-planar imaging, where FOV is 128×128 mm, Tr is 750 ms, Te is 18.4 or 20 ms, flip angle is $64°$, matrix is 64×64, slice thickness is 2 mm, and the interslice gap is 0.5 mm, with nine transverse slices. These parameters can be used for both anesthetized and conscious monkeys.

Data Analyses

Functional images are analyzed with SPM99 software package (http://www.fil.ion.ucl.ac.uk/spm/) running on MATLAB (Math Works, MA) with a Pentium-based PC. Functional images are first realigned, normalized spatially to a template with interpolation to a $1 \times 1 \times 1$-mm space, and then smoothed with a Gaussian kernel (FWHM 4–6 mm). The template is made from fine three-dimensional structural images of one monkey's whole brain (voxel size is $1 \times 1 \times 1$ mm) according to the standard procedure implemented in SPM99. The template is arranged in a bicommissural coordinate system where the origin is placed at the anterior commissure. This normalization procedure allows us to perform group analyses. In monkey experiments, because it is difficult to correct data from enough subjects for random effect analysis, conjunction analysis[24] is used for group analyses in our laboratory. Functional brain activation is modeled and analyzed statistically using a general linear model implemented in SPM99 software in both blocked and event-related designs.

Conclusion

fMRI in macaque monkeys is one of the most exciting tools for current and future neuroscience studies. We cannot only use this method alone, but we can also use it with other invasive methods in macaques. Moreover, we are now able to perform "parallel" researches into the mechanisms of primate brains using human and monkey fMRI. In conclusion, fMRI of macaque monkeys will open up new perspectives to cognitive neuroscience researchers and will provide insights into the brain mechanisms of human and nonhuman primates.

---

[24] K. J. Friston, A. P. Holmes, and K. J. Worsley, *Neuroimage* **10,** 1 (1999).

[6] Atlas Template Images for Nonhuman Primate Neuroimaging: Baboon and Macaque

By Kevin J. Black, Jonathan M. Koller,
Abraham Z. Snyder, and Joel S. Perlmutter

Introduction

Multisubject Studies in Atlas Space for Humans

Coregistration of functional brain images across many subjects is a technique that has been widely used for studies in humans. Some of the advantages to this multisubject approach include the ability to detect signals in regions not known *a priori,* reduced influence of individual anatomic variation, ease of analysis, and increased sensitivity to low-magnitude responses.[1–3] Combining data across human subjects is now most often accomplished using computerized, voxel-based image registration algorithms.[4,5] These programs rely on a high-quality template image that is usually related to a published atlas. Template images are readily available for human studies, but methods have lagged considerably for neuroimaging in other species.

Motivation for Neuroimaging Studies in Nonhuman Subjects

Neuroimaging in humans offers many advantages, but nonhuman species may be more appropriate for some studies, such as lesion models of human disease, drug development, pharmacologic investigations, and methods development. Baboons (*Papio* spp.) have been employed frequently in positron emission tomography (PET) studies due to their relatively large brain volume. Macaques of various species have long been used for functional neuroimaging studies[6] and have been trained to perform various tasks in the scanner.

Need for Three-Dimensional Atlas Template Image for Nonhuman Studies

Until recently, functional imaging studies in nonhuman species were analyzed as case reports or by combining numerical data extracted from

[1] P. T. Fox *et al., J. Comput. Assist. Tomogr.* **9,** 141 (1985).
[2] P. T. Fox *et al., J. Cereb. Blood Flow Metab.* **8,** 642 (1988).
[3] M. E. Raichle *et al., J. Cereb. Blood Flow Metab.* **11,** S364 (1991).
[4] M. A. Viergever *et al., in* "Image Processing" (M. H. Loew, ed.), Vol. 2434, p. 2. SPIE Proceedings, 1995.
[5] S. C. Strother *et al., J. Comput. Assist. Tomogr.* **18,** 954 (1994).
[6] J. S. Perlmutter *et al., J. Cereb. Blood Flow Metab.* **11,** 229 (1991).

functional images by reference to anatomic regions chosen *a priori.* Automated methods for combining nonhuman primate brain images across subjects not only convey an increased value to the images, but also lessen the need for physical restraint in images that can be acquired fast enough to minimize within-scan movement (e.g., [^{15}O]water images or echoplanar [EPI] images). These two advantages are favorable to animal welfare.

This Chapter describes the creation of template images for baboon brain using T$_1$-weighted magnetic resonance imaging (MRI) and PET [^{15}O]water blood flow images.[7] Similar templates were created for macaque brain.[8] Methods for creating the atlas template image have broader application and could be adapted for other species. In fact, similar methods were first used in generating a multisubject human template image conforming to the Talairach and Tournoux atlas.[9,10]

Our approach to testing our methods could also be applied to future work. Specifically, we examined the accuracy of the template vis-à-vis radiological landmarks and a photomicrographic atlas. Using two approaches to combining functional images across subjects, we confirmed both the accuracy of fit to the atlas and the accuracy of our voxel-based automatic image registration software. The various methods are discussed approximately in chronological order of development.

Methods

Development of the "b2k" Baboon Atlas

Subjects. The images used to create the MRI template derive from nine baboons, six male and three female. Ages for the animals ranged from approximately 4.5 to 20 years.

Image Acquisition. Using a 1.5-T Siemens scanner, we acquired sagittal magnetization prepared rapid gradient echo (MPRAGE) images with a voxel volume of ~1.25 mm^3 in sedated baboons. Higher image resolutions could be obtained by various techniques, but given the magnitude of residual anatomic variability after linear coregistration, the *de facto* resolution of commonly applied functional imaging methods, and the spatial smoothing used for cross-subject analysis, higher image resolution was felt to be superfluous. For some animals, two acquisitions were averaged to reduce noise.

---

[7] K. J. Black *et al., Neuroimage* **14,** 736 (2001).

[8] K. J. Black *et al., Neuroimage* **14,** 744 (2001).

[9] M. Corbetta *et al., Neuron* **21,** 761 (1998).

[10] J. Talairach and P. Tournoux, "Co-Planar Stereotaxic Atlas of the Human Brain." Theime Verlag, New York, 1988.

Image Registration. Automatic voxel-based registration of the individual images to the template was performed using in-house computer programs implemented on the Solaris operating system environment. Within-modality (individual MRI to template MRI) registration was performed by minimizing the variance of difference images, as described previously.[11] The result of each individual-to-template registration is a linear (12-parameter, or affine) transform of the individual image to the template image. No preparation is needed to run the registration programs unless an image is rotated by a significant amount ($>\sim30°$) relative to the template. In this case, the user must initialize the transformation manually to approximately account for the rotation before the automatic registration can be performed.

However, it is important to emphasize that other registration methods could be used, including the automatic image registration (AIR) method[12–14] or the approach of Friston and colleagues.[15,16] Laboratories familiar with one of these methods could use that approach.

Template Image Development. We created the template image by a "bootstrap" method that ensured registration of the template with the atlas, followed by iterative refinement of the averaged atlas image (Fig. 1).

The first step in creating the MRI template image is to transform each of the nine individual baboon MPRAGE images to atlas space. This "bootstrap" step is accomplished using a previously validated but labor-intensive method that requires expert identification of certain radiological landmarks.[17] The images are intensity scaled using a histogram method and averaged together voxelwise to create the initial template image, "template$_0$."

This image (top image in Fig. 2) demonstrates substantial spatial noise due to significant residual uncorrected spatial differences. This is addressed iteratively, with each iteration consisting of three steps. In the first iteration, each individual MPRAGE image is registered to template$_0$ using the within-modality registration method described in the previous section. These transformed images are then averaged, producing substantially improved uniformity of registration. This average is then registered to template$_0$ to prevent spatial drift and to ensure backward compatibility, yielding a new image, template$_1$ (center image in Fig. 2).

[11] A. Z. Snyder, *in* "Quantification of Brain Function Using PET" (R. Myers *et al.*, eds.). Academic Press, San Diego, 1996.
[12] R. P. Woods *et al.*, *J. Comput. Assist. Tomogr.* **16**, 620 (1992).
[13] R. P. Woods *et al.*, *J. Comput. Assist. Tomogr.* **14**, 536 (1993).
[14] R. P. Woods *et al.*, *J. Comput. Assist. Tomogr.* **22**, 153 (1998).
[15] K. J. Friston *et al.*, *Hum. Brain Mapp.* **2**, 165 (1995).
[16] J. Ashburner and K. J. Friston, *Hum. Brain Mapp.* **7**, 254 (1999).
[17] K. J. Black *et al.*, *J. Comput. Assist. Tomogr.* **21**, 881 (1997).

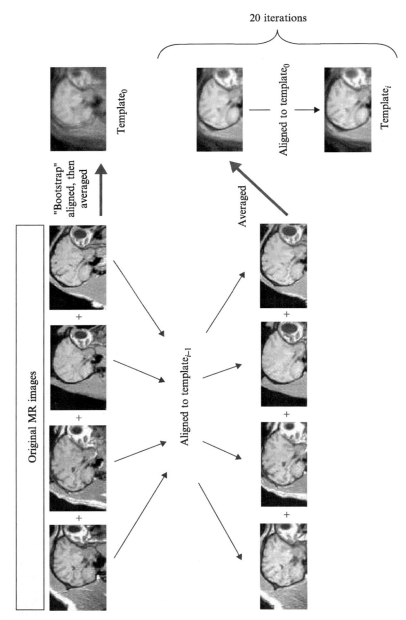

FIG. 1. Schematic diagram of the iterative process used to create b2k and n2k atlas templates.

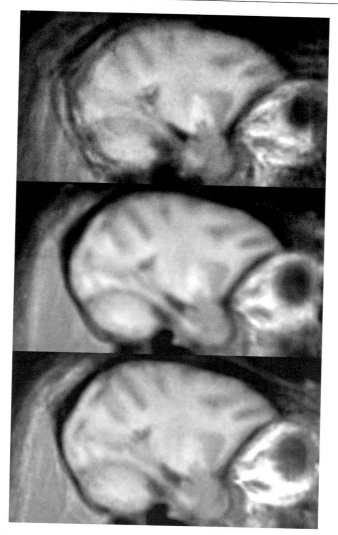

Fig. 2. Sagittal section from an image formed by averaging the transformed MPRAGE from each of nine normal baboons. (Top) Using our 1997 method (template$_0$). (Center) The first iteration using the new method (template$_1$). (Bottom) The final MRI template image (template$_{20}$ = "b2k"). For further details, see text. Note that the improvement in registration of extracranial structures is especially remarkable, as all transformations were computed after explicitly masking out these voxels in the target image. Reprinted from Black *et al.*,[7] with permission.

This process is iterated 20 times to create the final template image. Specifically, for $i = 1$ to 20, the individual MR images are registered to template$_{i-1}$, the most recent iteration of the average template image. These transformed individual images are averaged to create a (temporary) new average image. Third, this temporary image is registered to the template$_0$ image to ensure backward compatibility. At each step, template$_{i+1}$ is recreated from the original unfiltered MPRAGE images using matrix multiplication and a single resampling step, thus avoiding the accumulation of smoothing due to repeated interpolation. Because the coregistration algorithm matches the intensity of the image from each baboon, after averaging, each animal contributes approximately equally to the each iteration of the template. After the first few iterations, the transformation matrices describing registration of the individual images to the template image remain consistent. A C shell script automates this process. The final "b2k" template is template$_{20}$ (bottom image in Fig. 2).

Validation. Visual inspection of the new template image shows substantial preservation of the high-spatial-frequency image content (bottom image in Fig. 2).

For quantitative results, we tested absolute three-dimensional error (distance in mm) for 23 subcortical test points and 12 cortical sulcal landmarks. An expert rater identified these points in the published atlas[18] (for the subcortical landmarks) or in the final template atlas image (cortical landmarks). Later, the expert rater identified these same points on the individual images registered to the final template. The images were displayed without the atlas coordinates so that the expert rater was unaware of the "correct" answer. The mean error at each point (i.e., the Cartesian distance from the atlas coordinate to the point identified in the individual image, averaged across animals) ranged from 0.99 to 2.43 mm. The mean error averaged across all nine baboons and all subcortical test points was 1.53 mm, compared to a minimum possible mean error of 0.54 mm. The maximum error for any point in any animal was 3.96 mm. The mean error for the cortical test points across all nine baboons was 1.99 mm, with a maximum error of 4.43 mm.

We also tested absolute error in subcortical and cortical test points for four animals that were not used in creating the template atlas image. Testing these images for absolute error provides an estimate for the generalizability of our method. The average subcortical error in these images was 1.85 mm, with a maximum error of 5.24 mm. The average cortical error was 2.63 mm, with a maximum error of 6.61 mm.

---

[18] R. Davis and R. Huffman, "A Stereotaxic Atlas of the Brain of the Baboon." University of Texas Press, Austin, TX, 1968.

Development of the Baboon Blood Flow Template, "b2kf"

Subjects and Image Acquisition. The PET template image was created using 396 total individual PET images from seven of the nine normal baboons used to create the MRI template. Usually, 12 [^{15}O]water PET blood flow images were obtained from a single animal on a given day. Because the head of the subject was held in a fixed position relative to the scanner for all 12 scans, the 12 individual scans were averaged to form one image for each subject. We had confirmed previously that there was no meaningful movement of the brain among these individual scans.[19]

Registration. Within-modality image registration (PET to PET) was performed as described for the MRI template. Cross-modality (PET to MRI) image registration was implemented using a variant of the image intensity gradient correlation method.[20,21] Whereas in the original method the intensity gradient vector orientations were ignored, here the cost function depended on the relative orientations of these vector quantities. Again, other voxel-based image registration techniques could have been substituted. All cross-modality registrations were performed within a single subject and therefore were modeled as a rigid-body (six-parameter) transform.

Template Image Development. To create the PET template image, each subject's 12-scan average PET image was registered to that same subject's MPRAGE image. That subject's MPRAGE image was then registered to the MRI template image as described previously. The two transformation matrices resulting from these two registration steps were then multiplied to obtain a matrix that transformed the individual PET scans to MRI template atlas space (PET to MRI × MRI to template = PET to template).

Validation. The PET template was created by registering a 12-scan average PET image to that subject's MPRAGE image and then registering that MPRAGE image to the MRI template atlas. Ideally, these atlas-space 12-scan PET images would be identical to a transformation obtained by directly registering the 12-scan PET image to the PET template image. To test this, we compared the rotation, translation, and stretch (also called zoom) parameters computed from the transformation matrices from these two methods. The results were highly similar, thus confirming the accuracy of the two image registration methods, as well as of the two templates.

Macaque Template Images: The "n2k" and "n2kf" Templates

Using essentially identical methods, we created macaque template images by iteratively combining MPRAGE and [^{15}O]water PET blood flow

[19] K. J. Black *et al.*, *J. Comput. Assist. Tomogr.* **20**, 855 (1996).
[20] J. L. Andersson *et al.*, *J. Nucl. Med.* **36**, 1307 (1995).
[21] J. P. Pluim *et al.*, *IEEE Trans. Med. Imaging* **19**, 809 (2000).

images from 12 male, neurologically normal *M. nemestrina* monkeys (pigtail macaques).[8] Images were created in register with the baboon b2k template given our interest in comparing functional imaging data across species.[22,23] Additionally, we registered the monkey template to a published three-dimensional image set including MRI and labeled cryosection images from a single macaque.[24] The resulting "n2kc" atlas is also available at the n2k web site.

We validated the template accuracy using a subset of the points described earlier. Even when comparing macaque landmarks to the baboon template, the measured accuracy was 1.9 mm (mean error). This error measurement includes not only image registration error, but also all of the following: human error in identification of brain landmarks, true nonlinear morphologic differences among the individual macaque brains, morphologic differences between species, and the degree to which the baboon atlas is atypical of living baboon brains. This quite reasonable between-species fit may seem remarkable, but in fact, both we and others had shown previously that a macaque brain image can be aligned linearly to a baboon atlas with fairly good accuracy.[17,25]

Using the Templates in Analyzing Functional Imaging Data

Registering a PET Image to Atlas Space. The focus of this section is to illustrate some typical uses of the MRI and PET templates in neuroimaging studies based on our functional imaging studies of pharmacologic activation in baboons and macaques.[22,26–29] We usually had 5–12 [^{15}O] water PET blood flow images on a given day. We use an in-house program that registers each individual PET scan to each other PET scan and computes a best-fit transformation matrix for each scan to every other PET scan. We then create an average scan by transforming each scan to one of the scans (usually the first) and averaging them together. This 12-scan average image then serves as a source image for the rest of the process of transforming the scans to atlas space.

A different program then registers the 12-scan average image to atlas space. This can be done by registering the 12-scan average image directly

[22] T. Hershey *et al.*, *Exp. Neurol.* **166,** 342 (2000).
[23] J. A. Kaufman *et al.*, *Am. J. Phys. Anthropol.* **121,** 369 (2003).
[24] A. F. Cannestra *et al.*, *Brain Res. Bull.* **43,** 141 (1997).
[25] R. F. Martin and D. M. Bowden, *Neuroimage* **4,** 119 (1996).
[26] K. J. Black *et al.*, *J. Neurosci.* **17,** 3168 (1997).
[27] K. J. Black *et al.*, *J. Neurophysiol.* **84,** 549 (2000).
[28] K. J. Black *et al.*, *J. Neuropsychiatr. Clin. Neurosci.* **14,** 118 (2002).
[29] K. J. Black *et al.*, *Proc. Natl. Acad. Sci. USA* **99,** 17113 (2002).

to the PET template[30] or by registering to atlas space via that particular animal's MPRAGE. In the latter case, the 12-scan average PET image is registered via a rigid body model transformation to the MPRAGE, and the MPRAGE is registered to the MRI template image, as described earlier. The transformation from the 12-scan average image to PET template is then computed by matrix multiplication of the 12-scan to MPRAGE and MPRAGE to the MRI template. As shown in the previous section, either method is sound and produces very similar results. The direct PET to flow template method minimizes assumptions and is available if a structural MR image from the same time period is not available.

Use in SPM Analysis. The aforementioned steps result in functional images from a number of different baboons (or monkeys) registered to a common template with greatly reduced spatial differences between subjects. The images can now be analyzed for differences based on an independent variable of interest. Various approaches could be used, and the images could be analyzed for structural as well as functional images. We apply SPM99 (http://www.fil.ion.ucl.ac.uk/spm/spm99.html) to these images to generate voxelwise statistical maps and to correct for multiple comparisons. In doing so, we currently use the following protocol.

We copy a MATLAB .mat file to each template-registered image included in the analysis; this file identifies image orientation and the origin to SPM99 and facilitates use of our preferred image format in the preceding steps. The SPM99 "Full Monty" option allows explicit identification of a mask image, which we supply (available at purl.org/net/kbmd/b2k and purl.org/net/kbmd/n2k). For each image, we compute the average in-brain image intensity over the voxels included in this standard brain mask and include the list of global brain intensities as a linear covariate of no interest. Independent variables are identified for each scan. In several of our studies, this includes Boolean variables identifying each subject (to account for subject-specific effects) and the dose of drug given before each scan. An F image is computed to allow a single primary analysis that identifies either increases or decreases of regional blood flow with any of several independent variables of interest (e.g., drug dose). Subsequent analyses use pairwise contrasts to show unidirectional effects of a single variable. Our convention is to accept as significant clusters of voxels for which SPM99 computes a *corrected p* value of 0.05 or less. The clusters that enter this

---

[30] The atlas template images are freely available on the Internet (purl.org/net/kbmd/b2k and purl.org/net/kbmd/n2k). The template images are available in various formats, including the SPM format with atlas origin labeled; this allows SPM99 to automatically report results in atlas coordinates. Although we have used our own software for image alignment, the templates are suitable for use as targets with other registration methods, including automated image registration (AIR) or the methods included in the SPM suite of tools.

analysis are defined by contiguous voxels at which the F (or t) statistic exceeds the value corresponding to $p = 0.001$. To avoid type II error, all such clusters whose peak falls in the brain are reported for heuristic value, without a claim of statistical significance.

Anatomic Interpretation of Statistical Images

The final use of the templates is to provide anatomic identification for the results of the statistical analysis. In SPM99, activated clusters can be color coded readily for intensity and superimposed directly onto the grayscale atlas image for visual inspection (Fig. 3). For activation foci falling within the published atlas(es),[18,24] reference to those works provides quick anatomic identification. In the baboon atlas, cortical points can be displayed on a three-dimensional rendering of the brain; atlas points can be transformed into the coordinates of the display software by simple addition. We use the Volume Render program of ANALYZE, the template MR image, and the brain mask (loaded as a binary object) for this purpose.

Discussion

Similar Methods

Greer and colleagues[31] took a somewhat different approach by creating a baboon template based on a single brain and labeling anatomy based on their template image. Cross et al.[32] created a multisubject macaque template with reference to the bicommissural line. Another approach to data analysis that can provide valuable insights is to transform images to a flat map of the cortex.[33]

Future Applications

A surprisingly wide array of nonhuman primate brains can be registered linearly quite reasonably to each other.[8,34] Therefore, brain images from other macaque species, and perhaps other primate species, can likely be registered to the b2k or n2kc atlases. In order to work with data from species with substantially different brain volumes, however, the first step will likely be to estimate the approximate ratio, R, of the brain volume of the new species to the brain volume of the template (given on

[31] P. Greer et al., Brain Res. Bull. **58**, 429 (2002).

[32] D. J. Cross et al., J. Nucl. Med. **41**, 1879 (2000).

[33] D. C. Van Essen et al., Vision Res. **41**, 1359 (2001).

[34] D. M. Bowden and M. F. Dubach, in "Primate Brain Maps: Structure of the Macaque Brain" (R. F. Martin and D. M. Bowden, eds.), p. 38. Elsevier, New York, 2000.

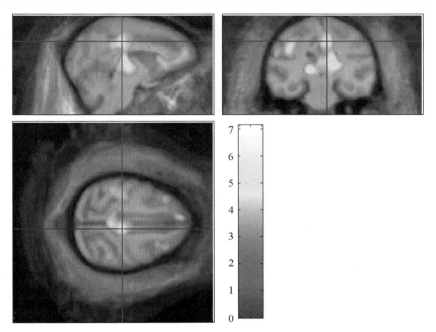

FIG. 3. An image consisting of voxel-by-voxel one-tailed *t* values (in color) from a comparison of regional cerebral blood flow (rCBF) at rest to rCBF after 0.5 mg/kg pramipexole IV in seven normal baboons (previously unpublished image from the study reported in Black *et al.*[29]) Prior to statistical analysis, individual positron emission tomographic images of rCBF were aligned to the b2k template, and hence to the Davis and Huffman[18] atlas of baboon brain. The *t* image is thresholded at 3.19, the value corresponding to $p = 0.001$ (87 df), and superimposed on slices from the b2k MRI atlas (in gray and white). Crosshairs are centered at atlas coordinate (4, 6, 26) (posterior cingulate cortex).

the b2k and n2k web pages). Then, before registering images to the template, one will need to initialize the registering software with stretch (zoom) factors of $R^{1/3}$ in each dimension. Alternatively, for some brain atlases in other species, digital images are available that could be useful in the first or "bootstrapping" step of creating a new template corresponding to that atlas.

Abstract

Coregistration of functional brain images across many subjects offers important experimental advantages and is widely used for studies in humans. Voxel-based image registration methods require a high-quality 3D template image, preferably one that correspondes to a published atlas. We created structural MR and blood flow PET images for neuroimaging studies in baboon and macaque (available at purl.org/net/kbmd/b2k and purl.org/net/kbmd/n2k). In this communication, we describe our methods in detail for creating such a template, for validating it with respect to a published atlas, and for using it to analyze functional imaging data. These methods can be applied to other image registration software programs and to other nonhuman primate species.

Acknowledgments

Preparation of this manuscript was supported in part by NINDS Grants NS044598 and NS39913.

[7] Direct Comparison of Visual Cortex Activation in Human and Nonhuman Primates Using Functional Magnetic Resonance Imaging

By DAVID J. DUBOWITZ

Introduction

The Need for Primate Functional Magnetic Resonance Imaging (fMRI)

Since its initial description[1,2] just over a decade ago, fMRI has revolutionized *in vivo* brain imaging allowing safe, noninvasive mapping of areas of functional neuronal activation in response to a specific task or stimulus. The signals measured with fMRI arise from a complex interaction of regional changes in local cerebral blood flow (CBF), blood volume, and oxygen saturation in response to changes in neural activity—the blood oxygenation Level-dependent (BOLD) effect. Despite the widespread use of fMRI for mapping brain activity in human subjects, the physiological foundations of the method remain poorly understood.

fMRI appears to be correlated to the spiking activity of large ensembles of neurons and to local field potentials.[3–5] The images generated

---

[1] S. Ogawa, T. M. Lee, A. R. Kay, and D. W. Tank, *Proc. Natl. Acad. Sci. USA* **87,** 9868 (1990).

[2] J. W. Belliveau, B. R. Rosen, H. L. Kantor, R. R. Rzedzian, D. N. Kennedy, R. C. McKinstry, J. M. Vevea, M. S. Cohen, I. L. Pykett, and T. J. Brady, *Magn. Reson. Med.* **14,** 538 (1990).

[3] N. K. Logothetis, J. Pauls, M. Augath, T. Trinath, and A. Oeltermann, *Nature* **412,** 150 (2001).

with fMRI are clearly spatially, temporally, and quantitatively related to these underlying neuronal events,[6–11] but exactly how neural activation leads to the vascular changes measured with fMRI images is still unknown.[12,13]

Evaluating the BOLD effect and correlating it with the underlying neural event is a complex process. Previous approaches have involved coregistration of fMRI with electroencephalography (EEG) and event-related potentials (ERPs)[8,14–16] or magnetoencephalography (MEG),[17,18] which helps elucidate the underlying temporal trends, but lacks spatial specificity. Another approach has been to compare neural activity in monkeys with fMRI in humans. This, however, often proves confusing because both interspecies and intertechnique variability need to be considered.[5,19,20] The most promising (and technically demanding) approach to correlating BOLD with neural activity comes from studies that measure fMRI and neural recordings simultaneously.[3,21]

[4] M. A. Paradiso, *Nat. Neurosci.* **2,** 491 (1999).

[5] G. Rees, K. Friston, and C. Koch, *Nat. Neurosci.* **3,** 716 (2000).

[6] N. K. Logothetis, *Philos. Trans. R. Soc. Lon. Ser. B Biol. Sci.* **357,** 1003 (2002).

[7] O. J. Arthurs, E. J. Williams, T. A. Carpenter, J. D. Pickard, and S. J. Boniface, *Neuroscience* **101,** 803 (2000).

[8] G. Bonmassar, K. Anami, J. Ives, and J. W. Belliveau, *Neuroreport* **10,** 1893 (1999).

[9] V. Menon, J. M. Ford, K. O. Lim, G. H. Glover, and A. Pfefferbaum, *Neuroreport* **8,** 3029 (1997).

[10] D. M. Rector, R. F. Rogers, J. S. Schwaber, R. M. Harper, and J. S. George, *Neuroimage* **14,** 977 (2001).

[11] E. A. Disbrow, D. A. Slutsky, T. P. Roberts, and L. A. Krubitzer, *Proc. Natl. Acad. Sci. USA* **97,** 9718 (2000).

[12] A. Villringer and U. Dirnagl, *Cerebrovasc. Brain Metab. Rev.* **7,** 240 (1995).

[13] M. Zonta, M. C. Angulo, S. Gobbo, B. Rosengarten, K. A. Hossmann, T. Pozzan, and G. Carmignoto, *Nat. Neurosci.* **6,** 43 (2003).

[14] A. Martinez, L. Anllo-Vento, M. I. Sereno, L. R. Frank, R. B. Buxton, D. J. Dubowitz, E. C. Wong, H. Hinrichs, H. J. Heinze, and S. A. Hillyard, *Nat. Neurosci.* **2,** 364 (1999).

[15] M. Khubchandani, H. N. Mallick, N. R. Jagannathan, and V. Mohan Kumar, *Magn. Reson. Med.* **49,** 962 (2003).

[16] C. Janz, S. P. Heinrich, J. Kornmayer, M. Bach, and J. Hennig, *Magn. Reson. Med.* **46,** 482 (2001).

[17] J. S. George, C. J. Aine, J. C. Mosher, D. M. Schmidt, D. M. Ranken, H. A. Schlitt, C. C. Wood, J. D. Lewine, J. A. Sanders, and J. W. Belliveau, *J. Clin. Neurophysiol.* **12,** 406 (1995).

[18] E. Disbrow, T. Roberts, D. Poeppel, and L. Krubitzer, *J. Neurophysiol.* **85,** 2236 (2001).

[19] R. B. Tootell, A. M. Dale, M. I. Sereno, and R. Malach, *Trends Neurosci.* **19,** 481 (1996).

[20] D. J. Heeger, A. C. Huk, W. S. Geisler, and D. G. Albrecht, *Nat. Neurosci.* **3,** 631 (2000).

[21] J. Van Audekerkea, R. Peeters, M. Verhoye, J. Sijbers, and A. Van der Linden, *Magn. Reson. Imaging* **18,** 887 (2000).

To interpret human functional MRI in terms of neuronal events, there is an urgent need to bridge this gap between our understanding of primate neurophysiology (based in large part on over half a century of studies in nonhuman primates) and the underlying physiological basis of fMRI.

Using a monkey model for fMRI provides this link in two ways. First, it allows comparison of behavioral, electrophysiological, and fMRI changes in the same animal (direct comparison of fMRI with electrophysiological recordings in humans being limited ethically to cases with preexisting pathology). Second, the ability to make direct comparisons between the human and the monkey brain using the same imaging technique, and while performing the same paradigm, has the potential to improve our understanding of the human brain greatly, allowing us to make direct inferences about human neurophysiology based on our existing wealth of knowledge from nonhuman primate neurophysiology. The value of nonhuman primate fMRI is summarized in Fig. 1.

Previous studies have addressed the potential neuroscience value of fMRI in nonhuman primates for investigating visual,[4,22–27] somatosensory,[28] cognitive,[29] olfactory,[30] and basal ganglia function,[31] as well as the underlying neurovascular physiology.[3,6,32] Because macaques currently represent the predominant nonhuman primate model for studying fMRI, this Chapter concentrates on methods related to fMRI in macaque monkeys.

[22] D. J. Dubowitz, D. Y. Chen, D. J. Atkinson, K. L. Grieve, B. Gillikin, W. G. Bradley, Jr., and R. A. Andersen, *Neuroreport* **9,** 2213 (1998).

[23] D. J. Dubowitz, D. Y. Chen, D. J. Atkinson, M. Scadeng, A. Martinez, M. B. Andersen, R. A. Andersen, and W. G. Bradley, Jr., *J. Neurosci. Methods* **107,** 71 (2001).

[24] N. K. Logothetis, H. Guggenberger, S. Peled, and J. Pauls, *Nat. Neurosci.* **2,** 555 (1999).

[25] L. Stefanacci, P. Reber, J. Costanza, E. Wong, R. Buxton, S. Zola, L. Squire, and T. Albright, *Neuron* **20,** 1051 (1998).

[26] W. Vanduffel, D. Fize, H. Peuskens, K. Denys, S. Sunaert, J. T. Todd, and G. A. Orban, *Science* **298,** 413 (2002).

[27] D. Y. Tsao, W. Vanduffel, Y. Sasaki, D. Fize, T. A. Knutsen, J. B. Mandeville, L. L. Wald, A. M. Dale, B. R. Rosen, D. C. Van Essen, M. S. Livingstone, G. A. Orban, and R. B. Tootell, *Neuron* **39,** 555 (2003).

[28] E. Disbrow, T. P. Roberts, D. Slutsky, and L. Krubitzer, *Brain Res.* **829,** 167 (1999).

[29] K. Nakahara, T. Hayashi, S. Konishi, and Y. Miyashita, *Science* **295,** 1532 (2002).

[30] C. F. Ferris, C. T. Snowdon, J. A. King, T. Q. Duong, T. E. Ziegler, K. Ugurbil, R. Ludwig, N. J. Schultz-Darken, Z. Wu, D. P. Olson, J. M. Sullivan, Jr., P. L. Tannenbaum, and J. T. Vaughan, *Neuroreport* **12,** 2231 (2001).

[31] Z. Zhang, A. H. Andersen, M. J. Avison, G. A. Gerhardt, and D. M. Gash, *Brain Res.* **852,** 290 (2000).

[32] D. J. Dubowitz, K. A. Bernheim, D. Y. Chen, W. G. Bradley, Jr., and R. A. Andersen, *Neuroreport* **12,** 2335 (2001).

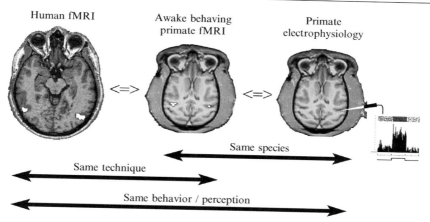

FIG. 1. Comparing data from half a century of invasive neuroscience investigation in nonhuman primates with the recent explosion in information from noninvasive studies of the human brain using fMRI often prove confusing because differences both in technique and between species need to be considered. Awake-behaving fMRI in nonhuman primates offers a unique opportunity to bridge this gap. Nonhuman primate fMRI provides insight into the organization of the human brain (same technique—comparing across primate species), as well as a model for understanding the nature of the fMRI activation with reference to the underlying neuronal events (same species—comparing different techniques).

Macaque monkeys are a natural choice as a model system for studying human neurophysiology. After gibbons and the great apes, Old World monkeys are the closest related primate group to humans (Fig. 2). Despite their apparent proximity on the evolutionary tree, macaques and humans last shared a common ancestor as far back as 30 million years,[33] and macaques have thus had considerable time to evolve independently from humans (as well as from other primates).

A number of differences have been described in the functional specialization of retinotopically organized visual areas between humans and macaques.[26,34,35] Such differences may represent specializations unique to macaques, or to Old World monkeys in general, or may be evolutionary developments specific to humans.[34] It is, however, because of the significant overlap in functional organization between the two species that macaques retain a prominent position as a model for investigating human neurophysiology. Macaque monkeys are readily trained to perform complex

[33] R. F. Kay, C. Ross, and B. A. Williams, *Science* **275**, 797 (1997).
[34] M. I. Sereno, *Curr. Opin. Neurobiol.* **8**, 188 (1998).
[35] R. B. Tootell, D. Tsao, and W. Vanduffel, *J. Neurosci.* **23**, 3981 (2003).

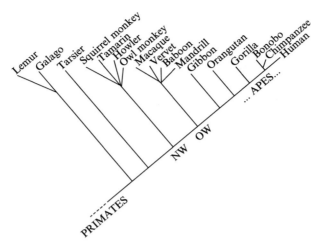

FIG. 2. A simplified tree of primate evolution. Macaques are the most commonly studied nonhuman primates for fMRI. Although there have been many millions of years of independent evolution between macaques and humans, there is still considerable overlap in structural and functional organization to make macaques a valuable model for studying the organization of the human brain (NW, New World monkeys; OW, Old World monkeys).

tasks and, with the aid of microelectrode mapping, radioisotope tracers, and anatomical studies, have formed the basis of much of our current understanding about primate visual pathways.[36–38]

To get a more complete picture, and thus a more comprehensive model of the human brain, there is still a need to study other primate groups, particularly the great apes, to determine which features are unique evolutionary specializations in humans and which are more likely to be developments common to the macaque and other primates. Determining the lineage of primates is continuously under review based on emerging DNA and genomic data.[39] Structural and functional MRI may also provide valuable data to aid with this. The safety, repeatability, and noninvasive nature of fMRI lends itself very well to studies in these other primate groups, and the general principles discussed here translate to fMRI studies in these species as well.

[36] J. H. Maunsell and W. T. Newsome, *Annu. Rev. Neurosci.* **10**, 363 (1987).

[37] R. B. Tootell, E. Switkes, M. S. Silverman, and S. L. Hamilton, *J. Neurosci.* **8**, 1531 (1988).

[38] D. J. Felleman and D. C. Van Essen, *Cereb. Cortex* **1**, 1 (1991).

[39] D. E. Wildman, M. Uddin, G. Liu, L. I. Grossman, and M. Goodman, *Proc. Natl. Acad. Sci. USA* **100**, 7181 (2003).

Optimizing MRI for Monkeys

Conventional BOLD

For the majority of fMRI studies of the brain, conventional BOLD functional imaging techniques are traditionally used as this provides high contrast-to-noise ratio (CNR). The measured MR signal for BOLD studies is complicated and represents a combination of physiological changes in CBF, cerebral blood volume (CBV), and metabolic rate of oxygen (CMRO$_2$), which can be modeled by Eq. (1) (for derivation, see Buxton[40] and Davis et al.[41]).

$$\frac{\Delta S}{\Delta S_0} = S_{max} \cdot \left(1 - \left(\frac{CBV}{CBV_0} \right) \cdot \left(\frac{\left(\frac{CMRO_2}{CMRO_{2,0}} \right)}{\left(\frac{CBF}{CBF_0} \right)} \right)^{\beta} \right) \tag{1}$$

S_{max} is the maximum BOLD signal change (at infinite CBF) and varies spatially within the brain with field strength. The parameter β expresses the relationship between the deoxyhemoglobin concentration in blood and the relaxation rate. At 1.5 T, it is calculated to be 1.5.[41] At higher fields (>4 T), β is close to 1 (from theoretical considerations.[40]) S_0, CBV_0, CBF_0, and $CMRO_{2,0}$ are baseline values or MR signal, cerebral blood volume, cerebral blood flow, and cerebral metabolic rate of oxygen, respectively.

At 1.5 T, measured values of S_{max} range from 0.08 to 0.22.[41,42] An imaging voxel is a mixture of large and small vessels and their immediately adjacent tissues, which contribute to this BOLD signal, and other tissues that do not. In practice, signal changes of 2–4% are seen at 1.5 T[43] (Fig. 3). This increases super-linearly with increasing B_0 field strength, where signal changes of 7% are seen in the visual cortex at 4.7 T.[6]

A stimulus is traditionally presented as a series of blocks or as an event-related paradigm. Imaging voxels that show changes in the MR signal that are temporally highly correlated with the presented stimulus are mapped as showing functional activation. When used for cortical mapping studies, the fMRI needs to be a good representation of underlying neuronal

[40] R. B. Buxton, "Introduction to Functional Magnetic Resonance Imaging: Principles and Techniques." Cambridge Univ. Press, New York, 2002.

[41] T. L. Davis, K. K. Kwong, R. M. Weisskoff, and B. R. Rosen, *Proc. Natl. Acad. Sci. USA* **95**, 1834 (1998).

[42] R. D. Hoge, J. Atkinson, B. Gill, G. R. Crelier, S. Marrett, and G. B. Pike, *Proc. Natl. Acad. Sci. USA* **96**, 9403 (1999).

[43] P. A. Bandettini, E. C. Wong, R. S. Hinks, R. S. Tikofsky, and J. S. Hyde, *Magn. Reson. Med.* **25**, 390 (1992).

Visual cortex activation – Macaque

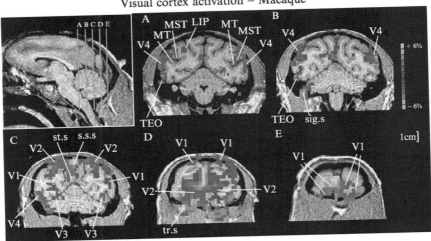

Visual cortex activation – Human

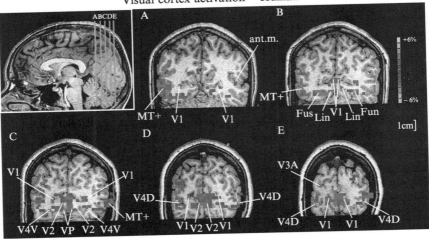

FIG. 3. Upper panel shows functional activation in macaque brain at 1.5 T. Visual paradigm is 25-s alternating blocks of video cartoon and darkness. Activated pixels are 2 × 2 mm superimposed on T_1-weighted MPRAGE images of 0.7 × 0.7 mm resolution (coronal images A–E are at 5-mm spacing ranging from 25 to 5 mm anterior to the occipital pole). Labeled areas are visual cortex (V1, V2, V3, V4), medial temporal area (MT), medial superior temporal area (MST), lateral intraparietal area (LIP), superior sagittal sinus (s.s.s), straight sinus (st.s), transverse sinus (tr.s), and sigmoid sinus (sig.s). Lower panel shows functional activation in a human visual cortex during the same stimulus. Activated pixels are 3.5 × 4 mm superimposed on T_1-weighted MPRAGE images of 1-mm isotropic resolution. Coronal images A–E are at 7-mm spacing ranging from 44 to 16 mm anterior to the occipital pole.

activity. Studies by Disbrow et al.[11] in anesthetized monkeys found correlation between neuronal activity and BOLD with imaging voxels of 9 mm³. There was better correlation in parenchyma, which decreased closest to large vessels. An important consideration for determining resolution relates to the interplay between neuronal activity and functional activation. For typical human fMRI with an imaging resolution of $3 \times 3 \times 3$ mm, the inclusion of small draining veins in the imaging voxel does not unduly skew the representation of neuronal activity. As resolution is increased, steps need to be taken to reduce the influence of large vessels. For fMRI in monkeys, where the smaller brain size necessitates higher-resolution imaging, it becomes more important that the MRI signal is appropriately spatially localized to parenchyma and not large draining veins. (See Fig. 3. The draining veins are very prominent during BOLD fMRI at 1.5 T in a monkey.) At higher field, the use of spin echo acquisition[44] or diffusion weighting during a BOLD experiment reduces the signal contribution from large veins. Alternatively, MR measures of other physiological parameters can be utilized to reduce the venous signal, such as cerebral blood flow (where the image is weighted toward tissue and the intravenous signal relaxes before being imaged) or cerebral blood volume (where the enhanced transverse relaxation in large veins from the use of T_2 contrast agents removes the venous signal).

Cerebral Blood Flow

Cerebral blood flow (CBF) studies have been used extensively for functional mapping with positron emission tomography (PET) using ^{15}O-labeled water as the contrast agent.[45] The magnetic resonance counterpart to this PET technique is to label arterial water magnetically with an applied radiofrequency (RF) pulse.[46] A difference image is produced (with and without spin labeling). The appearance of labeled spins in the imaging voxel is a reflection of the arterial blood flow into the voxel. Arterial spin labeling (ASL) can be used to measure the baseline cerebral blood

[44] S. P. Lee, A. C. Silva, and S. G. Kim, *Magn. Reson. Med.* **47**, 736 (2002).

[45] R. S. Frackowiak, G. L. Lenzi, T. Jones, and J. D. Heather, *J. Comput. Assist. Tomogr.* **4**, 727 (1980).

[46] J. A. Detre, J. S. Leigh, D. S. Williams, and A. P. Koretsky, *Magn. Reson. Med.* **23**, 37 (1992).

Green bars on the sagittal image indicate the position of coronal images A–E. Labeled areas are visual cortex (V1, V2, VP, V3A, V4-; dorsal and ventral), medial temporal complex (MT+), fusiform gyrus (Fus), lingual gyrus (Lin), and anterior motion area (ant.m). See Table I for imaging parameters (from Dubowitz et al.[23] with permission). (See color insert.)

flow, as well as cerebral blood flow changes, which are activation dependent. Most labeled spins exchange with water in the tissues and do not remain in the vascular compartment; thus the MR signal is seen in parenchyma rather than in draining veins. Images of CBV using the ASL technique result in improved spatial correlation with brain parenchyma compared to conventional BOLD fMRI in both human[47] and animal studies.[44] ASL techniques are capable of providing quantitative information about cerebral perfusion (expressed as volume of blood delivered to the capillary beds per unit volume of brain per unit time). In addition to better spatial localization, ASL may, in fact, be a more direct indicator of the underlying neuronal activity than BOLD.[48] ASL also provides a quantitative means to study the mechanisms underlying the BOLD technique itself.[49,50] ASL techniques can be broadly divided into continuous and pulsed tagging techniques. All have been used for measuring CBF changes in humans with neuronal activation. The duration of ASL and BOLD signals are similar[51] (Fig. 4), and the same stimulus design can be adapted for both BOLD and flow activation measurements.[40] One advantage of the ASL technique is the ability to measure both CBF and BOLD simultaneously. The difference between tag/no-tag conditions provides CBF data, and the sum of the conditions provides the BOLD signal.[47] From Eq. (1), the BOLD signal can be modeled in terms of changes in $CMRO_2$, CBV, and CBF. CBV and CBF are related empirically by the power relationship[52]:

$$CBV = (CBF)^{0.38} \qquad (2)$$

Using an addition experiment (e.g., altering CBF with increased pCO_2) allows the scaling constants to be calculated. Thus, this combined measure also allows estimates of the cerebral metabolic rate of oxygen ($CMRO_2$) to be made from fMRI activation studies. The limitation of this technique is that, currently, only a limited number of slices can be imaged at a time and the slices chosen should be as orthogonal as possible to the feeding artery for optimum spin tagging.[47] Although the magnitude of the physiological change is higher, the MR signal, and hence the available signal-to-noise ratio (SNR), is lower for quantitative CBF measures.[47]

[47] E. C. Wong, R. B. Buxton, and L. R. Frank, *NMR Biomed.* **10,** 237 (1997).
[48] R. B. Buxton and L. R. Frank, *J. Cereb. Blood Flow Metab.* **17,** 64 (1997).
[49] R. B. Buxton, L. R. Frank, E. C. Wong, B. Siewert, S. Warach, and R. R. Edelman, *Magn. Reson. Med.* **40,** 383 (1998).
[50] R. D. Hoge, J. Atkinson, B. Gill, G. R. Crelier, S. Marrett, and G. B. Pike, *Magn. Reson. Med.* **42,** 849 (1999).
[51] R. B. Buxton, E. C. Wong, and L. R. Frank, *Magn. Reson. Med.* **39,** 855 (1998).
[52] R. L. Grubb, Jr., M. E. Raichle, J. O. Eichling, and M. M. Ter-Pogossian, *Stroke* **5,** 630 (1974).

Cerebral Blood Volume

The earliest implementations of functional MRI used contrast based on cerebral blood volume. These utilized the intensity of signal change during the first-pass passage of an intravenous bolus of paramagnetic contrast agent through the brain and tracer kinetics analysis to estimate changes in CBV.[53] The passage takes approximately a minute to complete, and additional time is required for clearance of the contrast agent before the measure can be repeated. Temporal resolution is thus limited, and this approach provides only a "snapshot" measure of cerebral blood volume and is limited when studying dynamically changing cerebral physiology.[54] Greater temporal resolution can be achieved with intravascular contrast agents that maintain a prolonged steady-state concentration. Depending on the type of contrast agent used, the sensitivity to CBV can alter the T_1 or T_2 relaxation times, However, the use of T_2 agents appears to give better sensitivity to CBV changes than from T_1 agents.[54] Dextran-coated monocrystalline iron oxide nanoparticles (MION) affect T_2 relaxation and have a prolonged blood half-life (ranging from 3 to 15 h in monkeys[32,55]), making them very suitable as susceptibility agents for measuring CBV. Excretion half-lives as long as 96 h have also been reported.[56] At a sufficiently large dose the contribution to blood magnetization from the contrast agent far exceeds that from deoxyhemoglobin so T_2^*-weighted MR sequences are rendered insensitive to changes in the deoxyhemoglobin concentration. Flow changes also have no effect (the concentration of paramagnetic agent is the same in the arterial and venous circulation so increased flow does not affect susceptibility). The imaging contrast is thus only sensitive to the concentration of paramagnetic agent in the imaging voxel. As local cerebral blood volume increases, the vascular proportion of the voxel (and hence the observed amount of contrast agent) increases. Because T_2 agents shorten T_2 and T_2^*, the effect of increased blood volume (and increased iron concentration) is to reduce the signal in the voxel, thus neuronal activation is associated with a decrease in the MR signal for these CBV-weighted images.[32]

[53] J. W. Belliveau, D. N. Kennedy, R. C. McKinstry, B. R. Buchbinder, R. M. Weisskoff, M. S. Cohen, J. M. Vevea, T. J. Brady, and B. R. Rosen, *Science* **254,** 716 (1991).

[54] J. B. Mandeville, J. J. Marota, B. E. Kosofsky, J. R. Keltner, R. Weissleder, B. R. Rosen, and R. M. Weisskoff, *Magn. Reson. Med.* **39,** 615 (1998).

[55] F. P. Leite, D. Tsao, W. Vanduffel, D. Fize, Y. Sasaki, L. L. Wald, A. M. Dale, K. K. Kwong, G. A. Orban, B. R. Rosen, R. B. Tootell, and J. B. Mandeville, *Neuroimage* **16,** 283 (2002).

[56] D. J. Dubowitz, K. A. Bernheim, and R. A. Andersen, "Proceedings of Society for Neuroscience," Vol. 27. San Diego, 2001.

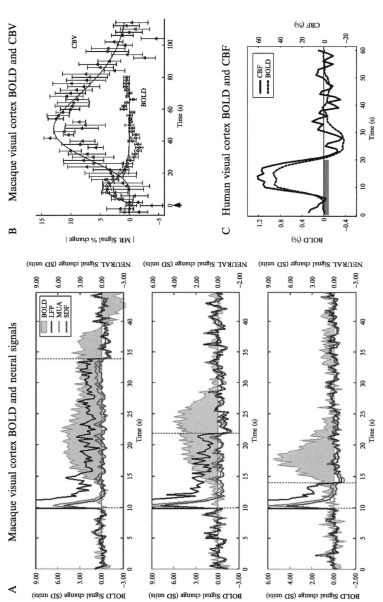

Fig. 4. Relationship of BOLD, CBF, CBV, and neural activity. (A) Simultaneous neural and hemodynamic recordings from a cortical site showing a transient neural response to pulse stimuli of 24 s (top), 12 s (middle), and 4 s (bottom). Curves show the time course of BOLD (gray shaded), local field potential (LFP) (black trace), multiunit activity (MUA) (light gray trace), and spike density function (SDF) (dark gray trace). The single and multiple unit responses adapt a couple of seconds after stimulus onset, with LFP remaining the only signal correlated with the

The signal advantage of the CBV measurement over BOLD for functional MRI is most evident at low B_0 field strengths. Several factors contribute to this. First, the effect of the BOLD signal (which increases with functional activation) is opposite to the CBV signal (which reduces with activation). The theoretical maximum contrast-to-noise (CNR) ratio $\approx$ %CBV/% BOLD at $T_E \approx T_2{}^*$. The percentage change in CBV is not field dependent, whereas the percentage change in BOLD does change with the applied magnetic field. Thus, the CNR decreases as the B_0 field increases for a given dose of contrast agent.[54] Second, as the effect of the contrast agent is to decrease the MR signal, the injected dose of the contrast agent needs to be reduced at large B_0 fields or the MR signal becomes too small to measure above the noise. Finally, the maximum CBV signal change is constrained by the actual physiological CBV change (which, unlike BOLD, is field independent). The signal-boosting effect of the contrast agent will be less when the BOLD effect is highest (at high B_0 field), and the maximum improvement in functional CNR is thus seen at lower field strengths.[54] Threefold increases in signal at 1.5 and 3 T have been reported compared to BOLD studies.[32,55] Because the BOLD signal and contrast-enhanced CBV signal have opposite signs, enough contrast agent must be used to overcome the BOLD signal. Using high-dose contrast and short echo times at 1.5 T, five-fold increases in functional CNR have been observed.[57] This increased signal allows higher-resolution imaging, and voxel dimensions of $2 \times 2 \times 2$ mm are achievable at 1.5 T with this method[57] (Table I). Because signals from large veins relax very quickly and do not contribute significantly to the image, CBV signals are better spatially correlated with brain parenchyma than BOLD fMRI.[54]

As discussed earlier, depending on the vessel size, the BOLD effect scales between linearly and quadratically with applied field strength. Thus, high-field imaging at 4.7 T could be expected to afford at least a threefold increase in SNR at 1.5 T. This theoretical gain needs to be offset by

[57] W. Vanduffel, D. Fize, J. B. Mandeville, K. Nelissen, P. Van Hecke, B. R. Rosen, R. B. Tootell, and G. A. Orban, *Neuron* **32**, 565 (2001).

BOLD response. Note also the poststimulus undershoot in neural and BOLD signals. Adapted from Logothetis *et al.*[3] with permission. (B) Time course of BOLD and CBV change in macaque primary visual cortex following a 6-s stimulus. Plot shows magnitude of response for comparison (the actual CBV signal is negative with increases in blood volume; see text). Note the approximately three-fold increase in signal change using CBV measurements, with prolonged recovery time and no poststimulus undershoot (reproduced from Dubowitz *et al.*[22] with permission). (C) Time course of BOLD and CBF signals in human primary visual cortex following a 20-s 8-Hz flickering checkerboard stimulus. Note similar time course, duration, and poststimulus undershoot (data courtesy of Richard Buxton).

TABLE I

SUMMARY OF MRI SEQUENCE PARAMETERS FOR BOLD fMRI AND ANATOMICAL MEASUREMENTS AT 1.5 T[a] AND FOR CBV FUNCTIONAL MRI[b]

	Anatomy (human)	Anatomy (monkey)	BOLD (human)	BOLD (monkey)	MION[c] (monkey)
Sequence	3D-MPRAGE	3D-MPRAGE	EPI-GE	EPI-GE	EPI-GE
Bandwidth (Hz)[d]	33,280 (130)	16,640 (130)	106,624 (833)	106,624 (833)	—
Echo time, Te (ms)	4.4	4.4	40	40	28
Repetition time, Tr (ms)	11.4	11.4	2000[e]	2000[e]	2400[e]
Inversion time, TI (ms)	20	250	—	—	—
Delay time, TD (ms)	0	600	—	—	—
Flip angle (degrees)	10	12	90	90	90
Number of averages	1	1	1	1	1
Field of view (mm)	256 × 256	90 × 90	448 × 448	256 × 256	128 × 128
Imaging matrix	256 × 256	128 × 128	128 × 128	128 × 128	64 × 64
Imaging plane	Sagittal	Axial	Axial	Coronal	Coronal
Slice (slab) thickness (mm)	140[f]	80[f]	4	5	2
Number of slices	140[g]	98[g]	—	—	—
Voxel resolution (mm)	1 × 1 × 1	0.7 × 0.7 × 0.8	3.5 × 3.5 × 4	2 × 2 × 5	2 × 2 × 2
Number of measurements[h]	—	—	84 × 2	84 × 4	240

[a] Modified from Dubowitz et al.[23] with permission.

[b] From Fize et al.[59]

[c] MION dose = 4–11 mg/kg, imaged used a 10-cm radial surface coil.

[d] Bandwidth per pixel shown in parentheses.

[e] Effective Tr determined by the fMRI experiment repetition time.

[f] Slab thickness is shown for 3D sequences.

[g] Indicates number of phase encode steps for 3D sequences.

[h] Number of time points measured × number of experiments averaged, or total number of volumes acquired.

increased bulk susceptibility artifacts at higher applied magnetic fields. Using CBV fMRI for primate imaging at 1.5 T allows a comparable increase in signal, but without the increase in bulk susceptibility distortions induced by a higher field.[32] SNR improvements from awake rather than sedated monkeys are additive. In comparing retinotopic mapping with fMRI,[58,59] Fize and colleagues reported better SNR measuring CBV in awake monkeys at 1.5 T than BOLD fMRI in anesthetized monkeys at 4.7 T. The CBV time course is slower than the BOLD change, which needs to be taken into consideration when designing CBV-based fMRI paradigms. Figure 4 shows the relative amplitude and time course for CBV and BOLD signal changes.

In addition to providing improved signal for fMRI studies, this method also allows the actual CBV change to be calculated, and, like CBF measures of functional activity, the CBV provides quantitative information. By measuring the iron concentration in venous blood, it is possible to calibrate the R_2^* relaxation rate, and thus get an absolute measure of CBV change with neuronal activation (the CBV can be calculated from the slope of R_2^* change with $[Fe]_{blood}$. For derivation, see Dubowitz et al.[32] ad Mandeville et al.[54])

$$K \cdot CBV(t) = \frac{R_2^*(t) - R_2^*(0)}{[Fe]_{blood}(t) - [Fe]_{blood}(0)} \tag{3}$$

The concentration of iron in blood is assumed to be constant (at least on the timescale of the activation experiment). The fractional change in CBV following a stimulus can be calculated from the CBV during rest and during activation.

$$\frac{\Delta CBV}{CBV(r)} = \frac{CBV(a) - CBV(r)}{CBV(r)} = \frac{R_2^*(a) - R_2^*(r)}{R_2^*(r) - R_2^*(0)} \tag{4}$$

The signal change is described by the exponential relationship:

$$S(TE) = S(0)e^{-T_E \cdot R_2^*} \tag{5}$$

So the relative CBV change can be written in terms of the signal change at rest and during activation:

$$\frac{\Delta CBV}{CBV(r)} = \frac{\frac{-1}{T_E} \cdot \ln\left(\frac{S(a)}{S(r)}\right)}{R_2^*(r) - R_2^*(0)} \tag{6}$$

[58] A. A. Brewer, W. A. Press, N. K. Logothetis, and B. A. Wandell, *J. Neurosci.* **22**, 10416 (2002).

[59] D. Fize, W. Vanduffel, K. Nelissen, K. Denys, C. Chef d'Hotel, O. Faugeras, and G. A. Orban, *J. Neurosci.* **23**, 7395 (2003).

R_2^* is the relaxation rate at rest (r), during activation (a), and without MION prior to the start of the experiment (0). T_E is the imaging echo time.

For photic stimulation from a 6-s duration 8-Hz flickering checkerboard, the CBV change in the macaque primary visual cortex is 32%. As expected, there is concordance in the responses in humans and macaques, and this value corresponds closely to 32% CBV changes measure in humans.[53] By comparison, CBV changes during activation in rodents appear lower than those in primates—20–24% CBV change in rats and mice.[54,60]

Iron oxide contrast media enter normal iron stores in the body and are excreted by the liver. The use of iron oxide formulations with a prolonged intravascular half-life improves the stability of the signal and reduces the need to redose the animal regularly. However, iron excretion is slow, and repeated imaging doses of 5–7 mg/kg have the potential for iron accumulation and the risk of morbidity associated with iron overload. Leite et al.[55] reported cumulative injected doses in macaque monkeys of 40–166 mg/kg without adverse effects. Following 1–3 weeks of study, the iron stores of the animal were returned to normal with 1 mg/day intramuscular deferoxamine for 4–6 days.[59]

Optimizing Pulse Sequences for Monkeys

For best signal to noise and contrast to noise in the MR image, it is important to optimize the pulse sequence for the relaxation parameters of the tissues of interest (T_1, T_2^*, and T_2). These parameters are dependent on a number of variables (water content, age of the animal, body temperature, B_0 field strength), and in monkeys they can be subtly different from relaxation parameters in humans. Given that smaller voxels are needed to resolve individual structures in the nonhuman primate brain, the available SNR can be quite small when imaging at lower magnetic fields such as 1.5 T. Thus, extra steps taken to optimize the sequences will help improve the available SNR. The optimum BOLD signal is with $T_E \approx T_2^*$ of the tissue.[61] Even at higher B_0 fields, sequence optimization remains an important step. The field strength dependency of the relaxation rates means that imaging sequence parameters should be optimized at a higher field to account for this. Suitable parameters for primate functional and anatomical MRI are detailed in Table I. T_1, T_2, and T_2^* values are given in Table II for macaque brain at different field strengths. Because T_2^* varies with field inhomogeneity and is dependent on the quality of the shim, this needs to be measured locally.

[60] T. Mueggler, D. Baumann, and M. Rudin, "Proc. of Int. Soc. Magn. Res. Med.," Vol. 1, p. 651, 2001.
[61] R. S. Menon, S. Ogawa, D. W. Tank, and K. Ugurbil, *Magn. Reson. Med.* **30**, 380 (1993).

TABLE II
T_1, T_2, AND T_2* VALUES FOR MACAQUE BRAIN AT 1.5 T FOR $n = 3$ MONKEYS EXPRESSED
AS MEAN ± STANDARD DEVIATION[a,b]

	Gray matter ($n = 3$) 1.5 T	White matter ($n = 3$) 1.5 T	Gray matter 4.7 T	White matter 4.7 T
T_1 (ms)	1010 ± 8.5 (920)	790 ± 4.0 (790)	1499	1097
T_2 (ms)	94 ± 0.8 (101)	92 ± 2.5 (92)	74	69
T_2* (ms)	49 ± 2.3	46 ± 6.5	36	—

[a] Modified from Dubowitz et al.[23] with permission.
[b] Values at 4.7 T are from Logothetis.[6] Typical values for human brain at 1.5 T are shown in parentheses [from M. L. Wood et al., J. Magn. Reson. Imaging **3** (1993)].

Spatial Resolution

The adult macaque brain is roughly 14 times smaller by volume than that of a human, so as a first approximation we would need a ~2.5-fold increase in resolution to make analogous comparisons with human MR studies. This requirement for improved resolution is somewhat offset by the simplified gyral folding in the macaque brain compared to humans. Additionally, the size differential is not reflected homogeneously across the brain, and the actual resolution improvement needed varies for different cortical structures. When measured on a flattened cortex, for the primary visual cortex, the difference in size is approximately a factor of 2 in area[62] (or 1.4-fold in linear resolution). For other areas of extrastriate cortex, the differential may be much greater (e.g., relative to V1, there is a 4-fold expansion in Vp between humans and macaques).[63] On average, to make inferences on visual areas in the macaque brain, at least a 2-fold improvement in spatial resolution is needed compared to human studies. Failure to achieve this level of resolution results in a poor definition of cortical structures. Such problems associated with inadequate resolution are magnified during postprocessing to expand or reconstruct the cortical surface.[64,65] The spatial resolution in MRI is limited by the available SNR per voxel. Measures that improve SNR will ultimately allow higher resolution

[62] D. C. Van Essen, J. W. Lewis, H. A. Drury, N. Hadjikhani, R. B. Tootell, M. Bakircioglu, and M. I. Miller, Vision Res. **41,** 1359 (2001).
[63] M. I. Sereno, A. M. Dale, J. B. Reppas, K. K. Kwong, J. W. Belliveau, T. J. Brady, B. R. Rosen, and R. B. Tootell, Science **268,** 889 (1995).
[64] S. A. Engel, D. E. Rumelhart, B. A. Wandell, A. T. Lee, G. H. Glover, E. J. Chichilnisky, and M. N. Shadlen, Nature **369,** 525 (1994).
[65] A. M. Dale, B. Fischl, and M. I. Sereno, Neuroimage **9,** 179 (1999).

imaging. For human BOLD fMRI studies, voxel sizes of 30–100 mm^3 are typical.[40] Conventional BOLD imaging of monkeys at 1.5 T does not achieve the required 2-fold improvement in resolution, and voxel sizes typically range from $2 \times 2 \times 5$ mm (20 mm^3) to $3.9 \times 3.9 \times 3$ mm (46 mm^3).[22,23,57] Further improvements in resolution require improvements in SNR. Three approaches to this have been described. (1) Intravascular contrast media weight the fMRI signal toward changes in cerebral blood volume rather than deoxyhemoglobin concentration.[22] This improves the functional CNR and can be used to increase imaging resolution at 1.5 T to $2 \times 2 \times 2$ mm.[57] (2) Increased field strength affords significant improvements in BOLD SNR. Numerical simulations show that the BOLD contribution to the transverse relaxation rate scales linearly with B_0 for large vessels and quadratically for small vessels[66]; thus, BOLD imaging with a resolution of 0.75×2 mm has been achieved at 4.7 T.[3] (3) Dramatic increases in SNR are achievable with the use of tailored RF surface coils (although with reduced spatial coverage). At 4.7 T, Logothetis *et al.*[67] were able to measure BOLD fMRI signals in sedated monkeys at $0.125 \times 0.125 \times 1$ mm resolution using implanted 22-mm surface coils.

Experimental Considerations for Monkey MRI

Animal Positioning and Compliance

For awake studies, it is important that the monkey is acclimatized and comfortable in the scanner and is rewarded regularly. The head of the monkey needs to be kept still within the MR scanner and positioned at the isocenter. The head is usually oriented with Reid's plane parallel to the horizontal plane of the MR scanner. This ensures that images are acquired in a standard stereotaxic orientation (without the need to reslice the images) and aids in the anatomical localization of structures. However, circumstances may mitigate alternate head orientation, such as space constraints[24] or tilting the head to optimize B_0 homogeneity (and thus improve SNR and image coregistration).[68]

Horizontal vs Vertical Orientation. Electrophysiology studies in awake-behaving, nonhuman primates have traditionally been done with the animal in an erect sitting position. This is considered a natural position for the animal and allows more scope for the animal to interact with its

[66] J. S. Gati, R. S. Menon, K. Ugurbil, and B. K. Rutt, *Magn. Reson. Med.* **38,** 296 (1997).
[67] N. Logothetis, H. Merkle, M. Augath, T. Trinath, and K. Ugurbil, *Neuron* **35,** 227 (2002).
[68] J. M. Tyszka and A. N. Mamelak, *J. Magn. Reson.* **159,** 213 (2002).

stimulus, particularly if the animal needs to indicate its perception or compliance with a task by use of a joystick or lever (the limbs are nonweight bearing and thus free to move). This also allows ample space around the animal to connect any recording devices, electrophysiology head stage, or head restraint. In the move from electrophysiology studies to MRI, it has thus seemed logical to keep the animal in the same erect seated position for MR studies. Indeed, there is already considerable expertise in training monkeys for studies in this orientation. A new line of large-bore (30–60 cm), vertically oriented MR scanners are now available specifically for imaging monkeys in an erect position. This allows some ease of transition to MR studies from physiology or other studies where the animal is erect and seated. For awake MRI studies, some researchers consider that, like erect physiology studies, this is the more natural position for the animal.[24] However, for sedated studies, having the monkey erect does introduce difficulties of maintaining vascular homeostasis and adds to the complexity and risks associated with anesthesia.[24]

In spite of this, MR scanners of the size needed for monkey imaging are much more commonly oriented horizontally (including the many MR scanners available from clinical MRI vendors that can also be used for nonhuman primate MRI). This allows more versatility in the choice of MR scanner. A number of research groups have successfully trained monkeys to sit semierect or to lie prone in a sphinx position inside a conventional horizontal bore scanner for MR studies.[22,25,29,57] This orientation is also more straightforward for sedated studies. Additionally, there is more space for the monkey to interact directly with a stimulus projected in front of him in a horizontal bore system (e.g., reaching or pointing at a screen), whereas space is more limited in current vertical bore systems. There are, at present, insufficient data comparing the two approaches to indicate that one orientation is clearly superior in all applications. Extensive training is required for a monkey to be fully compliant in either orientation. The decision to perform studies with the animal erect or prone depends on a number of factors: whether a dedicated monkey scanner will be used or if studies will be done on an existing large-bore research or clinical scanner, whether sedated studies are needed, and consideration of the nature of the stimulus to be used. Examples of setups for horizontal and vertical monkey imaging are illustrated in Fig. 5.

Anesthetized vs Awake. Successful fMRI requires that head motion be minimized or, ideally, completely eliminated. One approach to this has been to do studies with animals anesthetized. This allows prolonged studies, with negligible motion artifact, and is capable of generating extremely high-resolution studies.[24] Sedated or anesthetized studies also considerably reduce the chance of an animal escaping and remove the

A   Setup for horizontal primate MRI   B   Setup for vertical primate MRI

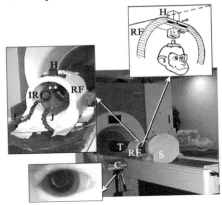

FIG. 5. An experimental setup for imaging macaque monkeys in (A) a conventional horizontal bore 1.5-T clinical MRI scanner (insert drawing adapted from Dubowitz *et al.*[22] with permission) and (B) a vertical bore 4.7-T primate MRI scanner. The monkey is confined to a plastic tube (T), which is inserted into the bore of the scanner. The head of the animal is secured within the radiofrequency coil (RF) by means of a head post (H), which marries up with a head cap on the head of the monkey. An infrared light source (IR) provides illumination for an eye-tracking camera (C), and the stimulus is projected onto a screen (S). The animal is rewarded with fruit juice via a flexible tube (J). In the vertical MRI scanner, video goggles (V) provide the stimulus and eye tracking.

need to train the animal or to provide a reward during the MRI study. Anesthesia leads to a number of difficulties; cerebral metabolism and blood flow are altered,[69–71] which attenuates the BOLD signal significantly.[69,72] Deeper anesthesia will also attenuate neuronal activation, thus further compromising the fMRI signal. While sedated studies permit motion-free

[69] F. Hyder, D. L. Rothman, and R. G. Shulman, *Proc. Natl. Acad. Sci. USA* **99,** 10771 (2002).

[70] M. T. Alkire, R. J. Haier, N. K. Shah, and C. T. Anderson, *Anesthesiology* **86,** 549 (1997).

[71] M. T. Alkire, C. J. Pomfrett, R. J. Haier, M. V. Gianzero, C. M. Chan, B. P. Jacobsen, and J. H. Fallon, *Anesthesiology* **90,** 701 (1999).

[72] E. Seifritz, D. Bilecen, D. Hanggi, R. Haselhorst, E. W. Radu, S. Wetzel, J. Seelig, and K. Scheffler, *Psychiatr. Res.* **99,** 1 (2000).

imaging, the loss in signal and departure from the normal physiological relationship between neuronal activity, blood flow, and metabolism all need to be taken into consideration. Anesthesia also limits studies that involve a behavioral component or knowledge of the actual percept of the animal. Anesthesia in rats has been shown to improve the spatial specificity of BOLD activation[73]; however, the effect of anesthesia is also heterogeneous across the brain,[70,74] and thus spatial variations in the fMRI signal can be generated by the choice of anesthesia alone. Despite the many advantages in imaging awake animals, there are many instances when it is preferable to study neural interactions in the absence of top-down modulation,[75] and thus less technically demanding sedated studies are preferable.

Anesthetic Regimens for Sedated Monkey fMRI. A number of anesthetic combinations have been proposed for primate fMRI. The optimal choice of agents depends on the area of cortex under investigation. Isoflurane (which is widely used as a gaseous anesthetic in veterinary work) attenuates the BOLD signal differentially in the brain. The BOLD activation signal is attenuated more severely in the visual cortex than in the lateral geniculate nuclei for any given dose, and within the visual cortex, there is more attenuation in the extrastriate visual cortex compared with the striate cortex (N. Logothetis, personal communication). Remifentanyl, a selective μ-opioid agonist, which decreases cerebral metabolism and intracranial pressure with minimal cerebral perfusion changes, appears to have much less effect on the BOLD signal in monkeys (N. Logothetis, personal communication). The same is not true in rats, where cerebral blood flow is better preserved with isoflurane than with fentanyl.[74] Thus, caution must be taken when extrapolating anesthetic protocols from one species to another. Logothetis *et al.*[24] recommend a balanced anesthetic regimen using isoflurane in air and fentanyl (3 mg/kg/h). Another alternative is ketamine, a dissociative anesthetic, which also appears to preserve both the neural activation and the BOLD signal in thalamocortical visual pathways when used at a low dose (1–5 mg/kg). Higher doses produce optokinetic nystagmus, dissociative stupor, and loss of the BOLD signal.[76]

Restraint. Because fMRI is critically sensitive to any gross motion, methods to prevent head motion are vital for both human and nonhuman primate imaging. Head restraint also helps ensure correct gaze and

[73] R. R. Peeters, I. Tindemans, E. De Schutter, and A. Van der Linden, *Magn. Reson. Imaging* **19,** 821 (2001).

[74] K. S. Hendrich, P. M. Kochanek, J. A. Melick, J. K. Schiding, K. D. Statler, D. S. Williams, D. W. Marion, and C. Ho, *Magn. Reson. Med.* **46,** 202 (2001).

[75] G. Rainer, M. Augath, T. Trinath, and N. K. Logothetis, *Curr. Biol.* **11,** 846 (2001).

[76] D. A. Leopold, H. K. Plettenberg, and N. K. Logothetis, *Exp. Brain Res.* **143,** 359 (2002).

interaction with a visual stimulus. Restraint may take the form of chemical restraint (sedation, neuromuscular blockers), rigid physical restraint, or a combination of the two. The traditional approach when rigid head restraint is needed for human imaging is to use a bite bar. Local experience shows that this can reduce head motion to the order of 1 mm in a well-trained and acclimatized subject (M. Sereno, personal communication). Bite bars are not practical for monkey MRI, particularly if the animal also needs to receive a juice reward during the study. A more practical approach to head fixation is a modification of the head restraints commonly used for electro-physiology studies, where the animal wears a small MR-compatible surgi-cally attached head cap. The head cap marries up to a restraining device within the scanner (see Fig. 5). Materials (usually plastics or ceramics) are chosen with good tensile rigidity, biocompatibility,[22] and susceptibility characteristics close to that of tissues.[77] Rigid restraint with a head cap made of dental acrylic, polyetherimide, and ceramic screws will reduce head motion to a few millimeters or less.[22] However, like a bite bar, the head cap provides a reference point in space rather than rigid immobilization. Even the strongest head restraint will not work well with an animal that is not co-operative and calm in the MRI scanner. Training and acclimatization to the MR environment are thus essential.

Training. The longest part of any functional MR study in awake-behaving, nonhuman primates is the preparatory training. Depending on the complexity of the task and the temperament of the monkey, this may take from 3 months to several years to accomplish. MRI time is costly, so most of the preliminary training can be done in a mock scanner environment. The behavior of the animal is shaped in a graded manner during daily training sessions, usually lasting 3 to 5 h. Eye movement re-cordings are made during training sessions, and the animal is rewarded with water or fruit juice, which is incorporated into the behavioral paradigm to provide continuous positive reinforcement of a correctly learned task. A flowchart summarizing how behavior is reinforced and rewarded is shown in Fig. 6. The entire paradigm is computer controlled to ensure con-sistency in the amount and timing of the reward. In cases of inappropriate eye movements, the trials are cancelled without reward, and thus the ani-mal quickly learns the correct oculomotor behavior. An example of a mock MRI is detailed in Fig. 7. This shows a simple setup for training a monkey in a horizontal orientation. The components are similar for a vertical orien-tation. The animal views a stimulus on the computer screen, and a video camera tracks the eye position of the animal (used to assess compliance

---

[77] J. Van Kylen, A. Jesmanowicz, E. C. Wong, and J. S. Hyde, "Proc. of Int. Soc. Magn. Res. Med." Vol. 1, p. 762, 1995.

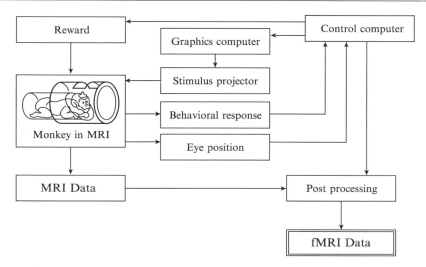

FIG. 6. Stimulus and behavioral control for awake nonhuman primate fMRI. The stimulus, reward, and timing are coordinated by the control computer. The monkey is presented with a stimulus, and if the task is performed correctly (monitored by a behavioral response or eye position), a reward is given. MRI data are acquired, and abnormal behavioral and eye position data are recorded for use in post processing (e.g., as a covariate of no interest in the statistical analysis).

and generate a reward for a correct task). The position of the animal within the containment tube and head restraint are the same as used in the MR scanner. A pair of loudspeakers are used to provide the same gradient noise to simulate the times when the scanner is imaging (and the monkey needs to remain still and attend to the stimulus) or when the scanner is quiet and the animal can break visual fixation and expect a reward. Further refinements (not shown in Fig. 7) are motion sensors on the limbs and head of the animal to remain motionless or pressure monitoring of the juice reward to discourage the animal from sucking on the tube while the experiment is running (i.e., while the gradient noises are audible).

Hygiene. Imaging nonhuman primates requires adherence with basic hygiene considerations to prevent zoonotic cross-transmission with animal handlers and between animals, particularly herpes simiae (herpes B), tuberculosis, and enteric bacteria. Additionally, awake animals need sufficient restraint to prevent them escaping from the MR scanner. Most groups involved in primate imaging use a dedicated primate chair, which serves both to confine the animal and to restrict the spread of body fluids and excretions (Fig. 5). Following imaging, the primate chair and scanner surfaces can be swabbed with diluted bleach or phenolic-based disinfectant. Toilet

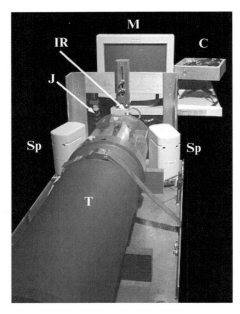

Fig. 7. Mock-up of horizontal MRI scanner used for training. The monkey lies prone in the same tube (T) used in the actual MRI scanner (see Fig. 3A) and observes the stimulus on a computer monitor (M). Eye tracking is recorded in complete darkness using an infrared source (IR) and video camera (C). The gradient noise of an EPI acquisition is played through the speakers (Sp) to simulate the actual noise in the scanner. The monkey is trained to keep still and only to expect reward from the juice tube (J) at the end of an imaging session (when the gradient noises cease). Behavioral control for training follows the same scheme used for data acquisition in the MRI scanner, as outlined in Fig. 6 (adapted from Dubowitz[78] with permission).

training the animal or using a disposable diaper further aids in maintaining hygiene.

MRI Hardware

Basic components of an MRI system are a superconducting magnet, gradient and shim coils, radiofrequency coils (and their respective amplifiers), and a control console. The more the hardware components are optimized for monkey imaging, the better the SNR and resolution that can be achieved. A dedicated gradient coil that closely approximates the size of the sample to be imaged (in this case a monkey head) will generally provide the best possible switching speeds and slew rates (and hence the fastest

[78] D. J. Dubowitz, Ph.D. Thesis, California Institute of Technology, Pasadena, 2002.

and highest-resolution imaging). The RF coil geometry and size have a profound effect on SNR (the increased SNR can be used to reduce imaging time and/or to increase resolution). High-fidelity, low-noise amplifiers also add to signal fidelity and SNR.

Experimental Setup

As described earlier, two approaches currently exist for awake nonhuman primate imaging: vertical and horizontal imaging. The exact choice will depend largely on the MR hardware available. For both approaches, the monkey is confined in a plastic tube, which provides both basic hygiene and confines the animal within the scanner, preventing escapes. Horizontally, the monkey sits on its haunches[57] or lies prone in a crouching sphinx position.[22] Vertically, the animal sits or squats upright inside the plastic tube.[24] For imaging the visual cortex, head fixation is required to reduce artifacts from gross movement and to ensure that the animal fixates appropriately on the stimulus. In a horizontal bore system, the stimulus can be presented via MRI-compatible video display goggles or projected onto a screen in front of the monkey (Fig. 5). Binocular video goggles are a necessity in the confined spaces within a vertical MR scanner. Goggles also allow more accurate control of the stimulus and background luminance and better control of visual disparity. For studies in which the animal needs to interact directly with the stimulus (e.g., pointing, reaching, grasping), the horizontal orientation with a rear projection screen is more versatile. The animal needs to be trained to fixate for prolonged periods (typically 30–60 s or longer) to allow adequate temporal averaging in the fMRI experiment. Correct behavior requires prompt rewards. Chewing and sucking need to be confined to times between actual data acquisition periods. Any task-related motion can cause false signal changes in the fMRI images.[79] Even small amounts of palatial movement during swallowing can cause significant stimulus-correlated motion artifacts in data (Fig. 8). For this reason, providing short duration drops of fluid as a reward is preferable to solid treats. Occasionally, unusual movements are seen that are not expected. Fig. 8 shows the effect of an animal wiggling its ears during a MRI study.

Experimental Design

Using awake nonhuman primates for fMRI imposes timing constraints on the choice of experimental paradigm, and these have to be taken into consideration when designing experiments for monkeys. A standard

---

[79] J. V. Hajnal, R. Myers, A. Oatridge, J. E. Schwieso, I. R. Young, and G. M. Bydder, *Magn. Reson. Med.* **31,** 283 (1994).

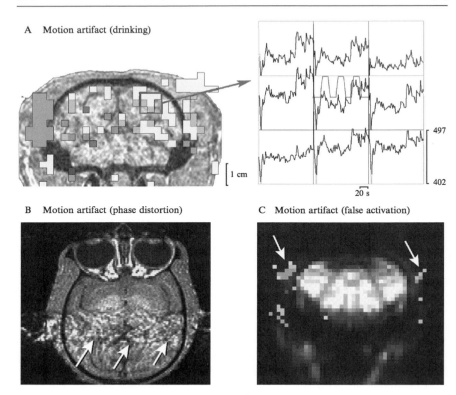

A   Motion artifact (drinking)

1 cm

497

402

20 s

B   Motion artifact (phase distortion)   C   Motion artifact (false activation)

FIG. 8. (A) MRI signal time course in visual cortex while receiving a juice reward. The monkey is sitting in complete darkness with no visual stimulus and receives a juice reward every 20 s. Artifactual signal changes of up to 10% are seen in the visual cortex due to palatial movement during sucking and swallowing. This stimulus-correlated noise would dominate any true BOLD signal. To prevent this, the monkey is trained not to expect a reward until the end of an imaging experiment (signaled by the gradient noise ceasing) (adapted from Dubowitz[78] with permission). (B) An unusual phase artifact due to the monkey moving his ears in the scanner (arrows). (C) The movement appears to be related to attention or interest in the visual task and is correlated with the stimulus paradigm. MR signal change simulating functional activation is seen over the ears in the functional images (arrows). This was prevented when the monkey wore an elasticated "skiing" headband and eventually disappeared with further training.

fMRI experiment in human subjects relies on several minutes of attending to the paradigm and compliance with the stimulus. During this period of attention, the stimulus is typically presented and modulated as a block design or event-related paradigm.[80–83] For awake monkeys, the time during

[80] A. M. Dale, *Hum. Brain Mapp.* **8,** 109 (1999).

which the animal will attend to the stimulus between rewards is considerably shorter than for human studies (typically 1 min or less). Monkeys receiving regular rewards will repeat a task with regularity many hundreds or even thousands of times. This is different from human studies, where such repetition leads to fatigue and loss of concentration. Imaging sessions in excess of 1–2 h are frequently poorly tolerated by human subjects, and attention to the paradigm is lost. However, a trained monkey that is engaged by the stimulus task and receives sufficient positive feedback and reward will "work" readily in the MRI scanner for 4–6 h at a time. It is thus not always practical to use the same experimental design for both human and monkey imaging studies.

Unlike human studies where an incomplete trial can be repeated, for studies in awake monkeys, the level of compliance with the task must be gauged in real time and a commensurate level of reward given. Thus, controlling the juice reward of the monkey also needs to be incorporated into the experimental design (Fig. 6). A timely reward is given for correct compliance with the visual task. A delay in delivering the reward is introduced if the monkey fails to comply with the task within certain parameters (e.g., maintaining accurate fixation to within a predefined radius of arc). If a trained monkey completely fails to do the task, it is important that the reward is not given for that particular trial. Ideally, the stimulus and data acquisition are also aborted and the experiment is restarted after a short delay.

For studies comparing visual cortex activation between the two species and requiring the same experimental paradigm in both species, some compromise is needed between the ideal paradigm for nonhuman and human subjects. A typical design for a functional imaging study in human subjects consists of imaging runs lasting several minutes, which are repeated within the confines of the subject's comfort. A more natural experimental design for nonhuman primate imaging is a shorter trial, which can be repeated many hundreds of times.

Postprocessing

Postprocessing of fMRI data is based on identifying areas of the image showing a statistically significant signal variation that is correlated with the stimulus. A large number of commerical and open-source software packages

[81] P. A. Bandettini, A. Jesmanowicz, E. C. Wong, and J. S. Hyde, *Magn. Reson. Med.* **30,** 161 (1993).

[82] T. T. Liu, E. C. Wong, L. R. Frank, and R. B. Buxton, *Neuroimage* **16,** 269 (2002).

[83] T. T. Liu, L. R. Frank, E. C. Wong, and R. B. Buxton, *Neuroimage* **13,** 759 (2001).

are now available to do this, which streamlines the process greatly (see the Internet Analysis Tools Registry at http://www.cma.mgh.harvard.edu/tools for a comprehensive list). However, a thorough inspection of the raw data and an understanding of the underlying analysis process are important when dealing with fMRI data. Most postprocessing schemes follow the same basic principle of modeling the MRI signal variation in terms of a scaled term related to the stimulus itself, a number of residuals, and noise. The more the residual terms can be reduced or at least incorporated into the model as covariates of no interest, the more accurate the result will be. Bulk motion is one of the biggest confounds that needs to be addressed during post processing, as this produces large signal changes at interfaces of structures with different MR contrast[79] (see Fig. 8). These signal changes due to motion may swamp the signal changes from hemodynamic changes. Motion that is highly correlated with the stimulus (e.g., a startle or head motion at the beginning of each stimulus presentation) can be very difficult to separate from true hemodynamic changes. Even motion outside the head (e.g., an animal moving its limbs or swallowing) may cause phase changes in the MRI signal measured in the head; thus, training an animal to minimize extraneous movements is as important as rigid head restraint. Additional variation in the MRI signal is also caused by physiological "noise" from respiratory and cardiac motion. This occurs due to phase changes induced by the movement of thoracic structures in the B_0 field, as well as actual moment of the brain due to directly transmitted pulsations through blood and CSF during the cardiac and respiratory cycles. Typical fMRI data are not sampled fast enough to remove cardiac and respiratory fluctuations by simple frequency domain filtering.[84] A number of post-processing strategies have been implemented to address physiological noise. By recording physiological and movement variables (head position, electrocardiogram, respiratory rate), they can be used to construct more accurate terms for the residuals when modeling the MRI signal.[85,86] Postprocessing strategies based on principal component analysis have also been used with some success to remove gross head movement.[87] As a rule of thumb, it pays to invest time minimizing bulk motion and other confounds *a priori* rather than addressing them only with postprocessing.

[84] L. R. Frank, R. B. Buxton, and E. C. Wong, *Magn. Reson. Med.* **45**, 635 (2001).

[85] T. H. Le and X. Hu, *Magn. Reson. Med.* **35**, 290 (1996).

[86] G. H. Glover, T. Q. Li, and D. Ress, *Magn. Reson. Med.* **44**, 162 (2000).

[87] A. H. Andersen, Z. Zhang, T. Barber, W. S. Rayens, J. Zhang, R. Grondin, P. Hardy, G. A. Gerhardt, and D. M. Gash, *J. Neurosci. Methods* **118**, 141 (2002).

Comparing Studies between and across Species

A Common Platform for Comparing Monkey and Human Neuroimaging

Much of the primate brain is composed of a thick, two-dimensional (2D) sheet of neurons. As the brain has expanded (with development and with evolution), these 2D sheets have formed into folds to allow the larger surface area to fit into the calvarium. In the human, much of the neocortex has become buried in these folds. This folding pattern complicates the task of mapping structural and functional regions in the large human brain.[34]

Many functional dimensions are mapped along the cortical surface (e.g., retinotopy, orientation tuning, ocular dominance), and the parameters describing these areas vary maximally and tangentially along the cortical sheet rather than in depth through it. Additionally, cortical areas are arranged in characteristic patterns across the cortical surface.[88] Despite this sheet-like geometry, the most commonly used coordinate systems are based on three-dimensional (3D) stereotaxic coordinates (e.g., Talairach)[89,90] rather than on position relative to the 2D cortical surface. Such stereotaxic maps do have advantages in terms of better applicability to subcortical structures and direct correspondence with surgical or histological approaches. However, the Talairach stereotaxic approach relies on rigid body rotation and linear scaling to fit to a standard brain, and the position of the cortical structure may actually be misrepresented by several centimeters.[91] These 3D stereotaxic methods also have significant drawbacks when assessing transcortical distances and cortical interconnections (particularly if structures lie on different banks of a sulcus). For comparing across subjects, representing the cortical surface as a flattened 2D map has a number of advantages, and subtle differences in structural and functional anatomy are identified more readily.[88,92] Comparative 2D cortical atlases are also available for different species.[62]

[88] B. Fischl, M. I. Sereno, R. B. Tootell, and A. M. Dale, *Hum. Brain Mapp.* **8**, 272 (1999).

[89] G. Paxinos, X.-F. Huang, and A. W. Toga, "The Rhesus Monkey Brain in Stereotaxic Coordinates." Academic Press, London, 1999.

[90] J. Talairach and P. Tournoux, "Co-planar Stereotaxic Atlas of the Human Brain." Thieme, New York, 1988.

[91] D. C. Van Essen and H. A. Drury, *J. Neurosci.* **17**, 7079 (1997).

[92] J. Dickson, H. Drury, and D. C. Van Essen, *Philos. Trans. R. Soc. Lond. B Biol. Sci.* **356**, 1277 (2001).

Representing neocortical anatomy as a surface is not new, and traditional anatomical studies have manually prepared sheets of cortical surface or histological slices for analysis.[93,94] Advances in automating surface reconstruction and unfolding[88] have the potential to make such surface representations as common in human and nonhuman primate fMRI studies as they have been in traditional anatomical studies of nonhuman primates.[34]

Unfolding of the sulcal architecture allows the neocortex to be represented as the smooth surface of an idealized balloon-like inflated brain. A further modification is to "cut" the stylized surface so that an entire hemicortex can be represented as a flat 2D map. There are advantages and disadvantages to both these approaches: the stylized inflated surface retains the characteristic relative geometry of a normal brain (frontal, parietal, temporal, and occipital areas retain their normal juxtaposition) so that the image still looks intuitively like a brain. The whole cortical surface can be visualized, and areas are no longer obscured inside sulci. The representation is still 3D, so it needs to be rotated to view an entire hemisphere. The cut 2D flat map presents the same information for an entire hemisphere in a single plane, providing a further reduction in the dimensionality of the image. The appearance, however, is less intuitively brain-like, and the lobar juxtaposition is distorted. For both these approaches, comparison between studies or across species relies on careful manual interrogation of the images by the investigator. For a more analytical comparison, the cortical surface needs to be mapped into a common representation. A method by Fischl et al.[95] defines a common 2D coordinate system by mapping the expanded cortex nonlinearly onto the surface of a sphere based on a measure of sulcal morphology. Geometric features on the cortical surface are used for anatomical registration. This ensures a more accurate structural and functional localization compared with 3D stereotaxic methods. The spherical coordinate system also produces less spatial blurring, which translates into increased statistical power for cross-subject averaging, This technique allows comparison across human subjects and holds promise as an approach for comparing across primate species as well (Fig. 9).

Retinotopic Mapping in Monkey and Human Visual Cortex

Visual cortical areas in monkeys are typically small and irregularly shaped. Additionally, they may vary in location, so defining their boundaries based on anatomical features alone is frequently difficult. One

[93] D. C. Van Essen and S. M. Zeki, *J. Physiol.* **277**, 193 (1978).
[94] D. C. Van Essen and J. H. Maunsell, *J. Comp. Neurol.* **191**, 255 (1980).
[95] B. Fischl, M. I. Sereno, and A. M. Dale, *Neuroimage* **9**, 195 (1999).

method for distinguishing visual areas relies on the fact that many are organized retinotopically, and that the retinotopic representations in adjacent visual areas are mirror images of each other.[38,96] By producing maps of receptive field eccentricity and polar angle, a spatial gradient (or fieldsign) can be calculated for a given area of the cortex. The gradient, or its sign, indicates if the representation on the cortex is a mirror or nonmirror image.[97] Expanding on conventional electrophysiology studies of receptive field orientation, this cortical organization of the cortex can also be studied with fMRI. Functional MRI for retinotopic mapping differs from traditional block or event-related paradigms. A Fourier-type paradigm is employed, thus the parameter of interest is the phase of the response rather than the amplitude. Additionally, at least two measured quantities are needed: the eccentricity and polar angle of the region of visual space that is represented at that cortical location. A flattened cortical representation of structural MRI data is needed on which to project the eccentricity and polar angle gradients.

The phase-encoded stimulus method of making these measurements was originated by Engel et al.[64] and used by Sereno et al.[63] and others.[98,99] Two continuously changing stimuli are swept repeatedly across the retina—expanding (or contracting) rings and clockwise (or counterclockwise) rotating wedges—while the subject fixates on a central point. Functional MRI data collected from each kind of stimulus are analyzed on a voxel-by-voxel basis by comparing the Fourier magnitude of the periodic response at the stimulus frequency to the response at other frequencies to determine which regions are modulated by the stimuli (regions that respond to all eccentricities or all angles will be "subtracted out" by this method). Then, the phase of the periodic signal at the stimulus frequency is determined. This phase angle (delay in response) corresponds to the position of the stimulus, in either eccentricity or angle, after correcting for common response delays.

These data are then sampled onto a high-resolution cortical surface reconstruction (folded, inflated, flattened). By taking the gradients in eccentricity and polar angle with respect to cortical x, y (on the flattened representation), it is possible to calculate whether each small patch of the retinotopically-organized cortex represents the visual field as a nonmirror

[96] M. I. Sereno and J. M. Allman, in "The Neural Basis of Visual Function" (A. G. Leventhal, ed.), p. 160. Macmillan, London, 1991.
[97] M. I. Sereno, C. T. McDonald, and J. M. Allman, Cereb. Cortex 4, 601 (1994).
[98] E. A. DeYoe, G. J. Carman, P. Bandettini, S. Glickman, J. Wieser, R. Cox, D. Miller, and J. Neitz, Proc. Natl. Acad. Sci. USA 93, 2382 (1996).
[99] S. A. Engel, G. H. Glover, and B. A. Wandell, Cereb. Cortex 7, 181 (1997).

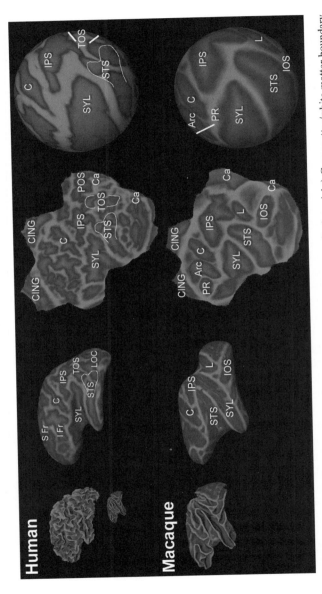

FIG. 9. Representations of the cortical sheet in macaques and humans. (Left to right) Gray matter/white matter boundary showing the sulcal and gyral pattern, inflated cortex, "cut" cortex mapped onto a 2D sheet, and cortex mapped onto the surface of a sphere. The small inset brain is a macaque brain at the same scale as the human brain for comparison. The sulcal pattern is shown in red; the gyral architecture is in green. Following inflation, the entire cortex is clearly seen and no areas are obscured inside the sulci. Each inflated brain is unique and is not bounded to a particular geometric shape. To compare across subjects, or across species, the flat and spherical representation provides a common reference. The 2D image is generated by "cutting" the cortical sheet and presenting it as a single sheet. The human brain is mapped onto an averaged brain in the spherical representation. The macaque brain is mapped onto the surface of a sphere and is rotated so that the view matches the human view, but is not constrained to any averaged sulcal representation. Note the close relationship of sulcal morphology

image or as a mirror image (fieldsign). Because many adjoining visual areas have an opposite visual field sign (e.g., V1 and V2), this makes it possible to automatically "color in" many visual areas. A detailed review of the technique can be found in Warnking et al.[100]

This method requires uninterrupted fixation of the stimulus for several minutes. Whereas this is quite tolerable for human subjects, this degree of fixation is much more challenging for awake nonhuman primate studies. One approach is to map the retinotopic organization in anesthetized animals.[58] As discussed earlier, some modification in the stimulus is needed for awake monkey studies. Fize et al.[59] addressed this by presenting discrete stimuli of a particular orientation or eccentricity as a randomized block design (rather than the traditional continuously varying Fourier stimulus) and also allowed the monkeys to break fixation for up to 20% of the studies. This approach provides discrete (rather than continuously variable) measures for the mirror image/no-mirror image fieldsign maps, but improves SNR for the chosen orientations and eccentricities.

Conclusions

Functional MRI in nonhuman primates offers a unique tool to link our understanding of primate neurophysiology from invasive techniques in nonhuman primate and our endeavors to understand our own brain from human fMRI studies. Primates have highly evolved visual systems and are a model organism for studying visual physiology. Additionally, studies of the visual system are particularly suited to investigation with MRI, as the visual cortex is superficial and in easy reach of RF imaging coils, and visual stimulus paradigms can be readily designed to minimize subject motion. Thus, fMRI of the visual system provides a valuable tool in our quest to understand the functional and anatomical organization of the human brain, as well as offering exceptional opportunities to investigate the underlying neurovascular coupling that is the basis of the fMRI signal. New tools

[100] J. Warnking, M. Dojat, A. Guerin-Dugue, C. Delon-Martin, S. Olympieff, N. Richard, A. Chehikian, and C. Segebarth, Neuroimage 17, 1665 (2002).

between the macaque and the human brain. Sulcal areas labeled in the human brain are sylvian (SYL), superior temporal (STS), transverse occipital (TOS, human "lunate" sulcus), intraparietal (IPS), central (C), cingulate (CING), calcarine (Ca), parietooccipital (POS), lateral occipital (LOC), inferior frontal (I fr), and superior frontal (S Fr). Additional sulcal areas labeled on the macaque brain are lunate (L), arcuate (Arc), principal (PR), and inferior occipital (IOS). Macaque MRI raw data kindly provided by Nikos Logothetis and Margaret Sereno; image reconstruction and postprocessing courtesy of Martin Sereno. (See color insert.)

that allow anatomical mapping of the brain across species can be combined with structural and functional MRI and hold promise as techniques for studying taxonomy and the functional interrelationship between primate species.

Acknowledgments

The author thanks Drs. Martin Sereno, Richard Buxton, Miriam Scadeng, Tom Liu, and Kamil Uludag for valuable discussion and help with the manuscript and figures. For Fig. 9, macaque MRI data used to generate the figure were provided by Nikos Logothetis and Margaret Sereno. Postprocessing was done by Martin Sereno using free surfer software http://surfer.nmr.mgh.harvard.edu/.

[8] Magnetic Resonance Imaging of Brain Function

By Stuart Clare

Introduction

The rapid development of methods for noninvasive brain mapping, particularly over the past decade, has led to exciting advances in our understanding of the human brain. Foremost in these methodologies is the technique of functional magnetic resonance imaging (fMRI). Utilizing the intrinsic magnetic properties of the blood, it is possible to identify the brain region associated with a specific sensory, motor, or even cognitive task to a high spatial precision. Unlike positron emission tomography (PET), MRI does not use radioactively labeled compounds and is essentially noninvasive and safe for repeat studies. Although fMRI does not share the temporal resolution of electroencephalography (EEG) or magnetoencephalography (MEG), it does have a spatial resolution of millimeters, and the most recent experiments suggest that it may be able to detect activations at the level of the cortical layers.[1]

While fMRI is a complex methodology, developments in recent years, particularly by the scanner manufacturers and other commercial and academic groups, have meant that the tools for fMRI, while expensive, are more commonly available. In particular, the available software for both fMRI stimulus presentation and image analysis are highly sophisticated and user-friendly. This means that human fMRI is now achievable by research groups without dedicated physics and image analysis support.

[1] A. C. Silva and A. P. Koretsky, *Proc. Natl. Acad. Sci. USA* **99**, 15182 (2002).

Experimental Procedures

Overview of Methods

Functional MRI relies on detecting the small changes in image brightness on MRI scans, associated with the hemodynamic changes in the brain, in response to a specific external stimulus or "internal" cognitive process.[2] Carrying out an fMRI experiment therefore consists of three primary components: presenting or otherwise cueing the stimulus, scanning the brain rapidly using MRI, and analyzing the MRI scans to detect changes in image intensity.

While the subject is being scanned repeatedly, ideally covering the whole brain every 3 s, a stimulus or cue is presented to them. This could be a simple visual stimulus, such as a flashing light, or a more complex stimulus, such as a list of numbers to remember and recall at some point. This stimulus is repeated a number of times to build up confidence in determining the brain regions that are truly responding to the stimulus, while averaging out other "random" brain processes. The resulting images are then analyzed using computer software to detect those regions in the images that show a significantly time-locked response to the stimulus. These regions are displayed as bright "activations" overlaid on a conventional brain scan or brain atlas, such as shown in Fig. 1.

Human fMRI can be performed using most modern MRI scanners found in radiology departments of many hospitals, operating at a field strength of 1.5 Tesla. However, in recent years, the desire to detect these small hemodynamic changes has led to the successful use of field strengths of 3 to 4 Tesla in research MRI systems. A small number of research sites worldwide are experimenting with the use of even higher field strengths for human imaging, the highest currently being 9 Tesla; however, the difficulties with producing high-quality human brain images diminish the benefits in signal detection offered at such high field strengths.[3]

Functional MRI of nonhuman primates and rodents is covered in more detail in other articles in this volume. For the rest of this Chapter, the use of human subjects is assumed; however, much of the underlying methodology is the same for human or animal subjects.

Blood Oxygenation and MRI

The oxygen carrier in blood, hemoglobin, gets its red color from the iron molecule, which forms the binding site of oxygen. This presence of iron in the molecule makes it magnetically sensitive. In its deoxygenated

[2] J. W. Belliveau, D. N. Kennedy, Jr., R. C. McKinstry, B. R. Buchbinder, R. M. Weisskoff, M. S. Cohen, J. M. Vevea, T. J. Brady, and B. R. Rosen, *Science* **254,** 716 (1991).
[3] S. Chan, *AJNR Am. J. Neuroradiol.* **23,** 1441 (2002).

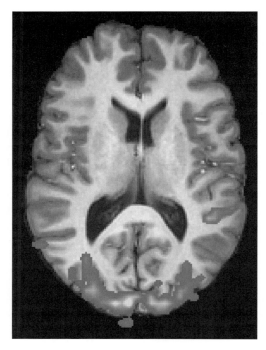

FIG. 1. Example of an fMRI result showing activation in the visual areas resulting from the subject looking at a contrast reversing checkerboard stimulus. (See color insert.)

state, hemoglobin displays paramagnetic properties, meaning that the local magnetic field is increased in the presence of an external magnetic field.[4] In contrast, oxygenated hemoglobin is diamagnetic and has little effect on the local magnetic field. The effect of these local changes in magnetic field can be detected in a type of MR scan that is said to be T_2^* weighted. On a T_2^*-weighted MR image, a pixel that contains predominantly oxygenated hemoglobin will appear brighter than a pixel that contains predominantly deoxygenated hemoglobin. This form of image contrast in MRI is termed blood oxygenation level-dependent (BOLD) contrast.[5]

Although it is possible to quantify the change in hemoglobin oxygenation using MRI, it is typical in fMRI to just detect the relative signal

[4] K. R. Thulborn, J. C. Waterton, P. M. Matthews, and G. K. Radda, *Biochim. Biophys. Acta* **714,** 265 (1982).
[5] S. Ogawa, T. M. Lee, A. R. Kay, and D. W. Tank, *Proc. Natl. Acad. Sci. USA* **87,** 9868 (1990).

changes. The exact link between neuronal firing and the BOLD signal change that is detected is complex and not entirely understood. Upon the metabolic demand that synaptic activity produces, oxygen is removed from the blood and the concentration of deoxyhemoglobin increases. This would result in a small dip in the MR image intensity, which is sometimes observed in fMRI experiments. However, the much stronger effect is the large increase in image intensity that follows, peaking at about 6 s after the neural activity. This represents a large increase in the concentration of oxyhemoglobin, far greater than its resting state level. This results from a large increase in the local blood flow rate and local blood volume due to capillary expansion. While this apparently excessive overcompensation in oxygen delivery was initially a puzzling result, recent physiological models have demonstrated the need for such increases to maintain the necessary oxygen delivery rate to the mitochondria.[6]

Following this peak in local oxygenation level, the signal returns back toward its baseline state, but is often observed to decrease still further (known as the undershoot), as the relative contribution of oxygen extraction, blood flow, and blood volume return to their baseline state. Most fMRI "activations" are detected as regions that display the large increase in signal, peaking several seconds after the stimulus. In fact, the presence of either the initial dip in signal or the poststimulus undershoot is not detected in many experiments, as it seems to vary by brain location and can be obscured by image noise, particularly at lower field strengths.[7]

The complex physiological processes that give rise to the signal changes observed in fMRI mean that there are a number of reasons for caution in interpreting the experimental results. First, and most obviously, is that the signal arises from hemodynamic effects and not directly from neural activity. The location of peak BOLD effects could indeed be some distance from the site of the activating neurons. This is particularly the case when imaging using methods that are more sensitive to the signal from the large draining veins, which could be centimeters from the actual activation site.[8] To guard against this particular problem it is advisable to interpret the spatial location of fMRI activations with reference to a map of veins (as can be acquired easily using MRI or from standard atlases).

Second, thought must be given to the time characteristics of the fMRI response. The timing of peak activation relative to the signal needs to be taken into account when analyzing the images, as it may vary over brain regions. Care must also be exercised in interpreting any differences in signal

[6] R. B. Buxton, E. C. Wong, and L. R. Frank, *Magn. Reson. Med.* **39,** 855 (1998).
[7] R. B. Buxton, *Neuroimage* **13,** 953 (2001).
[8] S. G. Kim, K. Hendrich, X. Hu, H. Merkle, and K. Ugurbil, *NMR Biomed.* **7,** 69 (1994).

timing as representing temporal differences in the onset of neural activation. While it is certainly possible to obtain an indication of neural timing from the BOLD response, a lag in signal in one region relative to another does not necessarily mean a difference in neural timing and may just represent a difference in blood supply in those regions.[9] The large delay after activation before the signal returns to baseline also has implications for sexperimental design and is discussed in more detail later.

It should also be noted here that BOLD contrast is not the only way to perform fMRI, although it is by far the easiest and most commonly used method. MR images can also be made sensitive to the blood flow rate alone. Such experiments suffer from low signal-to-noise ratio (SNR) and do not have the same temporal resolution as BOLD fMRI, but are very useful in interpreting the BOLD signal changes and may turn out to be more spatially specific than BOLD.[10]

Rapid MRI for fMRI

MR images are essentially maps of water content in the brain generated by the NMR phenomenon that certain atoms, when placed in a strong magnetic field, will absorb and emit radiofrequency energy at a specific frequency dictated by the strength of that applied magnetic field. However, it is straightforward in MRI to modulate this basic signal such that the intensity in a region of the image is not just dependent on water content, but also on the local structural environment or other physiological parameters. Examples of this are the T_1- and T_2-weighted images often used in clinical diagnosis, where, for example, the region of cell damage produced by a stroke can be seen very clearly. Another example of this is the so-called T_2^*-weighted image, which is highly sensitive to the local magnetic environment and is particularly sensitive to the BOLD effect.[11]

A typical diagnostic MR image is optimized for spatial resolution and contrast to detect the particular pathology of interest. This typically means a scan time of several minutes. Although it is possible to do fMRI with a scan lasting minutes, the requirement of needing to keep a discrete set of brain regions active for such a time makes this impractical for anything other than the simplest experiment.

Speeding up the scanning process requires not only very high-performance scanner hardware, but comes at a cost to image quality. However, with the

[9] F. M. Miezin, L. Maccotta, J. M. Ollinger, S. E. Petersen, and R. L. Buckner, *Neuroimage* **11,** 735 (2000).

[10] T. T. Liu, E. C. Wong, L. R. Frank, and R. B. Buxton, *Neuroimage* **16,** 269 (2002).

[11] M. E. Haacke, R. W. Brown, M. R. Thompson, and R. Venkatesan, "Magnetic Resonance Imaging." Wiley, New York, 1999.

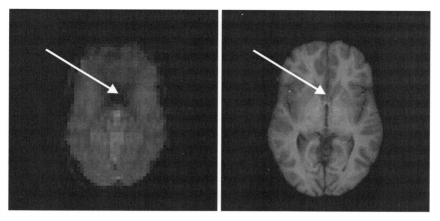

FIG. 2. An example of signal loss in the frontal lobes seen in echo planar imaging (EPI). (Left) A typical EPI scan used for fMRI and (right) the same slice as seen in a conventional MRI scan. Brain tissue that is clearly visible in the conventional MRI scan appears missing in the EPI scan, particularly in the region indicated by the arrow.

advent of fMRI, most modern scanners have the technological capability to run very fast imaging methods.

The most common fast imaging method used for fMRI is echo planar imaging (EPI). This method is able to collect data from a single "slice" through the brain in less than 100 ms, meaning that, at coarse resolution, it is possible to scan the whole brain in around 3 s.[12] EPI also has the benefit of being inherently a T_2^*-weighted sequence, so it is ideally suited to BOLD fMRI.

The largest drawback to using EPI is that the images often contain image distortion and signal loss. An example of this is shown in Fig. 2. The air-filled sinuses that sit below the frontal lobes of the brain cause a "hole" to appear in the EPI images, compared to the standard MRI, and the frontal lobes also appear smeared out and distorted.[13] Such effects mean that it is difficult to accurately detect activations in these regions of the brain. Several methods may be used to try and address this problem, such as correcting the distortions by using a "field map,"[14] by using specially designed mouthpieces containing graphite to compensate for the effect of air sinuses,[15] or by using imaging methods similar to EPI that do not suffer from distortion,[16] although these often have their own disadvantages.

[12] M. K. Stehling, R. Turner, and P. Mansfield, *Science* **254,** 43 (1991).
[13] P. Jezzard and S. Clare, *Hum. Brain Mapp.* **8,** 80 (1999).
[14] P. Jezzard and R. S. Balaban, *Magn. Reson. Med.* **34,** 65 (1995).
[15] J. L. Wilson, M. Jenkinson, and P. Jezzard, *Magn. Reson. Med.* **48,** 906 (2002).

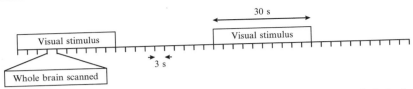

FIG. 3. A schematic diagram of a typical fMRI experiment where the whole brain is scanned every 3 s and alternating periods of visual stimulus and no stimulus are given every 30 s.

Unfortunately, the characteristics of the EPI method that make it susceptible to signal loss near sinuses are also those that make it sensitive to the BOLD effect. A typical fMRI experiment on a human subject using EPI would have image pixel sizes of 3×3 mm^2 and use a slice thickness of between 3 and 6 mm. Using thinner slices is one way of reducing signal loss near sinuses, but again this could reduce sensitivity to the BOLD effect.[17] Another parameter that affects signal loss is known as the "echo time" or TE. By reducing the echo time, signal loss is reduced, but again this comes at a cost to BOLD sensitivity. In practice, an echo time of 30–50 ms gives good results over the whole brain, but this is a parameter worth varying in initial pilot experiments to find the optimum balance in the brain regions of interest.[18]

One final note on rapid MRI is that these methods produce a high level of acoustic noise, often over 100 dB. This means that it is essential to provide the subject with adequate ear protection and warn them prior to the experiment.

Experimental Design

The simplest form of fMRI experiment consists of blocks of stimulus presentation and "rest," interleaved such as illustrated in Fig. 3 for a visual stimulus. Typical timings of such a block paradigm are 30 s of stimulus and 30 s of rest, both repeated four times, with the whole brain being scanned once every 3 s. Critical to the success of the experiment is that there is a single clear difference between the task and the rest condition. It is not conceivable that the brain is completely inactive during the "rest" period, but it needs to be assumed that such activity is equally likely to be present during the task period as the rest period.

[16] G. H. Glover and C. S. Law, *Magn. Reson. Med.* **46,** 515 (2001).
[17] K. D. Merboldt, J. Finsterbusch, and J. Frahm, *J. Magn. Reson.* **145,** 184 (2000).
[18] S. Clare, S. Francis, P. G. Morris, and R. Bowtell, *Magn. Reson. Med.* **45,** 930 (2001).

Because activations are detected by comparing serial MRI scans, any subject movement between each scan will reduce the ability to detect them. It is therefore important to minimize subject movement. For most short fMRI experiments, this is best done by using foam pads around the subject's head. Such pads or pillows not only provide support, but also act as points of reference for the subjects as they try and keep their own head still. Many experimenters advocate the use of thermoplastic masks or bite bars to keep the subject still. While these do minimize head movement very effectively, they are often uncomfortable for the subject and usually require some time to get used to using them.

Whatever stimulus is used, it is likely that some form of visual stimulus or cue will need to be given to the subject either to stimulate the visual cortex directly, to instruct the subject in the timing and pacing of a movement task, or present cognitive stimuli such as patterns of letters to compare or remember. The most versatile way of presenting such stimuli is to use a high-quality video projector connected to a computer. This can project text or images onto a screen near the end of the scanner, at the subject's feet. Then an angled mirror above the subject's face, or a prism arrangement, enables the subject to view the screen. There are a number of more specialized systems that deliver the picture directly to glasses worn by the subject; however, the complexity and expense of these systems only make their use justified for particular applications that need higher control on what the subject sees.

Visual presentation is not the only way to present stimuli to subjects. As indicated earlier, the high acoustic noise environment of the scanner is not ideal for using auditory stimuli; however, it is possible to use MRI-compatible headphones (either pneumatic or electrostatic). Successful auditory studies have been accomplished by scanning at a slower rate, with short gaps in the scanning during which the speech or sound is played.[19] Other devices, such as vibrotactile stimulators to stimulate the somato-sensory cortex, thermal devices to stimulate the pain network, or olfactometers to deliver smells, have all been used in the MRI environment. As with all equipment that comes in contact with a subject, it is essential that there is no chance of it causing any harm. The scanner environment adds additional constraints, both on safety and on the ability to get high-quality MR images, so it is important to check carefully before even taking a device into the magnet room.

As well as presenting a stimulus to a subject, it is often desirable to record some response from the subject. The most simple and versatile way to

[19] D. A. Hall, M. P. Haggard, M. A. Akeroyd, A. R. Palmer, A. Q. Summerfield, M. R. Elliott, E. M. Gurney, and R. W. Bowtell, *Hum. Brain Mapp.* **7,** 213 (1999).

do this is with an MRI-compatible, four button box. The subject can rest their four fingers of one hand on each of the buttons on the box, and the response can be fed back to the control computer for recording which button was pressed and the precise timing. Interface with the computer is usually best done via a dedicated analogue and digital interface card (such as from National Instruments, Austin, TX), which should be able to handle not only the signals from button boxes, but also from joysticks or other analogue devices, and be able to control other stimulus presentation modalities that require a digital or analogue signal. Getting the subject to verbalize a response is not advisable in general because the movement of the head caused by speaking can reduce the ability to detect small activations. Additionally, it can be hard to hear any response in the noisy environment of the scanner during the experiment.

Software for cueing the experiment and presenting the stimuli is available commercially (e.g., Presentation, Neurobehavioral Systems, Albany, CA) or from research groups (e.g., DMDX, www.u.arizona.edu/~kforster/dmdx/dmdx.htm; Cogent, www.vislab.ucl.ac.uk/Cogent). Such software enables a series of stimuli to be cued up then played out sequentially and will record the nature and timing of any responses by the subject. The pacing of an experiment relative to the acquisition of the scans is critically important, as it is the final images that contain the signal of interest. While this might seem trivial, it is often the case that neither the scanner nor the stimulus presentation computer can be relied upon to keep exact time. While this timing difference may only be one-hundredths of a second, over a long scan run this can make a significant difference. It is usually possible to arrange for the scanner to output a timing (TTL) trigger at the start of acquiring each scan. This trigger can then be detected by the stimulus presentation computer and be used to start the next stimulus.

A block design paradigm, where there are relatively long blocks of stimulation and rest, is suitable for many applications, but can suffer from a number of problems. First, it may not be possible to reliably get a brain region to be active for such a long period of time. Often habituation will occur as the subject easily learns a task or loses interest in the stimulus. There are also some stimuli that need to be presented for short periods of time, such as painful ones. Second, in some cases, particularly cognitive paradigms, it is not desirable to use the same type of stimuli continually. For example, if the experiment requires the subject to discriminate between stimuli, they may quickly learn that the response stays the same for 30 s and make the cognitive part of the paradigm invalid. Third, there are some cases where it is desirable to separate out brain regions that are involved in different aspects of a task. This is particularly the case in experiments on memory, where different brain networks may be involved in the storage of information to the retrieval. To deal with the second problem, it is possible to carry out a block design experiment where the majority of stimuli are of one type in the "task" period and of the other type in the "rest" period, but there are a few of the other type added to keep the novelty component of the experiment.

An alternative to block designs are "event-related" designs. This design type is similar to that used in evoked potential electroencephalography (EEG) recordings in the brain where a single stimulus is presented at some repetition rate. In EEG this is particularly suitable, as the electrical activity associated with the stimulus lasts less than a few hundred milliseconds. In fMRI, however, where the BOLD stimulus response takes 15–20 s to return to baseline following the stimulus, waiting such a long time between stimuli can be less suitable. This method is useful for looking for brain networks responding to different parts of a complex task (such as the memory illustration given earlier), but a more efficient use of event-related fMRI is to present the single stimuli at shorter intervals, often randomized in time. The analysis of such data is more complex, as it requires a model of how the overlapping BOLD responses to individual, closely spaced stimuli combine. However, such experiments can be highly efficient, particularly when stimuli need to be presented in some random order.[20]

Analysis of Images

In the simplest case, analysis of data consists of subtracting the average of all the images acquired during the "rest" phase of the experiment from the average of those acquired during the "task" period. While this gives a general qualitative indication of activated regions, in order to assign statistical significance to the result it is necessary to carry out a more detailed analysis. Also, this simple subtraction does not take account of the fact that the peak BOLD signal is delayed by around 6 s from the start of stimulation. In practice, typical fMRI analysis consists of three parts: preparation of fMRI data, model-based detection of the BOLD signal, and statistical inference and thresholding of the activations. If the results of a number of subjects are to be combined and compared, such as between a patient group and a control group, then an additional step of "group statistics" is required. Each of these areas is looked at in turn here. Most of these methods are available in commercial or freely available software packages (e.g., FSL, www.fmrib.ox.ac.uk/fsl; SPM, www.fil.ion.ucl.ac.uk/spm; AFNI, afni.nimh.nih.gov/afni; Brain Voyager, Brain Innovation, The Netherlands).

---

[20] K. J. Friston, E. Zarahn, O. Josephs, R. N. Henson, and A. M. Dale, *Neuroimage* **10,** 607 (1999).

Preparation of fMRI Data. As explained in the previous section, minimizing the head motion of the subject is vital to getting good results. However, even with the best restraint methods there is often some small residual motion that occurs in the images. This can be reduced by performing motion correction, in software, on the data. The motion correction algorithm compares each individual image with the first in the series and applies a mathematical transform to rotate or move the image until they look as similar to each other as possible.[21] Because the BOLD signal changes in the image are very small, they generally do not bias the motion correction. Next, the images are often spatially smoothed (blurred). The optimal detection of activations occurs when the spatial smoothness of the images is the same as the size of the region of activation. By applying spatial smoothing to the images, the ability to detect activations is often increased.[22] Finally, the time course of each pixel in the image is filtered to remove long-term drifts in the signal and is sometimes filtered to smooth the time course of the signal over time.

Model-Based Detection of the BOLD Signal. Figure 4 shows a representation of a number of scans from an fMRI time series. If a single pixel in an activated region is selected, and its intensity is plotted over time, it displays a clear delayed response with respect to the stimulus presentation. If a mathematical model for the amount of delay and smoothing that is seen in the theoretical BOLD response with respect to the stimulus pattern is assumed (such as shown in the bottom line of Fig. 4), then a statistical measure of how likely that pixel is truly activated in response to the stimulus can be obtained by calculating the correlation coefficient between this theoretical line and the actual time course. Critical to the success of this method is the choice of mathematical model that turns the stimulus time course into a theoretical BOLD response. All of the software packages have a range of choices for this function, but a commonly used one is a gamma function convolution of the stimulus time series. The mathematical framework for this correlation-based approach, known as the general linear model (GLM), is not the only one that can be taken, and it does indeed include assumptions that may not be fully appropriate for fMRI data; however, for the majority of fMRI experiments, it is sensitive and statistically reliable.[23] The GLM can also be used to analyze the "event-related" fMRI experiments described earlier, again producing a model for the theoretical

[21] K. J. Friston, S. Williams, R. Howard, R. S. Frackowiak, and R. Turner, *Magn. Reson. Med.* **35,** 346 (1996).

[22] M. E. Shaw, S. C. Strother, M. Gavrilescu, K. Podzebenko, A. Waites, J. Watson, J. Anderson, G. Jackson, and G. Egan, *Neuroimage* **19,** 988 (2003).

[23] K. J. Friston, A. P. Holmes, J. B. Poline, P. J. Grasby, S. C. Williams, R. S. Frackowiak, and R. Turner, *Neuroimage* **2,** 45 (1995).

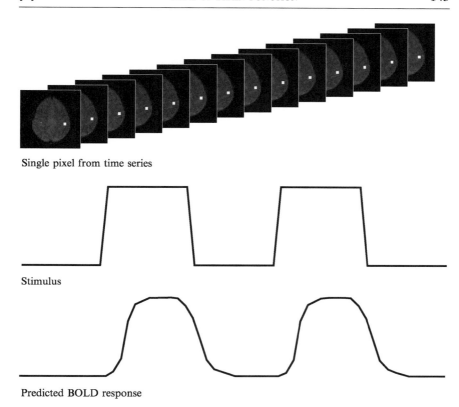

Single pixel from time series

Stimulus

Predicted BOLD response

FIG. 4. Representation of the predicted BOLD response pattern (bottom line) to the stimulus (middle line) in a single pixel that "activates" to that stimulus.

BOLD response given the stimulus timing pattern used. Analyzing the response through time of each pixel in the image results in a statistical "map" of the strength of correlation throughout the brain.

Statistical Inference and Thresholding of Activations. The output image produced by the general linear model looks something like Fig. 5A. It is clear that there is a high correlation at the bottom of the image, but it is not clear yet if this is significant. It is possible to threshold the image so that only those individual pixels that have a correlation coefficient that is significant to better than $p < 0.5\%$, but in an image of approximately 10,000 pixels, we know that 50 pixels would be labeled as "activated" purely by chance. An alternative is to threshold at a much lower significant threshold (such as 0.5% divided by 10,000), but this stringent threshold risks missing genuinely activated pixels. As an alternative it is common to use information

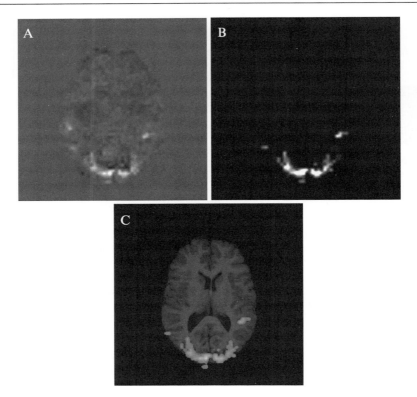

Fɪɢ. 5. (A) Statistical map obtained from the general linear model output. This map is then thresholded at an appropriate *p* score (B) and is then overlaid on a high-quality MRI scan of the same subject's brain (C).

on the number of pixels near each other to increase our confidence in the result. For example, if we see 10 pixels in a block together all showing high correlation coefficients, then we can be more sure that this represents a genuine activation than if we saw one pixel on its own with a similarly high correlation coefficient. The fMRI analysis software packages all contain the appropriate theory, such as that of Gaussian random fields[24] to threshold the activation images on the basis of both correlation coefficient (e.g., as reported as a "*Z* score") and pixel cluster size. What the software will output therefore is a list of pixel clusters that have a *Z* score of greater than say 2.3 (a good, if arbitrary, value to start with) and that have a pixel cluster size making the statistical significance (*p* value) less than say 0.5%. These

[24] K. J. Friston, K. J. Worsley, R. S. Frackowiak, J. C. Mazziotta, and A. C. Evans, *Hum. Brain Mapp.* **1,** 214 (1994).

thresholded activation images can then be overlaid on the subject's high-quality MRI scan (to which the fMRI images have been aligned) or on a standard brain atlas, such as shown in Fig. 5C.

Group Comparisons. If a typical result from a group of subjects is required or if a comparison between two groups needs to be made, then there are two additional steps that need to be performed. First, the MRI scans must be morphed to align to some standard brain template. Such brain templates have been made up of MRI scans from hundreds of individuals and are generally supplied with the software packages. Although not an ideal template, it is very common to report results with reference to the atlas of Talairach based on a dissection of a single brain. Software for morphing to template brain and alignment to the Talairach atlas is included in most fMRI analysis packages. Once all the subjects' scans from a single study are aligned to the same template, further statistical analysis can be performed on data to show regions that are significantly activated across all subjects or ones that are differentially activated in one group relative to another.

An Application of Functional MRI

The ability of the brain to reorganize functionally after injury is a fascinating but hard-to-study phenomenon. Functional MRI has been used to demonstrate the cortical changes that occur upon rehabilitation after stoke.[25] Here the experiment is described as an example of the integration of all the components described earlier to investigate brain function.

Patients with mild to moderate injury following stroke were scanned on four separate occasions, both before and after a movement therapy aimed at regaining motor control over their damaged side. In the scanner the subjects were cued visually to tap their hand, which was resting on a wooden board, at a rate that was either 25 or 75% of their maximum tapping rate. A 6-min series of EPI scans, with an echo time of 30 ms and a repetition rate of 3 s, was recorded during tapping of first the unaffected hand and then the affected hand. A simple block design was used, similar to that illustrated in Fig. 3, with the subject resting their hand for 30 s between 30 s of hand tapping. For all scanning sessions the hand-tapping rate was kept constant, even if the subject was able on a later occasion to tap at a faster rate. This was essential for determining whether the brain activation patterns associated with performing the movements changed over time. A recovery score for each subject was also determined based on their motor performance before and after therapy.

[25] H. Johansen-Berg, H. Dawes, C. Guy, S. M. Smith, D. T. Wade, and P. M. Matthews, *Brain* **125,** 2731 (2002).

Scans from the four sessions of each subject were analyzed together as one GLM analysis. This analysis was set up to detect not only correlations between the pixel time course and a predicted BOLD response to the stimulus, but also, in the same analysis, differences between the two sessions before therapy and the two sessions after therapy. This illustrates the strength of a well-designed GLM analysis; it is not necessary to individually analyze the results from each fMRI session and then look for differences in activation patterns, but it is possible to set up the analysis in such a way as to generate one statistical image with the particular result of interest, in this case the changes that have taken place after therapy.

The researchers then went on to perform a second level of analysis to obtain a summary result representing all the subjects. This was done by subtracting the pretherapy from the posttherapy activation image (unthresholded) for each subject and weighting this difference image by the recovery score for that subject. Combining these images across the group gave an image of areas where activation increases correlate with recovery across the group. The results of both the analysis of the individual subjects and the group analyses indicated that upon recovery, movement of the affected hand produced increased activity in the motor networks associated with both hands. That is to say that the motor networks of the unaffected hand were being recruited to compensate for the damage on the affected side.

Discussion and Conclusion

The advent of fMRI has made a huge impact in the way that the brain is studied, both in the pure neuroscience setting and in a clinical context. The increasing availability of MRI scanners in hospitals throughout the world means that although the technology is expensive, it is increasingly available to researchers. This has been coupled with a solid development of tools for stimulus presentation and, importantly, for data analysis that have the sophistication and statistical rigor required for solid inference, coupled with an ease of use suitable for general laboratory use.

The next big challenge in the development of fMRI is to more fully understand, quantitatively, the relationship between the signals observed in the MRI scans and the underlying neural activity of which it is a marker. This will require not only more sophisticated imaging methodology and more complex physiological models, but also ways to get a closer measure of the working of the neuron *in vivo*. Here, the experiments performed on animals, as described in other Chapters in this volume, will play a vital role.

[9] Laser-Polarized Xenon Nuclear Magnetic Resonance,
     a Potential Tool for Brain Perfusion Imaging:
     Measurement of the Xenon T$_1$ *In Vivo*

By Anne Ziegler, Jean-Noël Hyacinthe, Philippe
Choquet, Guillaume Duhamel, Emmanuelle Grillon,
Jean-Louis Leviel, and André Constantinesco

Introduction

Magnetic resonance imaging (MRI) in humans or animals uses mostly
^{1}H NMR because (1) ^{1}H is the most abundant and the most NMR-sensitive
nucleus in living beings and (2) when it belongs to the water molecules, it
generally presents a transverse relaxation time long enough to be detected
easily. However, even for this nucleus, at thermal equilibrium, in a magnetic field of a few Tesla and at room temperature, the resulting polarization is
quite low, in the order of 10^{-5} (from Boltzmann's law).

Different techniques have been explored to overcome this low signal
level. For instance, it follows from the Boltzmann equation that decreasing
temperature increases polarization. This approach has been explored to
polarize ^{3}He for rat lung imaging.[1] Such "brute force" polarization methods are currently being developed for the large-volume production of hyperpolarized gas.[2] For some privileged atoms, it is also possible to
dramatically increase the nuclear polarization by optical pumping. Such is
the case for ^{3}He and ^{129}Xe, which are nontoxic and have nonzero nuclear
spin (1/2) so that they can be used for *in vivo* NMR experiments. After
optical pumping, the polarization is routinely sufficient (20–40%) to
give a huge improvement in the NMR detection.[3] Gases with an enhanced
spin polarization are commonly designated as hyperpolarized or laser
polarized.

Since the first publication on the use of hyperpolarized gas in *in vivo*
MRI in the mid-1990s,[4] this research field has expanded rapidly. Due to
their biophysical properties, ^{3}He and ^{129}Xe are (almost) ideal for studying

[1] F. Kober, P.-E. Wolf, J.-L. Leviel, G. Vermeulen, G. Duhamel, A. Delon, J. Derouard,
M. Décorps, and A. Ziegler, *Magn. Reson. Med.* **41**, 1084 (1999).
[2] G. Frossati, *Nucl. Instr. Meth. Phys. Res. A* **402**, 479 (1998).
[3] W. Happer, E. Miron, S. R. Schaefer, D. Schreiber, W. A. Van Wijngaarden, and X. Zeng,
Phys. Rev. A **29**, 3092 (1984).
[4] M. S. Albert, G. D. Cates, B. Driehuys, W. Happer, B. T. Saam, C. S. Springer, Jr., and
A. Wishnia, *Nature* **370**, 199 (1994).

molecular structures, materials, porous media, proteins, living organisms, and so on.[5,6] The gyromagnetic ratio of ^{3}He is about 2.75 higher than that of ^{129}Xe, which results, for a given concentration and polarization level, in a signal gain of about 7.5; despite its high diffusion coefficient, ^{3}He is therefore preferable for studying lung, where there is a low concentration of atoms (for review, see Möller et al.[7]) In NMR spectroscopy, ^{129}Xe has been used for surfaces studies,[8] as its chemical shift is very sensitive to the environment.[9] Xenon has also been used as a probe for protein cavity through nuclear Overhauser effects between xenon and protons[10] or linked to proteins as a biosensor.[11] Unlike helium, xenon has a relatively high solubility in blood and tissue[12,13] and has a great affinity for lipid-rich tissue such as brain (for a review on hyperpolarized xenon in biology, see Cherubini and Bifone[14]). NMR images of hyperpolarized xenon in brain were performed on rats (Fig. 1)[15–17] and in humans.[18]

    The biophysical properties of xenon make it a good candidate for NMR tissue probing or perfusion studies. Indeed, xenon has been used for years in perfusion measurement studies, either by computed tomography(CT)[19] or in nuclear medicine. A γ-emitting xenon isotope, ^{133}Xe, has been widely used for brain perfusion studies following either bolus injection or inhalation.[20] Despite a rather poor spatial resolution of a few millimeters, this

[5] T. Pietrass, Magn. Reson. Med. **17**, 263 (2000).

[6] B. M. Goodson, J. Magn. Reson. **155**, 157 (2002).

[7] H. E. Möller, X. J. Chen, B. Saam, K. D. Hagspiel, G. A. Johnson, T. A. Altes, E. E. de Lange, and H.-U. Kauczor, Magn. Reson. Med. **47**, 1029 (2002).

[8] E. Brunner, Concepts Magn. Reson. **11**, 313 (1999).

[9] J.-L. Bonardet, J. Fraissard, A. Gédéon, and M.-A. Springuel-Huet, Catal. Rev.-Sci. Eng. **41**, 115 (1999).

[10] C. Landon, P. Berthault, F. Vovelle, and H. Desvaux, Protein Sci. **10**, 762 (2001).

[11] M. M. Spence, S. M. Rubin, I. E. Dimitrov, E. J. Ruiz, D. E. Wemmer, A. Pines, S. Q. Yao, F. Tian, and P. G. Schultz, Proc. Natl. Acad. Sci. USA **98**, 10654 (2001).

[12] R. Y. Z. Chen, F. C. Fan, S. Kim, K. M. Jan, S. Usami, and S. Chien, J. Appl. Physiol. **49**, 178 (1980).

[13] P. K. Weathersby and L. D. Homer, Undersea Biomed. Res. **7**, 277 (1980).

[14] A. Cherubini and A. Bifone, Prog. Nuclear Magn. Reson. Spectrosc. **42**, 1 (2003).

[15] S. D. Swanson, M. S. Rosen, B. W. Agranoff, K. P. Coulter, R. C. Welsh, and T. E. Chupp, Magn. Reson. Med. **38**, 695 (1997).

[16] G. Duhamel, P. Choquet, E. Grillon, L. Lamalle, J.-L. Leviel, A. Ziegler, and A. Constantinesco, Magn. Reson. Med. **46**, 208 (2001).

[17] G. Duhamel, P. Choquet, E. Grillon, J.-L. Leviel, A. Ziegler, and A. Constantinesco, C. R. Acad. Sci. IIC **4**, 789 (2001).

[18] W. Kilian, F. Seifert, and H. Rinneberg, "10th Scientific Meeting of the International Society for Magnetic Resonance in Medicine," p. 758. Honolulu, Hawaii, 2002.

[19] B. P. Drayer, S. K. J. Wolfson, and O. Reinmuth, Stroke **9**, 123 (1978).

[20] H. I. Glass and A. M. Harper, Brit. Med. J. **1**, 593 (1963).

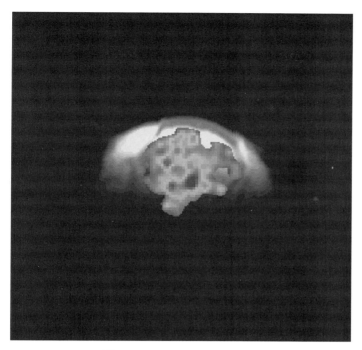

FIG. 1. NMR image of HP^{129}Xe in rat brain superimposed on a proton image. Xenon was hyperpolarized by optical pumping, dissolved in Intralipid 30%, and about 0.15 ml was injected in 1.5 s into the carotid. Acquisition was performed with a two-dimensional projection–reconstruction pulse sequence (FOV 55 mm, acquisition time 960 ms).

method remains a reference because it gives an absolute and quantitative measurement of perfusion.[21]

Quantification of organ perfusion, either globally or regionally, is of great interest. It allows alterations in perfusion to be detected and followed during pathologies, physiological transitions (exercise), artificial states (anesthesia), and evaluation of therapies. Quantitative measurements are needed to compare perfusion to a reference normal state or to evaluate variations in time. The reference methods for absolute cerebral perfusion measurement are based on the Kety–Schmidt method for a freely diffusible tracer.[22] Such tracers pass through the blood–brain barrier to the cerebral

[21] S. Holm, A. Andersen, N. A. Lassen, O. B. Paulson, and S. Vorstrup, *J. Cereb. Blood Flow Metab.* **7,** S565 (1987).
[22] S. S. Kety and C. F. Schmidt, *Am. J. Physiol.* **143,** 53 (1945).

tissue. The quantity of tracer in the tissue is linked directly to the blood flow, which delivers and removes it.

MRI combines the advantages of high spatial resolution and the absence of ionizing effects. However, contrast-enhanced MRI gives only relative values of cerebral blood flow and relies on sophisticated mathematical models for intravascular contrast agents. Arterial spin labeling, which is known to give absolute quantification of cerebral blood flow, is limited by a low signal-to-noise ratio (SNR) (for a review of these techniques, see Barbier et al.[23]). Using hyperpolarized xenon (HP^{129}Xe) MRI for perfusion measurements has two main advantages: the NMR signal is enhanced by the optical pumping process by a factor of about 10^5 and the biophysical behavior of xenon is well known (through studies on perfusion with ^{133}Xe). Indeed, both isotopes follow the same kinetic laws, and therefore the models and calculations developed for ^{133}Xe clearance can be applied to ^{129}Xe MRI. In addition, about 95% xenon is eliminated from the blood when it passes through the lungs[24] so that it might be used for perfusion studies as a first-passage technique.

After an intraarterial injection of a tight bolus of HP^{129}Xe dissolved in a lipid emulsion,[17] the NMR signal of HP^{129}Xe in the brain depends on three parameters: (1) the tissue concentration of ^{129}Xe, which can be described by the Kety–Schmidt method; (2) the radio frequency (RF) flip angle used for the NMR measurements, which is calibrated; and (3) the longitudinal relaxation time, T_1^{tissue}, of ^{129}Xe in the brain tissue. Determination of T_1^{tissue} is then necessary for absolute cerebral perfusion measurement.

Description of Method

Optical Pumping Process

Xenon is hyperpolarized by collisional spin exchange with rubidium vapor pumped optically at 795 nm.[3,25] Details of the theory and practical considerations can be found elsewhere.[5,7] The optical pumping device installed in our laboratory consists of three main modules (Fig. 2).

Vacuum Pumps and Gas Admission Module. The whole gas circuit has to be free of oxygen because (1) rubidium is highly reactive with oxygen, so care must be taken to protect it from any oxygen source, including air; (2) xenon, after laser polarization, must be removed from the proximity of relaxing sources, which could decrease its T_1 dramatically. Therefore,

[23] E. L. Barbier, L. Lamalle, and M. Décorps, *J. Magn. Reson. Imaging* **13**, 496 (2001).

[24] I. P. C. Murray and P. J. Ell, "Nuclear Medicine in Clinical Diagnosis and Treatment." Elsevier Science, London, 1999.

[25] M. A. Bouchiat, T. R. Carver, and C. M. Varnum, *Phys. Rev. Lett.* **5**, 373 (1960).

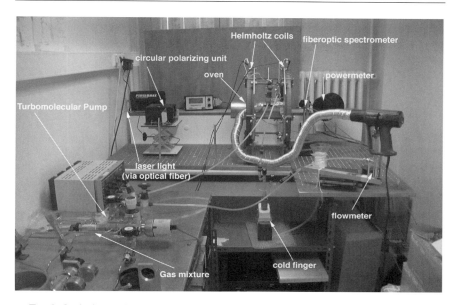

FIG. 2. Optical pumping device. It consists of three main modules: vacuum pumps and gas-mixing module, the optical pumping module to build up and monitor the polarization, and the storage module with the "cold finger" and the flowmeter.

the line is evacuated continuously and the high vacuum needed requires a two-stage system. A primary pump allows one to reach about 10^{-2} mbar and then a turbo pump allows one to reach 10^{-6} mbar. The 100-ml Pyrex cell, containing less than 0.1 g of rubidium, is then filled with the gas mixture for optical pumping. We use natural xenon (26% of ^{129}Xe), and the mixture proportions of 92.5% ^{4}He, 2.5% N_2, and 5% Xe are set to maximize the level of polarization reached with our system. After closing the cell outlet valve, the cell is filled with the gas mixture under 5.5 bar at room temperature, and the inlet valve is closed.

Optical Pumping Module. The cell is set in a 4.5-mT axial magnetic field generated by Helmholtz coils and is heated to 115° to vaporize the rubidium. A Coherent Fap system (Coherent, Santa Clara, CA) produces a circularly polarized laser beam tuned at 795 nm (P = 60 W), which induces Rb polarization in the cell. An optical spectrometer monitors the Rb absorption of the laser light (Fig. 3) in order to control some optical pumping process parameters (laser wavelength, cell temperature).

After 3 min, the cell is cooled rapidly to condense Rb. Then the outlet valve is opened and the gas mixture expands from the cell via a perfluoro-alkoxy (PFA) (Whitey, Swagelok, PFA tubing) line into the storage device.

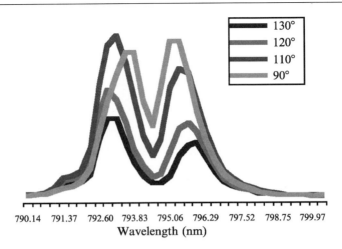

FIG. 3. Optical spectra of Rb absorption for different temperatures in the oven, measured to optimize the Rb vaporization.

HP^{129}Xe Storage Module. The storage device consists of a "cold finger" (Fig. 4), a cell immersed in liquid nitrogen and positioned in a permanent magnetic field (90 mT). Frozen xenon gets trapped, ^{4}He and N_2 remain gaseous, and, by temporary openings of the flowmeter, get sucked up by the vacuum pumps. In the "cold finger," the T_1 of ^{129}Xe is about 90 min, which enables transportation of HP^{129}Xe to the MRI magnet without a significant loss of polarization. These devices and procedures enable one to routinely obtain 20 ml, at 1 bar, of natural xenon (26% ^{129}Xe) polarized up to 18–20%.

Laser-Polarized ^{129}Xe NMR Specificities

The optical pumping process gives a nuclear polarization far above the polarization at thermal equilibrium. During the NMR experiments, each RF pulse destroys a part of this polarization, which cannot be recovered without returning the xenon to the optical pumping module. Therefore, the pulse sequences are devised to use the magnetization sparingly, generally with the use of small flip angles. This is the first source of NMR signal decrease.

The second source of signal decrease is the longitudinal relaxation time T_1: if pure xenon is known to have a long spin–lattice relaxation time at low temperature, which may reach several hours,[26] it is no longer the case when xenon is in contact with biological tissues, lipids, and so on. In conventional

[26] M. Gatzke, G. D. Cates, B. Driehuys, D. Fox, W. Happer, and B. T. Saam, *Phys. Rev. Lett.* **70,** 690 (1993).

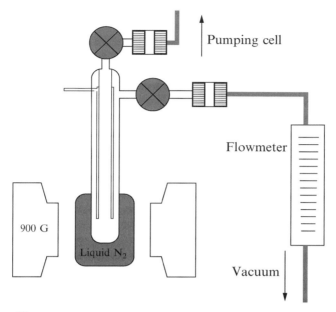

FIG. 4. HP^{129}Xe storage system. After optical pumping and cooling of the pumping cell to room temperature, the gas expands to the storage system and is then cooled down to nitrogen temperature, which makes xenon solid. The other gases are extracted through the flowmeter.

NMR, T_1 characterizes the return of the magnetization along the B_0 field axis to Boltzmann equilibrium. Hence, the shorter it is, the faster the magnetization is available for new NMR acquisitions. For laser-polarized noble gases, we are far from thermal equilibrium and the system tends naturally to return to such equilibrium, losing its polarization on the way to reach a level that is five orders less than the laser-polarized level. T_1 then becomes the time during which we could "use" the huge polarization of HP^{129}Xe. Hence, long T_1 is needed. Xenon T_1 is highly dependent on the atomic environment, and it is not only the components to which xenon interacts but also the state of these components (presence of oxygen, conformational state, etc.) that influence T_1. For instance, T_1 of HP^{129}Xe in blood has been investigated by several groups.[27–30] On the one hand, different values were

[27] M. S. Albert, V. D. Schepkin, and T. F. Budinger, *J. Comput. Assist. Tomogr.* **19,** 975 (1995).
[28] A. Bifone, Y.-Q. Song, R. Seydoux, R. E. Taylor, B. M. Goodson, T. Pietrass, T. F. Budinger, G. Navon, and A. Pines, *Proc. Natl. Acad. Sci. USA* **93,** 12932 (1996).
[29] J. Wolber, A. Cherubini, M. O. Leach, and A. Bifone, *Magn. Reson. Med.* **43,** 491 (2000) [erratum in *Magn. Reson. Med.* **47,** 213 (2002)].
[30] J. Wolber, A. Cherubini, M. O. Leach, and A. Bifone, *NMR Biomed.* **13,** 234 (2000).

found for oxygenated and deoxygenated blood: Wolber *et al.*[31] found 6.4 and 4.0 s, respectively, at 1.5 T, 37° (which opens promising perspectives of functional exploration of the brain using the blood oxygenation level-dependent effect). On the other hand, the quite short values of T_1 demonstrate the short life-time of $HP^{129}Xe$ in the blood and suggest some of the difficulties encountered in biological applications.

Delivering $HP^{129}Xe$ to the Brain

$HP^{129}Xe$ can be administered through inhalation,[15] but use of the route is limited because the relaxation time of xenon in blood is short and the solubility of xenon in blood is not high. Therefore, many studies have explored the feasibility of $HP^{129}Xe$ MRI after the injection of $HP^{129}Xe$ in biocompatible carriers. A large number of injectable carriers have been investigated.[32,33] $HP^{129}Xe$ can be either encapsulated (microbubbles, liposomes) or dissolved [lipid emulsion, perfluorooctylbromide (PFOB), a blood substitute]. In both cases, the T_1 of $HP^{129}Xe$ is crucial in that it limits the "lifetime" and level of the NMR signal and, hence, its ability to be detected in the organ. Furthermore, a compartmental model shows that, after injection of $HP^{129}Xe$, the signal in the brain depends critically on the solubility of xenon in the bolus,[34] and other parameters, such as chemical shift, droplet/bubble sizes, and experimental protocols, have also to be taken into account.

The saline solution of $HP^{129}Xe$ is characterized by a long T_1 (66 s[28]) and would have been a good candidate if the solubility were not so low (Ostwald coefficient $\sim$0.09). This solubility limitation led to the investigation of nonpolar environments. Dissolved in PFOB, $HP^{129}Xe$ presents long T_1 (94 s) and a high Ostwald coefficient ($\sim$1.2).[35] Encapsulated in liposomes, $HP^{129}Xe$ also has a long T_1 (116 s), but it seems that the exchange between the encapsulated xenon and the surrounding medium is low[33]; this carrier is thus better adapted for intravascular applications (cardiac imaging, angiography). We investigated other carriers, such as microbubbles,[36] but finally, for the *in vivo* T_1 measurement method, we chose an emulsion

[31] J. Wolber, A. Cherubini, A. S. Dzik-Jurasz, M. O. Leach, and A. Bifone, *Proc. Natl. Acad. Sci. USA* **96,** 3664 (1999).

[32] B. M. Goodson, *Concepts Magn. Reson.* **11,** 203 (1999).

[33] A. Venkatesh, L. Zhao, D. Balamore, F. A. Jolesz, and M. S. Albert, *NMR Biomed.* **13,** 245 (2000).

[34] C. Lavini, G. S. Payne, M. O. Leach, and A. Bifone, *NMR Biomed.* **13,** 238 (2000).

[35] J. Wolber, I. J. Rowland, M. O. Leach, and A. Bifone, *Magn. Reson. Med.* **41,** 442 (1999).

[36] G. Duhamel, P. Choquet, J.-L. Leviel, J. Steibel, J. Derouard, M. Décorps, A. Ziegler, and A. Constantinesco, "Gaz Hyperpolarisés en Résonance Magnétique: Explorations Biomédicales et Applications Cliniques." Les Houches, France, 1999.

of Intralipid 20% (Pharmacia and Upjohn, France) because xenon presents a good solubility (Ostwald coefficient $L^{\text{Intra}} = 0.4$), a T$_1$ long enough ($T_1^{\text{Intra}} = 23$ s) to perform the measurement, and only one resonance peak at 194.5 ppm from the gas peak reference (0 ppm).[17,37]

Rationale of the Method

Our method aims to determine the T$_1$ of hyperpolarized xenon *in vivo*. Due to the characteristics of the hyperpolarized magnetization, standard T$_1$ measurement methods such as inversion–recovery or saturation–recovery measurement of the magnetization are not applicable.

After an injection of hyperpolarized xenon dissolved in Intralipid into the rat brain, two xenon resonances (Fig. 5) can be detected in the rat brain by NMR spectroscopy.[16] The peak at 194.5 ppm from the gas peak (0 ppm) was assigned to ^{129}Xe dissolved in Intralipid and the peak at 199.0 ppm was assigned to ^{129}Xe dissolved in brain tissue.[16]

To determine the xenon T$_1$ in the brain tissue (T_1^{tissue}), we considered the kinetics of the NMR signal of xenon in brain tissue during a long (duration T^{inj}) injection of the HP tracer into the carotid (Fig. 6). We used a two-compartment model to interpret these dynamics.[38] Because xenon is known to diffuse freely in brain, we applied the Kety and Schmidt method.[22] The equation describing the evolution of the amount Q of xenon in the brain tissue is then

$$\frac{dQ(t)}{dt} = F\{C_a(t) - C_v(t)\} \qquad (1)$$

where F is the cerebral blood flow and $C_a(t)$ and $C_v(t)$ are the xenon concentrations in arterial and venous blood, respectively.

To go further, we need to make two assumptions. We assume that, during the injection, the amount $FC_v(t)$ of xenon leaving the brain tissue via the venous network can be neglected:

$$C_a(t) \gg C_v(t) \text{ for } t < T^{\text{inj}} \qquad (2)$$

We made a second assumption that the concentration of arterial xenon, $C_a(t)$, is constant during the hyperpolarized xenon injection; this means that the injection has to be done at a constant rate. Thus, Eq. (1) can be reduced to

[37] G. Duhamel, P. Choquet, J.-L. Leviel, J. Steibel, L. Lamalle, C. Julien, F. Kober, E. Grillon, J. Derouard, M. Décorps, A. Ziegler, and A. Constantinesco, *C. R. Acad. Sci. III-, Sci. Vie* **323**, 529 (2000).
[38] P. Choquet, J.-N. Hyacinthe, G. Duhamel, E. Grillon, J.-L. Leviel, A. Constantinesco, and A. Ziegler, *Magn. Reson. Med.* **49**, 1014 (2003).

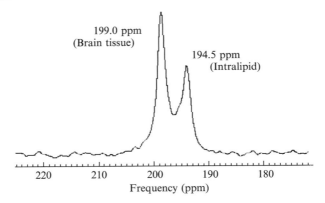

FIG. 5. ^{129}Xe NMR spectrum from a rat brain at 2.35 T after an intracarotid injection of hyperpolarized xenon dissolved in Intralipid.

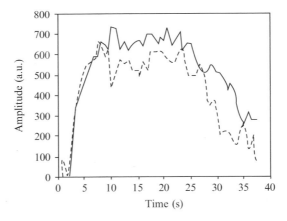

FIG. 6. Temporal behavior of the xenon signal in the blood pool (194.5 ppm, dashed line) and in brain tissue (199.0 ppm, solid line) resulting from 25-s injection. During the injection, the 199.0-ppm peak, corresponding to HP^{129}Xe in the brain tissue, increased rapidly for 5–15 s and was then approximately constant (or fell slightly) until the end of the injection, when it fell rapidly. Reproduced with permission from Choquet et al.,[38] copyright © 2003 Wiley-Liss, Inc.

$$\frac{dQ(t)}{dt} = FC_a \qquad (3)$$

The solution is

$$Q(t) = FC_a t \qquad (4)$$

Therefore, under the two assumptions mentioned previously, the quantity of xenon in the cerebral tissue increases linearly with time during the injection. The kinetics of the NMR signal of the xenon dissolved in the cerebral tissue then depends on three effects (Fig. 7): the continuing input of HP xenon by the injection (signal increase), the HP xenon relaxation (signal decrease), and the loss of signal due to RF pulses (signal decrease).

To evaluate the T_1 of hyperpolarized ^{3}He in lungs, magnetization is evaluated just before the nth RF pulse and immediately after.[39] We follow a similar approach, evaluating the magnetization of xenon in brain tissue before the nth RF pulse $[M_{Xe}(n^-)]$ and immediately after $[M_{Xe}(n^+)]$.

The available xenon magnetization at the beginning of the experiment relaxes first in the Intralipid (relaxation time T_1^{Intra}) during the time needed for the tracer to reach the brain. A constant fraction (η) of xenon contained in Intralipid diffuses in the brain tissue, giving rise to a magnetization, which relaxes with the constant T_1^{tissue}. To express the whole xenon magnetization in brain tissue M_{Xe} as a function of t, F, T_1^{Intra}, and T_1^{tissue}, let $\delta\tau$ be the time interval necessary for a volume $\eta F \delta\tau$ of xenon dissolved in the Intralipid and administered with the constant flow F to reach the brain tissue. During $\delta\tau$, the longitudinal relaxation of xenon takes place in Intralipid with the constant T_1^{Intra}. After time $\delta\tau$, the $\eta F \delta\tau$ volume of xenon that has diffused to the brain tissue relaxes with the T_1^{tissue} constant (Fig. 8). Then, at time t, the corresponding magnetization of this volume of xenon can be written as

$$\delta M_{\delta\tau} \propto \eta F \delta\tau \exp\left(-\frac{\delta\tau}{T_1^{Intra}}\right)\exp\left(-\frac{t-\delta\tau}{T_1^{tissue}}\right) \quad (5)$$

The whole magnetization of xenon in brain tissue is thus obtained by summing the elementary magnetizations $\delta M_{i\delta\tau}$ along all the time intervals:

$$M_{Xe} = \sum_{i=0}^{N} \delta M_{i\delta\tau} \propto \sum_{i=0}^{N} \eta F i \delta\tau \exp\left(-\frac{i\delta\tau}{T_1^{Intra}}\right)\exp\left(-\frac{t-i\delta\tau}{T_1^{tissue}}\right) \quad (6)$$

with $N\delta\tau = t$.

Allowing $\delta\tau$ to tend to 0, we can rewrite Eq. (6) as

$$M_{Xe} \propto \int_0^t \eta F d\tau \exp\left(-\frac{\tau}{T_1^{Intra}}\right)\exp\left(-\frac{t-\tau}{T_1^{tissue}}\right) \quad (7)$$

Finally, by integration of Eq. (7), assuming that $T_1^{Intra} \neq T_1^{tissue}$ and that this exchange occurs only during Tr, the magnetization of xenon in Intralipid

[39] H. E. Möller, X. J. Chen, M. S. Chawla, B. Driehuys, L. W. Hedlund, and G. A. Johnson, *J. Magn. Reson.* **135,** 133 (1998).

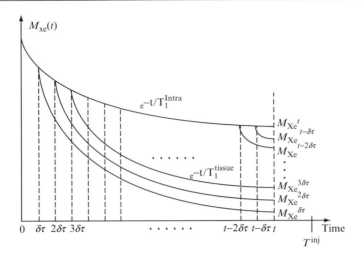

FIG. 7. Schematic description of xenon magnetization evolution in brain tissue in the absence of RF pulses. The injection duration T^{inj} is divided in small delays $\delta\tau$, where an amount $\eta F \delta\tau$ passes through the blood–brain barrier to the cerebral tissue and therefore decreases with T_1^{tissue} constant instead of T_1^{Intra}.

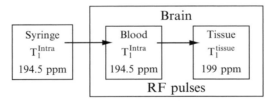

FIG. 8. Compartmental modeling of the $HP^{129}Xe$ experiment. The brain is described by two compartments (blood and tissue), and xenon behavior is characterized by different chemical shifts and T_1 values, depending on the compartment where the xenon is. Xenon in brain is also submitted to the RF-pulse influence. Before the injection, the xenon relaxes with T_1^{Intra} and is not submitted to the RF pulses (the syringe is outside the RF coil).

diffusing from the vascular compartment to the brain tissue during Tr is $M_{I \to T}$:

$$M_{I \to T} = \eta F \frac{T_1^{Intra} \cdot T_1^{tissue}}{T_1^{Intra} - T_1^{tissue}} \left[\exp\left(-\frac{Tr}{T_1^{Intra}}\right) - \exp\left(-\frac{Tr}{T_1^{tissue}}\right) \right] \quad (8)$$

Before the first RF pulse, the magnetization $M_{Xe}(1^-)$ given by the magnetization of xenon in Intralipid diffusing from the vascular compartment to the brain tissue during Tr is $M_{I \to T}$:

$$M_{\mathrm{Xe}}(1^-) = M_{I \to T} \propto \frac{T_1^{\mathrm{Intra}} T_1^{\mathrm{tissue}}}{T_1^{\mathrm{Intra}} - T_1^{\mathrm{tissue}}} \left(E_1^{\mathrm{Intra}} - E_1^{\mathrm{tissue}} \right) \tag{9}$$

where $E_1^{\mathrm{Intra}} = \exp(-\mathrm{Tr}/T_1^{\mathrm{Intra}})$ and $E_1^{\mathrm{tissue}} = \exp(-\mathrm{Tr}/T_1^{\mathrm{tissue}})$

Immediately after the first RF pulse,

$$M_{\mathrm{Xe}}(1^+) = M_{\mathrm{Xe}}(1^-)\cos(\alpha) \tag{10}$$

During the next Tr, xenon coming from the syringe diffuses as described previously. The magnetization of this amount of xenon has relaxed, during Tr, with the constant T_1^{Intra}. In the same interval, xenon that has reached the brain tissue during the first Tr has relaxed with the constant T_1^{tissue}. Therefore, before the second RF pulse,

$$\begin{aligned} M_{\mathrm{Xe}}(2^-) = M_{\mathrm{Xe}}(1^+)E_1^{\mathrm{tissue}} + M_{I \to T}E_1^{\mathrm{Intra}} &= M_{I \to T}\cos(\alpha)E_1^{\mathrm{tissue}} \\ &\quad + M_{I \to T}E_1^{\mathrm{Intra}} \end{aligned} \tag{11}$$

The general expression is then

$$M_{\mathrm{Xe}}(n^-) = M_{I \to T}\left(E_1^{\mathrm{Intra}} \right)^{n-1} \left[1 - \left(\frac{E_1^{\mathrm{tissue}}\cos(\alpha)}{E_1^{\mathrm{Intra}}} \right)^n \right] \left[1 - \left(\frac{E_1^{\mathrm{tissue}}\cos(\alpha)}{E_1^{\mathrm{Intra}}} \right) \right]^{-1} \tag{12}$$

The NMR signal acquired after the nth RF pulse is given by

$$S(n) = S_0 \sin(\alpha) \frac{T_1^{\mathrm{Intra}} T_1^{\mathrm{tissue}}}{T_1^{\mathrm{Intra}} - T_1^{\mathrm{tissue}}} \left(E_1^{\mathrm{Intra}} - E_1^{\mathrm{tissue}} \right) \left(E_1^{\mathrm{Intra}} \right)^{n-1}$$
$$\left[1 - \left(\frac{E_1^{\mathrm{tissue}}\cos(\alpha)}{E_1^{\mathrm{Intra}}} \right)^n \right] \left[1 - \left(\frac{E_1^{\mathrm{tissue}}\cos(\alpha)}{E_1^{\mathrm{Intra}}} \right) \right]^{-1} \tag{13}$$

As the α angle and T_1^{Intra} are supposedly known, there are two unknown variables in Eq. (13): the proportionality constant S_0 (which includes the initial hyperpolarization level of the xenon, the cerebral blood flow, etc.) and T_1^{tissue}. Fitting the NMR signal from the dissolved xenon in the cerebral tissue using the model function leads to T_1^{tissue} determination.

Animal Preparation Procedures

Male Sprague–Dawley rats weighing 250–350 g are used. All operative procedures and animal care conform strictly to the French government guidelines (decree No. 87–848 of October 19, 1987). Each animal is anesthetized by intraperitoneal injections of chloral hydrate. After ligation of the external and common carotid arteries of one hemisphere, a catheter is inserted into the ipsilateral internal carotid artery. The rats are then placed in the NMR probe and introduced into the magnet for NMR experiments. At the end of the experiments, the rats are killed by an excess of chloral hydrate.

HP Tracer Injection

The HP^{129}Xe is dissolved in the lipid emulsion in a ratio of 1 ml lipid emulsion:1 ml of gaseous xenon. This resulting emulsion is called in the following HP tracer, and a volume V of HP tracer contains a volume $V \times L^{\text{Intra}}$ of xenon.

The syringe containing the HP tracer is connected to the intracarotid catheter. For each experiment, 0.55 ± 0.10 ml of HP tracer is injected at a constant flow into the carotid for 25 to 35 s.

For the model to be validated, two assumptions have been done, leading to constraints on the experimental conditions of the tracer injection. They were devised to ensure that both assumptions are verified. First, despite the manual injection procedure, we can be sure that the HP tracer is administered to the brain at an approximately constant rate during the first 20 s after the beginning of the injection. Therefore, the fitting process is performed on data acquired during that time. Second, the nonsaturation of the brain tissue in xenon is based on the injected volume of xenon and the anatomic and physiologic characteristics of the rat brain. The injected volume of xenon in 20 s was 0.15 ml. The part that has diffused into the brain is deduced from the NMR spectrum: neglecting the relaxation rate, the ratio of the 199.0 ppm peak area to the 194.5 ppm one corresponds to the ratio of the number of xenon atoms having diffused from Intralipid to the brain tissue over the number of xenon atoms remaining in the Intralipid. From Fig. 6, this ratio is evaluated to $r = 1.15$, resulting in $r \times 0.15$ ml$/(1 + r) = 0.08$ ml of xenon that has diffused into the brain tissue during injection, and the remaining 0.07 ml is still dissolved in Intralipid. Because the rat brain weighs about 1.2 g and is 97% tissue and because the Ostwald solubility coefficient of xenon in the brain tissue is $L^{\text{tissue}} \sim 0.14$ ml g^{-1} (estimated from Steward *et al.*[40]) assuming that, for rat brain, 16% of the volume is white matter and about 80% is gray matter,[41] the maximum volume of xenon that can be dissolved in the rat brain tissue is

[40] A. Steward, P. R. Allott, A. L. Cowles, and W. W. Mapleson, *Brit. J. Anaesth.* **45,** 282 (1973).

[41] P. D. Lyden, L. M. Lonzo, S. Y. Nunez, T. Dockstader, O. Mathieu-Costello, and J. A. Zivin, *Behav. Brain Res.* **87,** 59 (1997).

$(0.97) \times (1.2 \text{ g}) \times (0.14 \text{ ml g}^{-1}) \sim 0.16$ ml. The volume of xenon having diffused into the brain tissue ($\sim$0.08 ml) is smaller than the maximum xenon volume soluble in the brain tissue ($\sim$0.16 ml); it seems justified to assume nonsaturation of brain tissue.

NMR Acquisitions

All experiments are performed on a SMIS console (Surrey Medical Imaging Systems Ltd., Guildford, UK) interfaced to a 2.35-T, 40-cm bore, Bruker magnet (Bruker Spectrospin, Wissembourg, France) equipped with actively shielded Magnex gradient coils (Magnex Scientific Ltd., Abingdon, UK). A 25-mm, three-turn, home-built surface coil, tunable to the Larmor frequencies at 2.35 T of either ^{129}Xe (27.67 MHz) or ^{23}Na (26.45 MHz), is used. Shimming and RF pulse calibration are carried out using the ^{23}Na signal of the sensitive volume in the rat brain. Without moving the animal and keeping the coil matched during the frequency switching between the ^{23}Na and the ^{129}Xe frequencies, the ^{129}Xe RF pulse angle is derived from the ^{23}Na one using the following equation:

$$\alpha_{Xe} = \frac{\gamma_{Xe}}{\gamma_{Na}} \alpha_{Na} \qquad (14)$$

with α being the flip angle and γ the gyromagnetic ratio. A "90° pulse" is considered to be that which gives the maximum signal. The pulse duration is then reduced to obtain the desired flip angle (about 15°). For each experiment, ^{129}Xe spectra are acquired using a simple one-pulse/acquisition sequence. The acquisition of ^{129}Xe spectra starts simultaneously with injection of the HP tracer. The number of spectra (N) and repetition times (Tr) are, respectively, 40 and 760 ms for rats 1 to 4 and 120 and 250 ms for rats 5 and 6. The acquisition bandwidth is 5 kHz, sampled with 2048 complex data points. Time-domain data are filtered exponentially (20-Hz line-broadening factor) and zero filled to 8192 points before Fourier transformation and phase correction. Spectra are then fitted in the frequency domain by two Lorentzian line shapes in order to calculate the area of each peak (Levenberg–Marquardt algorithm). The fit of the areas of the ^{129}Xe dissolved in brain tissue peak to the model function described by Eq. (13) is performed using the nonlinear Newton–Raphson iterative procedure (JMP software; SAS Institute, Inc., Cary, NC).

Results

Results of the fitting process (Fig. 9) give an interanimal T_1^{tissue} estimation of 3.6 ± 2.1 s ($\pm$ standard deviation, $n = 6$). The fit was performed with the nominal α flip angle, as well as with $\alpha \pm 2°$, to evaluate the sensitivity of

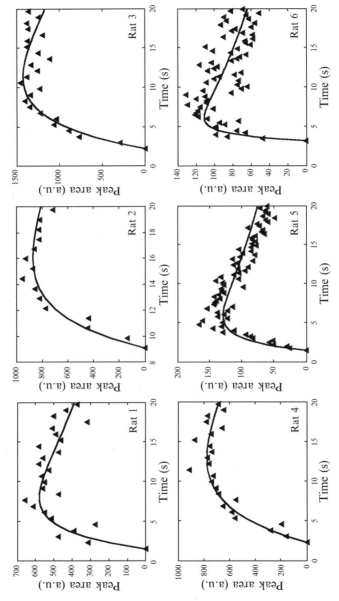

FIG. 9. Determination of T_1^{tissue}. Areas of the 199-ppm peaks (filled triangles), and the curves defined by Eq. (13) and fitted to these data for the six rats (solid lines). Reproduced with permission from Choquet *et al.*,[38] copyright © 2003 Wiley-Liss, Inc.

the T$_1^{\text{tissue}}$ estimate to the uncertainty on the flip angle. The calculation with $\alpha \pm 2°$ resulted in a 5.5 ± 1.9% error ($n = 12$) on the T$_1^{\text{tissue}}$ value with the nominal flip angle.

Concluding Remarks

Absolute perfusion measurements with HP^{129}Xe require the knowledge of xenon T$_1$. The proposed method to measure the xenon T$_1$ in the brain tissue has been validated through the first measurements.[38] However, the results showed some variability, which could be explained, for example, by RF flip angle distribution over the sensitive volume or by the fact that T$_1$ can be different in different types of cerebral tissues. Rat temperature and physiological parameters could also influence the T$_1$. Some improvements could be brought through a better flip angle determination (B$_1$ mapping) and a mapping of xenon T$_1$ by combining the method we described with a fast spectroscopic imaging method for instance. Because xenon T$_1$ is linked to the oxygenation level,[42] T$_1$ mapping could also provide useful information on brain oxygenation.

Acknowledgments

The authors particularly thank J.-M. Castejon for his expert glass blowing, M. Décorps, C. Segebarth, and J. Coles for helpful discussions, and H. Reutenauer and O. Montigon for maintaining the NMR system. This work was supported by the Région Rhône-Alpes and a BQR grant from the Université Joseph Fourier, Grenoble. G. D. and J.-N. H. benefit from Ph.D. grants from the French Ministry of Research and Technology. The NMR facilities were supported in part by the "Programme Interdisciplinaire Imagerie du Petit Animal" (CNRS–INSERM).

[42] G. J. Wilson, G. E. Santyr, M. E. Anderson, and P. M. J. DeLucas, *Magn. Reson. Med.* **41,** 933 (1999).

Section II

Imaging of Receptors, Small Molecules, and Protein–Protein Interactions

[10]   Positron Emission Tomography Receptor Assay with Multiple Ligand Concentrations: An Equilibrium Approach

By JAMES E. HOLDEN and DORIS J. DOUDET

Introduction

Among the many current modern imaging modalities, positron emission tomography (PET) presents the advantage of allowing the medical researcher to open a window into the mechanisms controlling a variety of physiological processes in health and disease. Although PET is widely used as a therapeutic and diagnostic tool in cardiology and oncology, many of these uses still rely on its more qualitative aspects. Although its diagnostic role in neurological and psychiatric disorders is limited, the true power of PET is to allow insights into the physiology and neurochemistry of the brain, where its quantitative capabilities are an asset and are being widely developed and used in human subjects, both normal controls and patients. One of the most studied aspects to date is the role of a variety of neurotransmitters and their receptors in health, aging, and disease. Although a field still in infancy, an assortment of tracers has been or is being developed for various forms of dopamine (DA), serotonin, acetylcholine, opiate, glutamate, and GABA receptors. Some receptor ligands may also be used as surrogate markers of the extent of release of their presynaptic neurotransmitters, with the implied assumption that the changes in binding of the tracer to the receptor in a given subject or a given population result from a change in the competition with the endogenous ligand. Issues and questions raised by this approach have been reviewed in detail by Laruelle,[1] and although some are relevant to the subject of this Chapter, they will not be reiterated here.

A challenge facing the study of living neuroreceptors is to be able to reconcile *in vivo* PET findings with those from *in vitro* postmortem studies in patients or in animal models of disorders. *In Vitro* studies (or *ex vivo* studies in isolated but intact tissues) present the advantage of having direct access to the tissue, allowing not only complete control of the conditions in which the measurements are obtained, but also the use of techniques leading to fully quantitative answers, such as the Scatchard's[2] methodology,

[1] M. Laruelle, *J. Cereb. Blood Flow Metab.* **20,** 243 (2000).
[2] G. Scatchard, *Ann. N. Y. Acad. Sci.* **51,** 660 (1949).

which permits separate evaluation of the density and the affinity of the receptors of interest in a given condition. Knowledge of the number of receptors and of their affinity in specific brain regions, specific pathological conditions, or after acute or long-term drug therapy would greatly enhance our understanding of the pathophysiological mechanisms and the induced compensatory effects involved in idiopathic diseases or in response to injuries (such as stroke) and drug exposure. A number of attempts have been made to evaluate *in vivo* neuroreceptor characteristics. However, working with a living organism poses a number of constraints on the type of studies and the type of data that can be acquired safely and in a less invasive manner.

This Chapter summarizes some of the issues associated with obtaining adequate and accurate data *in vivo* and how and why they may differ from published *in vitro* findings. We review some of the pertinent literature briefly and present the results of our own efforts in the development of an *in vivo* multiple ligand concentration receptor assay (MLCRA) method that could be applied to human patient studies with the best reproducibility, reliability, and minimal burden on the patient. All the studies reported here have been obtained using [^{11}C]raclopride, a ligand of the dopamine $D_{2/3}$ receptors. Raclopride is known to bind to $D_{2/3}$ with a single affinity and represents the best example currently available. Thus, studies using other ligands with different characteristics may have to be validated using similar principles as described later.

Overview of *In Vivo* Receptor Assay

Basic Principles

The routine outcome of most neuroreceptor PET studies is a binding potential (BP) or an equivalent. The concept of BP was introduced in the early days of the application of PET to the study of neuroreceptors by Mintum *et al.*[3] and continues to be the most prevalent experimental endpoint, primarily because its determination from measured data is remarkably reliable and robust, regardless of the particular data acquisition and reduction methods chosen. However, it is, by definition (BP $= B_{max}/K_d$), a reflection of both the density of receptors B_{max} in the target tissue and the apparent affinity, represented inversely by the dissociation constant K_d, between those receptors and the tracer. Thus, although it is a useful and widely used indicator of altered receptor function, a change

[3] M. A. Mintun, M. E. Raichle, M. R. Kilbourn, G. F. Wooten, and M. J. Welch, *Ann. Neurol.* **15**, 217 (1984).

in BP does not provide information on the specific nature of the change, and a lack of change does not necessarily represent maintenance of normal function. In an effort to resolve this ambiguity, Farde *et al.*[4] developed an elegant method to assess independently, *in vivo,* receptor density and affinity. In analogy with the *in vitro* scatchard method, Farde's original method calls for the sequential use of multiple concentrations of test ligand [i.e., two PET scans at a very minimum, a first one at high specific activity of the tracer followed by at least one (or more) with coinjection of the specific ligand with its unlabeled form to induce a significant occupancy of the receptors]. The main assumptions in the *in vivo* studies are that the tracer binds to a single receptor site and that its rates of binding and release are compatible with the length of the experiment [i.e., equilibrium in all the region of interest occurs within four to five half lives of the positron emitter with which the tracer is labeled (maximum of 90–100 min in the case of most ^{11}C-labeled compounds)].

Variations on this theme have been reported, with the required changes in receptor occupancy in the sequential scans being induced either by prior or conjoint administration of a competitor[5–7] or, in our own studies, straight administration of the tracer at low specific activity.[8] This design was introduced in analogy with similar approaches in postmortem ligand-binding studies in preparations of dilute concentrations of membranes isolated from the tissue of interest. Despite the apparent common principle, there are many important differences between the *in vivo* MLCRA and the *in vitro* assay applied in membrane preparations. These methodological differences were addressed in detail in an earlier publication.[8]

Distinctions Between In Vivo *and* In Vitro *Methods*

Both *in vivo* and *in vitro* approaches attempt to determine the density of receptors per given unit of tissue and the apparent affinity between the receptors and the test ligand, expressed inversely as the apparent dissociation constant K_d. Although the experimental endpoints in *in vitro* MLCRA are the density of receptor molecules per mass of isolated membrane and the apparent affinity between those receptors and the labeled

[4] L. Farde, H. Hall, E. Ehrin, and G. Sedvall, *Science* **231,** 258 (1986).
[5] D. F. Wong, A. Gjedde, and H. N. Wagner, *J. Cereb. Blood Flow Metab.* **6,** 137 (1986).
[6] J. Delforge, S. Pappata, P. Millet, Y. Samson, B. Bendriem, A. Jobert, C. Crouzel, and A. Syrota, *J. Cereb. Blood Flow Metab.* **15,** 284 (1995).
[7] J. Delforge, M. Bottlaender, C. Loc'h, I. Guenther, C. Fuseau, B. Bendriem, A. Syrota, and B. Maziere, *J. Cereb. Blood Flow Metab.* **19,** 533 (1999).
[8] J. E. Holden, S. Jivan, T. J. Ruth, and D. J. Doudet, *J. Cereb. Blood Flow Metab.* **22,** 1132 (2002).

ligand, the determination of both of these parameters is strongly dependent on the details of the methods used to isolate and purify the membranes from the postmortem tissue, and the literature shows abundantly that both density and apparent ligand–receptor affinity values determined by *in vitro* MLCRA can vary by orders of magnitude.[9,10] The main advantage of the *in vivo* MLCRA is the avoidance of the confounding effects of the membrane preparation; the bound and free concentrations of the tracer are assumed to be distinguished unambiguously on the basis of the time-course data in target and reference tissues. The upper limiting value of the kinetically determined bound component represents an accurate estimate of the true total concentration of receptors in a tissue.[11,12] The relationship between the kinetically identified free ligand concentration[12] and the concentration of ligand actually available for binding is at the outset unknown, but its effects can be absorbed into the value of the apparent affinity between receptor and ligand. As the primary goal in both experimental contexts is to distinguish receptor density from affinity effects in measured data, this issue does not disqualify the *in vivo* compared to the *in vitro* MLCRA.

Potential Confounds in In Vivo Methods

The greatest challenge of *in vivo* MLCRA arises because the receptors are embedded in the membranes of intact, fully functioning cells in an intact, fully functioning central nervous system; the potential confounds from this lack of isolation must be accounted for to the greatest degree possible. In both *in vitro* and *in vivo* MLCRA, the measured relationship between the multiple values of the bound ligand concentrations and the free ligand concentrations with which they are equilibrated is used to evaluate single values of the total concentration of receptors and the apparent receptor–ligand affinity. The assumption is made that these values are equally valid at all ligand concentrations. The most obvious potential confound in the *in vivo* situation is that increasing antagonist concentrations may induce changes in the concentration of endogenous ligand,[13] which could result in each data point being characterized by a different apparent affinity. Furthermore, circadian rhythms have been shown to alter the

[9] W. H. Riffee, R. E. Wilcox, D. M. Vaughn, and R. V. Smith, *Psychopharmacology* **77,** 146 (1982).
[10] P. Seeman, C. Ulpian, K. A. Wreggett, and S. Wenger, *J. Neurochem.* **43,** 2211 (1984).
[11] J. Delforge, M. Bottlaender, S. Pappata, C. Loc'h, and A. Syrota, *J. Cereb. Blood Flow Metab.* **21,** 613 (2001).
[12] J. Delforge, A. Syrota, and B. Bendriem, *J. Nucl. Med.* **37,** 118 (1996).
[13] A. Gjedde and D. F. Wong, *J. Cereb. Blood Flow Metab.* **21,** 982 (2001).

concentration of endogenous ligand,[14] which could result in compounding or opposing effects on the tracer affinity throughout the course of the day. These issues have been addressed in previous reports.[8,15] In particular, an approach to testing for the first of these confounds was described previously.[8] We believe that such a test must be performed each time a new ligand is introduced or when a previously established ligand is applied in a new experimental context.

Mathematical Considerations

Both *in vivo* and *in vitro* MLCRA share the controversy about the best approach for evaluating the parameters of interest from measured data. We present here two approaches, one using the linear analysis used most routinely for PET studies and another using a nonlinear approach favored by practitioners of *in vitro* MLCRA. Details of the advantages and disadvantages of each, and the considerations taken into account in our choices of analyses, were reported previously.[8]

Other Considerations

Radiation Dosimetry. The radiation dose has to remain within the guidelines of acceptable dosimetry, which limits the number of studies permissible within a day or even a quarter. For that reason, many PET researchers have limited their studies to two scans, one with practically null receptor occupancy from the high specific activity tracer and one at a significant receptor occupancy by the unlabeled analog.

Receptor Saturation. In living subjects it is impossible to perform total or near-total receptor saturation studies without potentially affecting the health of the subject. The degree of saturation may, however, be marginally adapted to the particular condition studied. For example, performing a study inducing 80–90% D_2 receptor occupancy by a neuroleptic such as raclopride or spiperone in a schizophrenic subject would not have the same impact as inducing the same occupancy in a parkinsonian patient. In most *in vivo* MLCRA, one usually attempts to reach about 50–60% saturation.

Pharmacological Interventions. In studies of acute pharmacological interventions, the performance of multiple studies requires that attention be paid to the stability of the drug effect over time, particularly the stability of the changes in endogenous ligand induced by the challenge. Changes in the synaptic concentration of endogenous ligand during the course of the two

[14] A. D. Smith, R. J. Olson, and J. B. Justice, Jr., *J. Neurosci. Methods* **44,** 33 (1992).
[15] D. J. Doudet, S. Jivan, and J. E. Holden, *J. Cereb. Blood Flow Metab.* **23,** 280 (2003).

to three PET studies (acquired over 6–8 h) necessary to obtain data may interfere with the binding of the tracer by altering the apparent affinity of the tracer and thus its BP.

Avoidance of Blood Sampling. We developed our design and performed validation experiments in anesthetized nonhuman primates but with further studies in human subjects in mind. For this reason, we employ a data analysis method that uses the activity in a reference region devoid of the receptor of interest (cerebellum in the case of raclopride) to obtain information of the concentration of free ligand, whereas activity in the target region is assumed to provide information on the bound concentrations. Graphical methods of analysis, such as the Logan method[16] or the reference tissue method[17] provide good data without the need for a metabolite-corrected input function. In addition to avoiding arterial puncture, this approach obviates the need for metabolite analysis, which often introduces large amounts of noise and variability in the determination of the input function.

Equilibrium Administration. These graphical methods, however, can be sensitive to changes in blood flow brought about by disease or drug administration. Thus, we use the combination of bolus plus constant infusion administration of the PET tracer to produce a stable equilibrium between bound and free concentrations at each administered concentration.[18] With raclopride, we found that this equilibrium is reached within 25–30 min of the start of the tracer administration and thus remains within the time frame of the C-11 study.

Nonsequential Studies. Other concerns were the capability of the subjects, especially patients, to withstand two to three scans throughout the course of a single day and whether consecutive doses of a receptor antagonist would affect both data and the subject. We performed a simple study in normal monkeys and monkeys with drug-induced parkinsonism, comparing data obtained using a routine sequential MLCRA design (three to four consecutive scans in 1 single day) versus a nonsequential MLCRA design in which the individual scans were obtained on separate days, days to weeks apart, in random order of specific activity, but at the same time of day. This validation experiment also allowed us to assess both the potential effects of circadian rhythms and the effects of prior sequential administration of pharmacological doses of the unlabeled tracer. Results of this study

[16] J. Logan, J. S. Fowler, N. D. Volkow, G. J. Wang, Y. S. Ding, and D. L. Alexoff, *J. Cereb. Blood Flow Metab.* **16**, 834 (1996).

[17] A. A. Lammertsma and S. P. Hume, *Neuroimage* **4**, 153 (1996).

[18] R. E. Carson, A. Breier, A. de Bartolomeis, R. C. Saunders, T. P. Su, B. Schmall, M. G. Der, D. Pickar, and W. C. Eckelman, *J. Cereb. Blood Flow Metab.* **17**, 437 (1997).

demonstrated that baseline MLCRA studies, at least with raclopride, may be performed over several days if necessary and allow more flexibility for the patient and the physician.

Other Ligands. One should, however, keep in mind that each tracer has different characteristics and may present different sets of challenges. Although the methods presented here have been found sound for studies using raclopride in the striatum, further validation may be necessary when starting work with other ligands, which may have different binding and dissociation characteristics, have affinity to more than one class of receptor in a given brain region, and so forth. This Chapter aims only at presenting the use of an almost ideal ligand, a blueprint of the type of studies that need to be performed before *in vivo* MLCRA can be conducted with confidence in living subjects.

Methods

Radiochemistry and Determination of Specific Activity

One of the most important aspects of our study is that we know the specific activity (SA) of the ligand before we inject it into the subject. This allows improved reproducibility and less variability not only between subjects, but also within subjects when repeated studies are being performed. This may represent a crucial aspect, as the concentrations of antagonist, unlabeled analog or competitor, may be a determinant in the measured K_d^{app}.[13] The chemists are being told in advance of the required SA before synthesis begins.

Raclopride was synthesized as described previously.[19] High specific activity [^{11}C]methyl iodide is produced via the gas-phase reaction of [^{11}C]methane with I_2.[20,21] The high SA for raclopride was greater than 1000 Ci/mmol.

Specific Activity Adjustment. Raclopride stock solutions are prepared by dissolving raclopride tartrate in USP water to make a 0.5 mM solution. This solution is used to prepare a 0.1 mM stock solution. Stock solution concentrations are determined precisely from standard calibration curves on analytical HPLC prior to use. The tracer production is initiated such that the final formulation will be complete 20 min (one half-life) before the anticipated time of injection. The yield in mCi is measured, and the

[19] M. Namavari, A. Bishop, N. Satyamurthy, G. T. Bida, and J. R. Barrio, *Appl. Radiat. Isot.* **43,** 989 (1992).

[20] P. Larsen, J. Ulin, K. Dahlstrom, and M. Jensen, *Appl. Radiat. Isot.* **48,** 153 (1997).

[21] J. M. Links, K. A. Krohn, and J. C. Clark, *Nucl. Med. Biol.* **24,** 93 (1997).

amount of stock solution to be added is calculated using the following formula:

$$\text{ml of stock solution} = X/(2YZ)$$

where X is yield at the time of formulation (mCi), Y is desired specific activity in Ci/mmol at time of injection, and Z is stock solution concentration (0.1 mM). The factor of two in the denominator accounts for the decay occurring between the time of formulation and the time of injection.

The carrier supplied from tracer production of >1000 Ci/mmol is considered to be insignificant compared to the concentration added. The requisite volume of stock solution is added, and the time is noted. The vial is then weighed, and an aliquot for precise measurement of the specific activity is drawn. The remainder is sent to the PET center. An exact volume of the quality control sample is injected onto the previously prepared analytical HPLC system. The SA of the adjusted solution is confirmed, and the investigators are notified of any small adjustment of the injection time required to attain the desired exact specific activity.

General Experimental Procedure

All the examples presented in this Chapter are obtained in healthy rhesus monkeys, as well as animals with MPTP-induced parkinsonism, as part of a number of studies. Detailed reports have been published.[8,15,22,23] Some animals received intraperitoneal or intravenous adminstration of methamphetamine (2 mg/kg ip)[24] and/or a dopamine transporter inhibitor, methylphenidate (0.5 mg/kg IV) or brasofensine (0.5 mg/kg ip).[25] All animal procedures are approved by the Committee on Animal Care of the University of British Columbia. Animal procedures, anesthesia regimens, and PET scan acquisition and analysis are described in detail elsewhere.[8,22] Briefly, the animal is positioned prone in a stereotactic head holder. A Siemens ECAT 953-31B allows the simultaneous acquisition of 31 coronal slices through the head and brain of the monkey (in-plane resolution: 6 mm FWHM; axial resolution: 5 mm). Scan data are acquired in two-dimensional mode over 1 h. For all scans, raclopride is administered as a bolus (2.5 mCi in 1 min in 10 ml saline) followed by constant infusion (2.5 mCi in 59 min in 30 ml saline) to create a true equilibrium condition.[18] Scanning starts with the start of bolus injection. The scanning sequence

[22] D. J. Doudet, J. E. Holden, S. Jivan, E. G. McGeer, and R. J. Wyatt, *Synapse* **38,** 105 (2000).

[23] D. J. Doudet, S. Jivan, T. J. Ruth, and J. E. Holden, *Synapse* **44,** 198 (2002).

[24] D. J. Doudet and J. E. Holden, *Biol. Psychiatr* **53,** 1193 (2003).

[25] D. J. Doudet, S. Jivan, C. English, and J. E. Holden, *Neuroimage* **11**(6), S5 (2000).

consists of six 30-s scans, two 1-min scans, five 5-min scans, and four 7.5-min scans for a total duration of 60 min.

For the sequential acquisition method, three or four bolus/infusion injections of raclopride (specific activity: SA1 > 1000 Ci/mmol; 40 < SA2 < 20 Ci/mmol, 12 < SA3 < 9 Ci/mmol, 6 < SA4 < 3 Ci/mmol) are performed throughout the day. The scans are separated by a minimum of 2 h (SA1–SA2) and up to 3.5 h (SA3–SA4). For nonsequential studies, the SA are similar, but each raclopride scan is performed on a separate day, 1–3 weeks apart, all at the same time of day. The order of scan acquisition is random.

Data Analysis

Data Reduction. As described previously, regions of interest (ROIs) are placed over the left and right striatum (circular ROIs: 37 pixels; pixel size: 4 mm^2) in four consecutive slices and four ROIs (16 pixels each) are positioned over an area of nonspecific ^{11}C accumulation on the cerebellum in two consecutive slices (for location, see Doudet *et al.*[22]). Time–activity curves are obtained for each ROI and averaged for each animal into left and right striatum and cerebellum (Fig. 1).

The two MLCRA methods of parameter estimation presented in the next section, linear and nonlinear, require the equilibrium ratio *B/F* between the concentrations of bound and free ligand in the target tissue. This value was derived from the tissue-input Logan graphical method for the evaluation of equilibrium distribution volume ratios (DVRs) of reversible

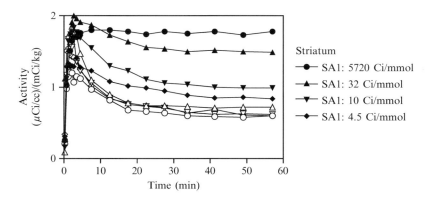

FIG. 1. Representative time courses of radioactivity concentration in the striatum (filled symbols) and cerebellum (open symbols) in a normal monkey at 4 decreasing SA. Constancy over time was obtained consistently in both regions during the four to five last frames, at all specific activities.

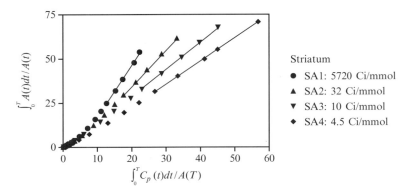

FIG. 2. Plots from the application of the tissue-input graphical estimation of DVR to data shown in Fig. 1. Linearity was observed consistently over the last four to five frames at all specific activities.

tracers[16] using the time–activity course in the cerebellum as the input function, between 30 and 60 min postinjection, by which time equilibrium was reached at all SA used (Fig. 2). The distribution volume ratio minus one (DVR-1) was interpreted as an estimate of the equilibrium ratio of bound and free ligand (B/F) in striatal regions. This graphical estimate of the bound/free ratio is represented by a variable surrounded by brackets and with a subscript g in the following equations to distinguish it from the same ratio estimated from individual measurements of B and F themselves. The other variables required by the fitting routines are these equilibrium concentrations B and F themselves. These were derived from the radioactivity concentrations in cerebellum $C(t)$ averaged over the final 30 min of the study, the graphical B/F estimate, and the SA values at injection time. B was estimated as $(B/F)C/SA$ and F as C/SA, where C represents the time average of $C(t)$. In summary, the measured time courses in the striatum and cerebellum are reduced to provide the estimates $(B/F)_g$ and C, which are used to compute the values provided to the parameter optimization process.

Parameter Optimization. We have previously reported a thorough investigation of four distinct approaches to evaluating the desired parameters BP, B_{max}, and K_d^{app} from reduced data.[8] We report here the two that had optimal properties. Both are based on the relationship between the bound concentration B and the free ligand concentration F with which it is equilibrated:

$$B = B_{max}F/(K_d^{app} + F) \tag{1}$$

B thus increases asymptotically to the saturated concentration B_{max} as F increases without limit, with half the receptors occupied when F reaches the value K_d^{app}. The superscript app has been added to this equilibrium dissociation constant to signify that it is the apparant value seen *in vivo* and to distinguish it from the idealized value derived from *in vitro* studies of the same ligand–receptor pair.

Both fitting equations are rearrangements of this saturation equation. The first is equivalent to the original linearization of the equation by Scatchard:

$$\left(\frac{B}{F}\right)_g = \text{BP} - \frac{1}{K_d^{app}} \left(\frac{B}{F}\right)_g F \tag{2}$$

Thus, $(B/F)_g$ is fitted against the bound value $(B/F)_g F$ (Fig. 3, top) with BP and K_d^{app} as optimized parameters (y axis intercept and negative inverse of the slope, respectively). The third outcome parameter B_{max} is the x axis intercept of the resulting straight line.

For the nonlinear method, the saturation equation is rearranged into

$$\left(\frac{B}{F}\right)_g = \text{BP} \; K_d^{app}/(K_d^{app} + F) \tag{3}$$

Thus, $(B/F)_g$ is fitted against the free concentration F (Fig. 3, bottom), again with BP and K_d^{app} as optimized parameters. BP is again the y axis intercept, and K_d^{app} is reflected in the rate of decline of the nonlinear curve. Similar to the growth of B in the saturation equation, the prediction for $(B/F)_g$ falls to half the BP value when F equals K_d^{app}. The receptor density is not represented in the graphical presentation of the curve; the value B_{max} is estimated separately as the product of the two fitted parameters.

This nonlinear parameter optimization approach was determined in our previous study to have optimal characteristics from the perspectives of the independence of the two sides of the fitting equation, and the covariance observed between the two fitted parameters. However, the resulting parameters from the two approaches are nearly always statistically indistinguishable. Thus, the linear method is used in the presentation of the following example applications to illustrate changes in density or affinity, as these changes are more easily perceived from the changes in a straight line graph. Agreement between the two approaches reflects the perfect conformity of data to the saturation equation from which the fitting equations are derived. Routine comparison of the results from the two methods thus provides a simple tool for discovering systematic deviations from this conformity when the method is applied in a new experimental context.

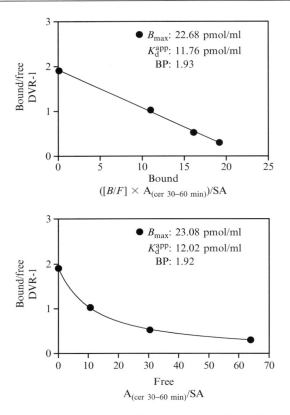

FIG. 3. Example fits from the application of two parameter-optimization approaches, linear (top) and nonlinear (bottom), to the four-point data set shown in Figs. 1 and 2. Solid curves are the model predictions of the two respective equations (see text) with the model parameters set to the optimized values shown.

Example Applications

Distinction of Density from Affinity Effects

Representative data sets from two experimental contexts are shown in Fig. 4. In both contexts the BP was altered significantly from the baseline value by the experimental intervention being studied, with a significant increase seen in animals that had been rendered parkinsonian by MPTP (a selective toxin of the dopaminergic nigrostriatal neurons) and a significant decrease seen in otherwise normal subjects following the acute administration of methamphetamine. Application of MLCRA showed that the

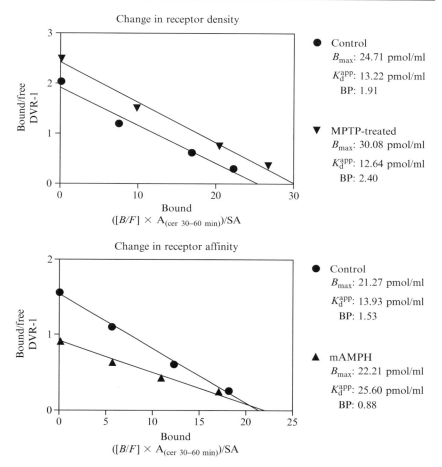

FIG. 4. Examples of the use of the MLCRA analysis. *Top:* Example of a change in receptor density. Straight-line plots were obtained using the fitted parameters in the striatum of a normal monkey (●) and in the striatum of a MPTP-treated monkey with clear clinical signs (▼). Note that in the MPTP-treated monkey, only BP and B_{max} are increased while K_d^{app} remains unchanged. *Bottom:* Example of a change in receptor affinity. Straight line plots of data in a monkey at baseline (●) and after methamphetamine (mAMPH: ▲). After methamphetamine, there is an increase in K_d^{app} and a decrease in BP in the striatum without significant changes in B_{max}.

increase seen in the MPTP parkinsonian model is due primarily to a change in receptor density, with no significant change in affinity,[23] whereas following methamphetamine the decrease is due primarily to the change in affinity brought about by competition between raclopride and endogenous dopamine mobilized by the drug, with no significant change in receptor density.[26]

Sequential vs Nonsequential Studies

As noted earlier in the overview section, our comparison of results from sequential studies performed on the same day with those from multiple studies performed nonsequentially on separate days has been motivated by several different questions. In the absence of an acute pharmacological intervention, the comparison provides assurance about the potential interference between studies and the effects of diurnal changes in the sequential case, and the effects of changes of the dopamine system over days or weeks in the nonsequential case. The comparison also bears on the future use of the method in a clinical setting in patients that are incapable of undergoing the rigors of sequential studies. Figure 5 shows a representative comparison under baseline conditions.[15] Similar results were observed in studies in the MPTP parkinsonian model. In the absence of acute pharmacological interventions, the two approaches can be regarded as equivalent.

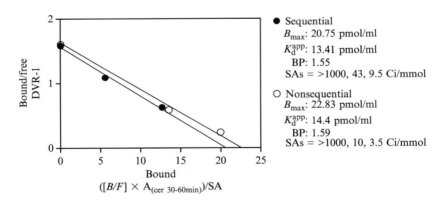

• Sequential
B_{max}: 20.75 pmol/ml
K_d^{app}: 13.41 pmol/ml
BP: 1.55
SAs = >1000, 43, 9.5 Ci/mmol

○ Nonsequential
B_{max}: 22.83 pmol/ml
K_d^{app}: 14.4 pmol/ml
BP: 1.59
SAs = >1000, 10, 3.5 Ci/mmol

FIG. 5. Comparison of the straight-line optimization procedure used to estimate BP, B_{max}, and K_d^{app} from data measured with sequential (●) and nonsequential (○) methods in the same animal. Note the good agreement between the measures of interest, although the specific activities used were different.

[26] D. J. Doudet and J. E. Holden, *J. Cereb. Blood Flow Metab* **23**, 1489 (2003).

Following drug intervention, the comparison also provides critically important information about the evolution over time of the drug effects on the system. Examples are not presented here, as these comparisons are a work in progress. We have observed cases such that the sequential and nonsequential results were the same, implying that the single drug administration prior to the sequential studies effected a change that was stable over time.[26] However, we have also seen very significant differences between sequential and nonsequential results in other cases.[25] We interpret these to mean that the effect of those drugs on the receptor system is strongly dependent on time after drug intervention. Our validation of nonsequential studies in the absence of drug intervention supports the validity of MLCRA for the study of such drug effects if care is taken to make the multiple drug administrations in nonsequential studies as equivalent to each other as possible, particularly with regard to their timing relative to the tracer studies. However, caution must then be exercised about long-term changes in the system induced by multiple, rather than single, administrations of the drug under study.

Two-Point Studies

Another work in progress is our study of the use of only two SA values. We provisionally suggest that two-point studies, possibly of critical importance in clinical applications of MLCRA, may conform adequately to the assumptions of the approach to justify their ease and efficiency. Investigation of Figs. 4 and 5 reveals that the combination of the high SA point with any other data point yields a two-point plot that would not be in strong disagreement with results from a full three- or four-point analysis. However, the dependence of the measured value of $(B/F)_g$ on the desired endpoints B_{max} and K_d^{app} varies strongly with the degree of saturation, with the dependence on B_{max} increasing, and that on K_d^{app} decreasing, as saturation is increased. Our work to date strongly suggests that two-point linear analyses may require receptor saturations of 80% to be reliable, particularly in studies of acute drug interventions. In that event, the y and x axis intercepts would be interpreted as the true BP and B_{max} values, respectively, and the K_d^{app} value estimated from the ratio B_{max}/BP would be interpreted as the affinity measure associated with the high SA data point.

Conclusions

PET studies with specific antagonists of brain neuroreceptors can distinguish between the influences of receptor density and receptor–ligand affinity by the performance of studies at multiple concentrations of the test ligand in exact analogy with the well-established methods of the postmortem receptor assay *in vitro*. This Chapter attempted to convey that such *in vivo* studies have exactly the same goal as their *in vitro* counterparts, to distinguish receptor density from receptor–ligand affinity as determinants of the binding process. While considerable care and caution must be exercised in the performance and interpretation of such studies, our work to date strongly supports the claim that the approach can serve as a reliable and useful tool in both basic and clinical settings.

Acknowledgments

This work was supported by the CIHR (formerly Medical Research Council of Canada) MPO 14535. We thank Astra Research Centre for their gift of the raclopride precursor. The authors thank the staff of the UBC/TRIUMF PET program for their assistance and contribution to this work. TRIUMF is funded by a contribution by the National Research Council of Canada. The authors are grateful to Dr. T. J. Ruth (Head, PET program) and J. A. Stoessl (Director, Pacific Parkinson Research Centre). These studies would not have been possible without the assistance of Ms. S. Jivan, M. Pronk (chemists), C. English, and C. Williams (technologists). We are especially indebted to J. Grant (AHT). Special thanks are due to Dr. J. Love and Mr. M. Boyd and the personnel of the UBC Animal Care Facilities for their outstanding care of the animals.

[11]   Estimation of Local Receptor Density, B'_{max}, and Other Parameters via Multiple-Injection Positron Emission Tomography Experiments

By Evan D. Morris, Bradley T. Christian, Karmen K. Yoder, and Raymond F. Muzic, Jr.

Introduction

Positron emission tomography (PET) is a functional imaging technique that allows an investigator to probe the biochemistry of an organism noninvasively. Because every PET study involves the injection of a radioactive molecule—a radiotracer—the practice of quantitative PET is tightly intertwined with the theory of tracer kinetics. The mission of tracer kinetics, in turn, is to estimate physiologically relevant parameters (e.g., blood flow rate, local cerebral metabolic rate, binding or dissociation rate constants) by modeling the uptake of a labeled molecule that mimics ("traces") the behavior of an endogenous or physiologically relevant exogenous chemical substance. Tracer kinetics merges experimentation and modeling. The experimental process involves both injection of a radioactive tracer and observation of local concentrations of said tracer over a period of time. (In PET, "observations" take the form of individual pixels in a time sequence of images, but this is secondary to our discussion.) Mathematically, what is needed is a model of the tracer uptake and its sequestration into various species and/or compartments. A formal comparison of model predictions with experimental observations yields estimates of model parameters; these parameters often represent the speed or magnitude of a physiological process.

Mere construction of a mathematical model, however, does not assure that each and every parameter of a model can be estimated from data. This goes to the identifiability of the parameters. Identifiability is hindered by noise in data or by ambiguity in the model structure. Identifiability is achieved through a combination of modeling parsimony and experimental design optimization. For example, no amount of optimization of an experiment could help identify the parameters n and m in the model, $Y = n*m*X$, where observed values of Y are related linearly to measured values of X; X is an independent and Y a dependent variable. However, as discussed later, if the model is something akin to $Y = n(m - X_1)X_2$, where both X_1 and X_2 are independent variables, then by suitably varying each of them, it may be possible to collect data that will enable an investigator to identify the parameter m as distinct from the parameter n. Roughly speaking, an appropriate experiment design would modulate the value of X_1 over a sufficient operating range, which is significant compared to m, so that the $(m - X_1)$ term did not behave effectively like the constant m.

The beauty of multiple-injection (M-I) PET experiments is that they are used to methodically perturb—and then observe—the system in question over a range of operating points so that the resultant data contain information that will differentiate the effects of parameters from one another. These complicated but elegant experiments (first conceived by Delforge et al.[1,1a]) can enable the investigator to dissect out effects of otherwise highly correlated kinetic parameters such as those that describe the kinetics of a PET tracer.

What parameters do we want to distinguish? Let us focus on PET studies with receptor–ligand tracers. One can imagine that the amount of ligand bound to target receptors (specific binding) will be dependent on at least two physiological factors that have direct correlates in parameters of a kinetic model. Net receptor-mediated uptake of a tracer will be

[1] J. Delforge, A. Syrota, and B. M. Mazoyer, *Phys. Med. Biol.* **34,** 419 (1989).
[1a] J. Delforge, A. Syrota, and B. M. Mazoyer, *IEEE Trans. Biomed. Eng.* **37,** 653 (1990).

dependent on the speed of interaction of the tracer with the receptor (association and dissociation rate constants), as well as the number of the receptors in a given volume of tissue (receptor density). Not surprisingly, in the standard single-injection experimental design, the respective parameters that represent speed and number of binding sites are highly correlated. In many PET studies, it may not be necessary to distinguish the binding rate constant from the receptor density. In such cases, it would not be necessary to mount the demanding experiments or data analysis described in this Chapter. However, in those specialized situations where it is important to identify the receptor density as distinct from binding rate constants for a tracer, M-I PET studies are the only way to do so.

Multiple-injection PET experiments have been carried out to closely examine the kinetics of various ligands that bind to receptors in the brain and heart.[2–12] What specialized situations might require identification of individual rate constants and receptor densities from PET data? Experiments that are intended for any of the following would be appropriate specialized applications of the techniques described herein:

- Fully evaluate the kinetics of a new tracer; determine if new tracer is different because it binds to different populations of receptors or binds to same population more avidly than established tracers.
- Accurately determine the regional variation in receptor density.

[2] B. T. Christian, T. Narayanan, S. Bing, E. D. Morris, J. Mantil, and J. Mukherjee, *J. Cereb. Blood Flow Metab.* **24,** 309 (2004).

[3] N. Costes, I. Merlet, L. Zimmer, F. Lavenne, L. Cinotti, J. Delforge, A. Luxen, J. F. Pujol, and D. Le Bars, *J. Cereb. Blood Flow Metab.* **22,** 753 (2002).

[4] J. Delforge, M. Bottlaender, C. Loc'h, I. Guenther, C. Fuseau, B. Bendriem, A. Syrota, and B. Maziere, *J. Cereb. Blood Flow Metab.* **19,** 533 (1999).

[5] J. Delforge, C. Loc'h, P. Hantraye, O. Stulzaft, M. Khalili-Varasteh, M. Maziere, A. Syrota, and B. Maziere, *J. Cereb. Blood Flow Metab.* **11,** 914 (1991).

[6] J. Delforge, S. Pappata, P. Millet, Y. Samson, B. Bendriem, A. Jobert, C. Crouzel, and A. Syrota, *J. Cereb. Blood Flow Metab.* **15,** 284 (1995).

[7] J. Delforge, A. Syrota, M. Bottlaender, M. Varastet, C. Loc'h, B. Bendriem, C. Crouzel, E. Brouillet, and M. Maziere, *J. Cereb. Blood Flow Metab.* **13,** 454 (1993).

[8] M. C. Gregoire, L. Cinotti, L. Veyre, F. Lavenne, G. Galy, P. Landais, D. Comar, and J. Delforge, *Eur. J. Nucl. Med.* **8,** PS(2000).

[9] E. D. Morris, J. W. Babich, N. M. Alpert, A. A. Bonab, E. Livni, S. Weise, H. Hsu, B. T. Christian, B. K. Madras, and A. J. Fischman, *Synapse* **24,** 262 (1996).

[10] R. F. Muzic, Jr., A. D. Nelson, G. M. Saidel, and F. Miraldi, *IEEE Trans. Biomed. Eng.* **15,** 2 (1996).

[11] R. F. Muzic, Jr., G. M. Saidel, N. Zhu, A. D. Nelson, L. Zheng, and M. S. Berridge, *Med. Biol. Eng. Comput.* **38,** 593 (2000).

[12] T. Poyot, F. Conde, M. C. Gregoire, V. Frouin, C. Coulon, C. Fuseau, F. Hinnen, F. Dolle, P. Hantraye, and M. Bottlaender, *J. Cereb. Blood Flow Metab.* **21,** 782 (2001).

- Assess the validity of using a particular brain region as a reference region (i.e, test the assumption of no receptors but otherwise identical kinetics as a target region).
- Differentiate possible diseases of receptor (or neuronal) loss from diseases of receptor dysfunction.

This Chapter is intended as a guide to graduate students, postdocs, and principal investigators who want to quickly get up to speed on key theoretical and experimental aspects of the M-I PET technique and begin to appreciate the attendant sensitivity analysis that gives the technique its power. The following discussion covers (1) the basics of the theory behind M-I PET studies, (2) practical considerations in planning and executing a successful M-I study, (3) key elements of the numerical implementation of the kinetic model and data-fitting algorithms, (4) an approach to interpretation of the parameter estimates once data are fitted, and (5) an examination of the sensitivity of PET data to the parameters and how that information can be used to improve the design of subsequent experiments.

Theory

Need for Models

PET is an imaging technique that measures radioactivity indiscriminately. No distinction can be made at measurement time between radioactivity (actually the detection of two simultaneously emitted photons) that emanates from a tracer molecule flowing with the blood, free in the extracellular space, bound to a cell protein in the intracellular space, or even from radioactivity that comes from a radionuclide attached to a metabolic product of the injected tracer. All of the aforementioned sources of radioactivity are detected and logged by the PET scanner, and all contribute to the reconstruction of a PET image. However, not all of these sources of detected signal are of equal importance to the investigator. In fact, in the case of a receptor-binding tracer molecule, the primary signal of interest is the radioactivity associated with tracer bound to a target molecule—typically a receptor or enzyme. To discern the wheat of the bound tracer signal from the chaff of the free and metabolized sources of radioactivity, the investigator must rely on a mathematical model.

Compartmental Models

The models used to describe PET data are usually compartmental. That is, they do not take account of spatial gradients in tracer concentration but rather assume that tissue concentrations can be properly described as well-mixed compartments. In fact, compartment models have been compared rigorously to distributed models and have been found to be satisfactory to describe PET data.[13] In PET, the volume of the compartment might correspond to the volume of the voxel (if the model is being applied on a pixel-by-pixel basis) or to a larger region of interest (ROI) (it applied on the ROI level). Compartmental models are described mathematically by a series of ordinary differential equations (ODEs); one ODE is required for each compartment. Compartments typically correspond to distinct kinetic states taken on by the radiolabeled tracer. These compartments can be distinct entities physically (e.g., intra- vs extracellular pools), distinct kinetically (e.g., bound to enzyme and bound to cell surface receptor), or distinct chemically (e.g., native vs metabolized tracer). As long as the states represent radioactive species, they must be included in the model of the PET measurements. Sometimes, as shown later, it is necessary to model nonradioactive species as well. Usually, the tracer is introduced into the organism via bolus injection(s) and so the uptake, retention, and eventual efflux of tracer from the tissue region of interest are transient phenomena that never reach steady states. That is, the concentrations of tracer in tissue or plasma do not achieve a constant level. If the system (tracer and tissues of interest) were to reach equilibrium, the system of ODEs would reduce to a set of algebraic equations that could be solved analytically. Since this is often not the case, and certainly not true with M-I experiments, the differential equations must be solved—either analytically or numerically—to solve the model for the predicted PET activity over time in a given region (details of this procedure are given in the section on models and data fitting).

In most PET models of receptor–ligand interactions, we hypothesize three kinetically distinct compartments and an arterial plasma pool of tracer—all of which contribute to the measured PET radioactivity. The arterial pool is not a compartment in the mathematical sense, although it is physically distinct from tracer in tissue. Because the arterial plasma concentration in most PET studies is a measured (i.e., applied to the model as a known) quantity, its depiction does not require a differential equation. In fact, the plasma concentration (or some other input function) must exist to drive the model. If no activity is introduced into the plasma, none is ever taken up into the tissue of interest. A version of the compartmental model corresponding to free, specifically (i.e., receptor-) bound, and nonspecifically bound tracer is shown schematically in Fig. 1 (arrows between compartments connote rate constants). K_1 and k_2 are first-order constants

[13] R. F. Muzic, Jr. and G. M. Saidel, *IEEE Trans. Med. Imaging* **22,** 11 (2003).

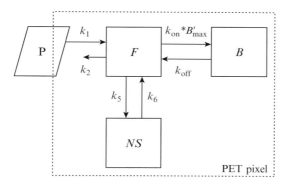

Fig. 1. Standard compartmental model used to describe dynamic PET data. The PET pixel is indicated to show that the measured quantity is a weighted sum of radioactivity in the compartments F (free), B (bound), NS (nonspecific), and some amount in the blood. P indicates that the metabolite-corrected plasma concentration is not a compartment because it is measured separately from the PET images and is assumed to be known.

that are related to blood flow. The terms k_{on}, B'_{max}, and k_{off} are apparent first-order rate constants related to association and dissociation of tracer and receptor. Terms k_5 and k_6 are first-order constants that measure the rates of forward and reverse nonspecific binding.

Standard Model Equations

In the language of mathematics, the "boxes" in Fig. 1 represent three unknown concentrations whose time-varying functions are encoded in their respective mass balances. The mass balances state that the change in concentration with time of species x (where $x = F$, B, or NS) is a function of those processes that contribute to an increase of x minus those processes that cause a loss of x.

$$\frac{dF}{dt} = K_1 C_p - (k_2 + k_5)F - k_{on}\left[B'_{max} - (B + B_c)\right]F + k_{off}B + k_6 NS \quad (1)$$

$$\frac{dB}{dt} = k_{on}\left[B'_{max} - (B + B_c)\right]F - k_{off}B \quad (2)$$

$$\frac{dNS}{dt} = k_5 F - k_6 NS \quad (3)$$

where C_p is the time-varying plasma radioactivity associated only with labeled native tracer. C_p is measured via blood samples (see later). The state variables of the model, F, B, and NS, represent the time-varying concentrations of tracer (in pmol/ml) in free, bound, and nonspecifically bound states, respectively. B_c is the concentration of unlabeled (or "cold") tracer bound to receptors.

The part of the model of greatest interest to investigators of receptor binding is $k_{on}[B'_{max} - (B + B_c)]F$. This expression describes binding of free tracer to available receptors. It states that the concentration of available receptors is the difference between available receptors at steady state, B'_{max} (a constant), and receptors bound to either labeled, B, or unlabeled tracer, B_c (time-varying functions). *Note:* In a single-injection experiment, there is always a known relationship between labeled and unlabeled bound ligand. The ratio of the labeled to the unlabeled is the specific activity (SA is given in μCi/pmol or Bq/pmol, ratios of radioactivity to mass of ligand). Thus, the expression for available receptors is often written as $(B'_{max} - B/SA)$. It will be clear why this is not adequate for modeling M-I PET data.[14]

Because the binding of a ligand to a receptor is a bimolecular process, it depends on available receptors, the presence of free ligand, F, and a bimolecular rate constant, k_{on}. In conventional single-injection experiments, which are predicated on injecting only a tiny ("trace") amount of radioligand, the amount of bound tracer (labeled or unlabeled) never rivals the available sites at steady state and so the term of interest, $k_{on}[B'_{max} - (B + B_c)]$, reduces to $k_{on}B'_{max}$. In this case, the model is analogous to the example described in the introduction. Namely, the parameters k_{on} and B'_{max} are not uniquely identifiable and the parameter estimation problem is reduced, of necessity, to finding an effective first-order rate constant k_3 ($=k_{on}B'_{max}$).

The raison d'etre of M-I PET experiments is specifically to overcome the problem of k_{on} and B'_{max} being irretrievably correlated (i.e., unidentifiable). Why do we want to identify these parameters separately? For one, the equilibrium dissociation constant (affinity constant) K_D is the ratio of the rate constants k_{off} and k_{on}. Thus, estimation of the *in vivo* K_D for a PET ligand is effectively dependent on the estimation of k_{off} and k_{on}. Estimation of these two constants—or their ratio—is not possible from a single-injection PET study. However, if the injected mass of tracer is modulated sufficiently in the course of multiple bolus injections of tracer such that the occupancy of receptors varies over a large enough range, then the term $(B'_{max} - B - B_c)$ must be retained explicitly in the model. It then becomes possible to identify the unique roles of the association rate constant and the concentration of available receptors in the uptake and binding of a tracer. In doing so, we move toward being able to estimate the receptor number and the affinity constant separately and possibly

[14] E. D. Morris, N. M. Alpert, and A. J. Fischman, *J. Cereb. Blood Flow Metab.* **16,** 841 (1996).

toward using PET to distinguish a defect of receptor function from a defect of receptor number.

M-I Model Equations

How do we adapt the standard model equations to accommodate the description of M-I data? One approach is to treat the separate injections as separate species that compete for the same receptor sites. Figure 2 diagrams the case of three separate bolus injections of tracer. The important thing to recognize about this extension for M-I data is that the specific activity of each injection is intentionally different. Therefore, it will be necessary to somehow track the individual inputs over the entire course of the study. This is one of the subtleties of the M-I PET technique. A blood sample taken shortly after a third bolus injection will contain radioactivity that originates with each of the three injections (assuming that all three injections contain radioactivity). Figure 3 depicts the multiple injections in terms of measured radioactivity and in terms of molar quantities needed

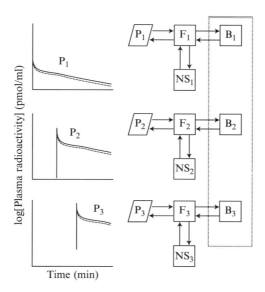

Fig. 2. Three parallel, coupled models with distinct input functions, P_1, P_2, and P_3, used to describe the dynamic time–activity curves generated from different regions of interest by a multiple-injection PET study. [*Note:* P_1 in the figure corresponds to the individual input functions $C_p^j(t - T^j)U(t - T^j)$ in the text.] All parameters are assumed to be identical across parallel models. The models are coupled because they share a common pool of receptors, B'_{max} (indicated by the dotted box surrounding all bound compartments), initially available to the tracer, regardless of injection. The injections, offset in time, that correspond to each respective subcompartmental model are illustrated to the left.

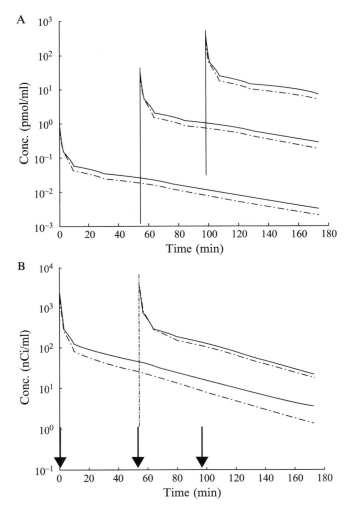

FIG. 3. (A) Input functions for each injection in terms of total ligand concentration in pmol/ml [all species (solid) and metabolite-corrected (dotted)]. Metabolite-corrected molar concentrations are used to construct the input function (see text for details). (B) Input functions in terms of radioactivity concentration (nCi/ml). The third injection consisted of unlabeled ligand only; therefore, there is no peak of radioactivity at the time of the third injection. There is input to the system, however, that must be measured somehow or modeled based on the shape of the other input functions as described in the text. Injection times are indicated by vertical arrows.

for solving the model. One approach to a posteriori dissection of the measured blood radioactivity is discussed at length by Morris et al.[9] In brief, we might assume that all input functions have the same shape but different scales. Thus, the observed plasma radioactivity can be described as

$$C_p(t) = \sum_j S^j C_p^j(t - T^j) U(t - T^j) \tag{4}$$

where S^j is a scale factor related to injected dose, $C_p(t)$ is an analytical expression of exponentials, and $U(t-T)$ is the unit step function at time T (i.e., $U = 0, t < T; U = 1, t \geq T$). From this equation, it is possible to recover separate C_p^j curves for each injection from the measured plasma radioactivity.

The general model equations for multiple injections take on the following form:

$$\frac{dF^j}{dt} = K_1 C_p^j - (k_2 + k_5) F^j - k_{on}(B'_{max} - \sum_l B_l^j) F^j + k_{off} B^j + k_6 NS^j \tag{5}$$

$$\frac{dB^j}{dt} = k_{on}(B'_{max} - \sum_l B_l^j) F^j - k_{off} B^j \tag{6}$$

$$\frac{dNS^j}{dt} = k_5 F^j - k_6 NS^j \tag{7}$$

where j is the index over injection number; B_l^j is either B^j or B_c^j.

The assumption of Eqs. (5)–(7) is that the kinetic parameters (K_1 k_2 k_{on} k_{off} B'_{max} k_5 k_6) are unaffected by the injection of either a high or a low specific activity tracer.

From the mass balance equations in Eqs. (5)–(7), we can construct the instantaneous output equation to describe the total radioactivity, $T(t)$, measured in the tissue at any moment in time as a result of one or more injections:

$$T(t) = \sum_j SA^j(t)(1 - F_v)[F^j(t) + B^j(t) + NS^j(t)] + F_v C_{wb}^j(t) \tag{8}$$

where SA^j converts the concentration associated with each injection in the tissue (F, B, and NS are in pmol/ml) into radioactivity, F_v is the blood volume fraction, and C_{wb} is the radioactivity concentration (nCi/ml) in whole blood. *Note:* The concentration of tracer in the arterial plasma (pmol/ml) is the driving force for uptake of the tracer into the tissue and, hence, the appropriate input function. However, any radioactivity in the microvasculature in the ROI contributes to the PET signal, so C_{wb} is the appropriate term for the output equation, which is in the units of the PET measurement (nCi/ml).

Experimental Protocol and Considerations

Multiple-injection PET studies are sophisticated experiments, both in design and in implementation. The appropriate duration for the experiment is dependent on both the kinetics of the ligand and the half-life of the positron-emitting nuclide. In nuclear-counting experiments, the noise in the data and hence, the parameter precision are determined by the amount of radioactivity (i.e., photons) collected. From this point of view, it would be desirable to acquire PET data for as long as possible. Unfortunately, due to the expenses associated with reserving PET scanner time, veterinarian staff, and anesthetization of the animal, one of the primary considerations when designing an M-I experiment is to minimize the duration of the experiment. On the other hand, because we use these experiments to maximize precision of the parameter estimates, there is a trade-off between convenience and precision. Successful experimental optimization can help balance this and other trade-offs and achieve a desired level of parameter precision.

Animal Preparation

Conducting M-I studies requires a small team of personnel to ensure a successful experiment. Input from all of the team members is needed to carefully plan and design an experiment that follows the guidelines of the Institutional Animal Care and Use Committee (IACUC). A skilled veterinary staff is needed to anesthetize the animal and to insert the catheter lines for ligand injection (venous) and blood withdrawal (arterial). In general, the choice of anesthetic is determined by the investigator and the veterinary staff based on their familiarity with the anesthetic agent, ease of use, and animal safety considerations. Because M-I PET experiments measure tiny (subnanomolar) concentrations, care must be taken that any biochemical effects of the anesthetic drugs do not perturb the biochemical system under study. For example, ketamine is a widely used preanesthetic known to interact with the dopaminergic system of the brain[15]; therefore, a M-I study targeting the dopaminergic system should allow adequate time (>1 h) for the effects of ketamine to subside before administration of the PET ligand. Most M-I studies require a minimum

---

[15] G. S. Smith, R. Schloesser, J. D. Brodie, S. L. Dewey, J. Logan, S. A. Vitkun, P. Simkowitz, A. Hurley, T. Cooper, N. D. Volkow, and R. Cancro, *Neuropsychopharmacology* **18**, 18 (1998).

TABLE I

EXAMPLE OF A TIMELINE FOR A MULTIPLE-INJECTION PET EXPERIMENT[a]

Sample experiment measurement of D2/D3 receptor density in the
thalamus with fallypride

00:00:00	Pre-anesthesia with glycopyrolate (0.01 mg/kg)
00:30:00	Anesthesia with ketamine (10 mg/kg)/xylazine (0.5 mg/kg)
00:45:00	Intubate monkey and maintain with 1–2% isoflurane
00:60:00	Insert venous (saphenous) and arterial (femoral) catheters
01:30:00	Position monkey in PET scanner, monitor vitals
01:45:00	Acquire 10-min transmission scan for attenuation correction
02:00:00	Begin PET data acquisition
02:00:00	Injection #1, "tracer study" with high specific activity injectate (SA_1 ∼2000 mCi/μmol); withdraw blood samples periodically for analysis (∼1 ml each)
02:54:00	Injection #2, "partial saturation" with low SA injectate (SA_2 ∼ 100 mCi/μmol); take blood samples
03:38:00	Injection #3, "saturation" with unlabeled fallypride only (SA_3 = 0 mCi/μmol); take blood samples
05:00:00	Terminate PET acquisition; remove anesthesia
05:15:00	Remove intubation when gag reflex is recovered
06:00:00	Monitor monkey during recovery from anesthesia

[a] The protocol was designed to elicit a precise estimate of B'_{max}, available receptors in the thalamus that bind [^{18}F]fallypride. In this particular design, the last injection contains only unlabeled fallypride. Note the absence of a "hot" peak at the corresponding third injection in Fig. 3B.

of 3 h of animal anesthetization, including animal preparation (see example timeline in Table I). The entire experiment can last up to 12 h. The anesthetic must provide a stable physiological system throughout this time course, it must minimize changes in regional blood flow (which affects ligand delivery), and it must withstand possible drug-induced stimulation. For the safety of the animals, typically 1–2 weeks must be allowed between experiments for the animal to recover from the effects of the anesthesia.

The insertion of two catheters is needed for M-I studies: a venous port for the administration of ligand and an arterial port for the temporal sampling of plasma radioactivity. The ligand is generally administered into a vein as a bolus infusion (5–30 s in duration) in several milliliters of saline. A bolus infusion is needed to accurately identify the ligand delivery parameter (K_1). As nearly as possible, all injections of ligand for each experiment should be given in an identical fashion so that ligand delivery is consistent throughout each epoch of the experiment (see section on constructing input curves).

Measuring Blood Activity and Constructing Input Curves

Accurate measurement of the radioligand concentration over time in the arterial plasma is essential to precise estimation of model parameters. The input function provides the essential time-varying details of radioligand delivery to the tissue of interest. An example of three arterial plasma input functions [in terms of both molar concentration (Fig. 3A) and radioactivity (Fig. 3B)] and the plasma radioactivity curves from which they are derived are depicted in Fig. 3. The graph in Fig. 3B shows measurable quantities of *radioactivity* (in nCi/ml) in the plasma. The plot displays two different measured quantities: total radioactivity in arterial plasma (solid curve) and metabolite-corrected arterial radioactivity (dash–dot line). The latter, corrected plasma concentration, is data needed for each of three input functions that drive the model, whereas whole blood radioactivity measurement (not shown) is needed for solution of the output equation (see the section on theory). With knowledge of the specific activity, blood data corresponding to each of the injections can be converted to input functions for the total (molar) ligand concentration, as shown in Fig. 3A. In the case of an experiment that includes a "saturation" component, the contribution of the third injection to the total ligand input function, $C_p(t)$, cannot be measured directly from the (radioactivity) blood curve, as no additional radioactivity is injected (see earlier discussion). Instead, the *shape* of the curve must be inferred from the previous injections. The scale factor, S, for the injection of unlabeled material can be determined from the ratio of the doses.

The shape of the input function is, in part, determined by the speed of the venous injection [e.g., a rapid bolus injection will result in a sharply peaked input function (blurred by dispersion as the bolus travels through the vasculature)]. The blood curve is measured by withdrawing blood samples from the arterial port. The frequency of withdrawal must be matched to the anticipated shape of the input function. Blood samples are usually drawn every 5–10 s for the first several minutes following ligand injection. The samples can be drawn less frequently as the ligand begins to equilibrate between plasma and tissue(s).

As a general rule of thumb, the total volume of blood withdrawn for an experiment should not exceed 10% of the blood volume of the animal. Sampling of a 10-kg rhesus monkey would be limited to roughly 70 ml (assuming 7% of body weight is blood volume). This volume should be replaced by an iv drip of saline over the course of the experiment. The arterial blood samples are centrifuged to separate the plasma from the red blood cells. The plasma samples can be assayed further to separate the native ligand from the radiolabeled metabolic by-products. The volume of each arterial plasma sample must be large enough to yield an accurate measurement of radioactivity in the final plasma fraction (as gauged by radioactive counting statistics). In the case of primates, the volume of each arterial sample is typically 1 ml.

Generation of Regional Time–Activity Curves (TACs) from PET Images

As in most dynamic PET experiments, the M-I model is fitted to tissue TACs derived from PET scans yielding estimates of the kinetic parameters of interest (see section on data fitting). Data at each time point are based on investigator-defined regions of interest (ROIs) placed on the PET images at each time. Although single-injection PET studies are sometimes analyzed in a pixel-by-pixel manner (to generate parametric maps), fitting the M-I model to data would be too demanding computationally to do so.

In most cases, ROI analysis requires that high-resolution structural image (typically a T_1-weighted MRI) is acquired and coregistered to a PET image for each PET subject. The preferred PET image (for use in registration only) is usually an image of the summed (or averaged) radioactivity over the entire duration of the PET study. The MRI allows the investigator to precisely define the exact anatomical location of ROIs, and the coregistration then gives coordinates that allow the ROIs to be applied in proper spatial orientation to each frame of PET data. Taking the average radioactivity in the ROI at each time point generates the desired TAC. Typically, PET images are not suitable as the basis for ROI templates for two reasons: (1) specific anatomy may not be resolved easily and (2) hot spots in the PET image may induce unintentional bias in the ROI placement by the investigator. Several intermodality registration algorithms [e.g., automated image registration (AIR) by Woods et al.[16,17]] are available both in commercial medical image analysis packages (e.g., MEDx, Sensor Systems, Inc.) and as stand-alone code for free download (http://bishopw.loni.ucla.edu/AIR5).

The ROIs selected will probably depend on the characteristics of the radiotracer used. [^{18}F]Fallypride is a highly selective D_2/D_3 receptor ligand with a high PET signal-to-noise ratio and excellent resolution in areas with low-to-moderate concentrations of D_2 receptors (e.g., cortex, thalamus).[18]

[16] R. P. Woods, S. R. Cherry, and J. C. Mazziotta, *J. Comput. Assist. Tomogr.* **16,** 620 (1992).
[17] R. P. Woods, J. C. Mazziota, and S. R. Cherry, *J. Comput. Assist. Tomogr.* **17,** 536 (1993).
[18] J. Mukherjee, Z. Y. Yang, M. K. Das, and T. Brown, *Nucl. Med. Biol.* **22,** 283 (1995).

In investigations with this ligand, several brain regions are available for analysis [(in contrast, other ligands may not provide reliable data outside of brain regions with extremely high numbers of D_2/D_3 receptors (e.g., striatum) and ROIs of interest may range from very large volumes (e.g., whole striatum, whole thalamus) to smaller, more specific volumes (cortical areas, nucleus accumbens, individual thalamic areas)]. An important point to consider in selecting ROIs is their volume. Larger volumes (composed of many voxels) result in TACs with higher signal-to-noise ratios. These low-noise curves typically lead to successful data fitting and thus precise parameter estimation. Unfortunately, larger ROIs are also more prone to be heterogeneous in tissue composition and therefore lead to data that reflect an average of kinetically different regions. The investigator must also be aware that small ROIs may suffer from partial volume (PV) effect error if the structures they circumscribe are small relative to the resolution of the scanner. Partial volume error will lead to nonlinear underestimation of the true radioactivity in the ROI and, subsequently, bias in the parameter estimates. For a review of PV error and for various approaches to correcting for it, see Kessler et al.,[19] Meltzer et al.,[20] Morris et al.,[21] Muller-Gartner et al.,[22] Muzic et al.,[23] Rousset et al.,[24–26] and Strul and Bendriem.[27]

The resulting TACs from three brain regions generated by placing ROIs on images made in the example protocol given earlier (Table I) are shown in Fig. 4. Specific binding is highest in the striatum, moderate in the thalamus, and nearly absent in the cerebellum. We can tell this, in part, from observing (1) similarity of the decline in tracer concentration following the first and second injections and (2) the absence of deflection from the descending curve of the cerebellum at the time of the third injection. That

[19] R. M. Kessler, J. R. Ellis, Jr., and M. Eden, *J. Comput. Assist. Tomogr.* **8,** 514 (1984).
[20] C. C. Meltzer, J. K. Zubieta, J. M. Links, P. Brakeman, M. J. Stumpf, and J. J. Frost, *J. Cereb. Blood Flow Metab.* **16,** 650 (1996).
[21] E. D. Morris, S. I. Chefer, M. A. Lane, R. F. Muzic, Jr., D. F. Wong, R. F. Dannals, J. A. Matochik, A. A. Bonab, V. L. Villemagne, S. J. Grant, D. K. Ingram, G. S. Roth, and E. D. London, *J. Cereb. Blood Flow Metab.* **19,** 218 (1999).
[22] H. W. Muller-Gartner, J. M. Links, J. L. Prince, R. N. Bryan, E. McVeigh, J. P. Leal, C. Davatzikos, and J. J. Frost, *J. Cereb. Blood Flow Metab.* **12,** 571 (1992).
[23] R. F. Muzic, Jr., C. H. Chen, and A. D. Nelson, *IEEE Trans. Med. Imaging* **17,** 202 (1998).
[24] O. Rousset, Y. Ma, M. Kamber, and A. C. Evans, *Comput. Med. Imaging Graph.* **17,** 373 (1993).
[25] O. Rousset, Y. Ma, S. Marenco, D. F. Wong, and A. C. Evans, *in* "Quantification of Brain Function Using PET" (R. Myers, V. Cunningham, D. Bailey, and T. Jones, eds.), p. 158. Academic Press, San Diego, 1996.
[26] O. G. Rousset, Y. Ma, and A. C. Evans, *J. Nucl. Med.* **39,** 904 (1998).
[27] D. Strul and B. Bendriem, *J. Cereb. Blood Flow Metab.* **19,** 547 (1999).

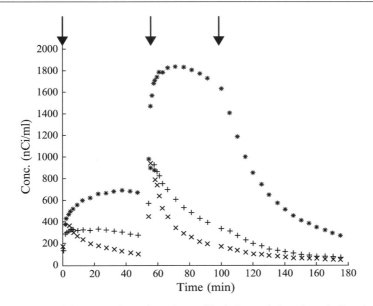

Fig. 4. Time–activity curves from the striatum (*), thalamus (+), and cerebellum (×) from a multiple-injection study of D_2/D_3 dopamine receptors with [^{18}F]fallypride. All data are from the same animal. Injection times are indicated by vertical arrows.

is, injection of high or low specific activity does not change the shape of the curve because of (1) the absence of receptors in the region, and (2) the injection of cold tracer does not accentuate displacement because there is no specific binding to be displaced. The next sections address the fitting of models to data and present a resulting fit to one of the curves in Fig. 4.

Models and Data Fitting

Implementing Model Equations

This section describes the framework for relating the theoretical models (described earlier) to data that are collected (as described earlier). Since a PET scanner measures radioactivity concentration, the output equation, Eq. (8), is used to relate the radioactivity concentration (e.g., Bq/ml or, μCi/ml) to the molar concentration (e.g., pmol/ml) in various compartments. However, a modification to Eq. (8) should be used in practice, as the PET scanner does not measure instantaneous radioactivity concentration. Rather, it measures concentration averaged over acquisition time intervals commonly referred to as frames. Thus we define model$_i$ as the time-averaged concentration over frame i and compute this quantity as

$$\text{model}_i = \frac{1}{d_i} \int_{t_i}^{t_i+d_i} T(\tau) d\tau \tag{9}$$

where t_i and d_i are the start time and durations of frame i and T is from Eq. (8).

To implement such an equation, it is convenient to express it in a form that can be solved with an ordinary differential equation (ODE) solver such as is used to solve the state equations [Eqs. (5) to (7)]. Accordingly, we introduce a new expression for the integrand in the previous equation,

$$\frac{dh}{dt} = T \tag{10}$$

When this differential equation, Eq. (10), is solved with initial condition $h(0) = 0$ (no radioactivity in the system at time zero), the expression for the model-predicted PET signal in time frame i is simply

$$\text{model}_i = \frac{1}{d_i} [h(t_i + d_i) - h(t_i)] \tag{11}$$

Parameter Estimation

In M-I studies, estimating values of model parameters as precisely as possible is often the primary goal. Parameter values provide quantitative assessments of receptor concentration, affinity, blood flow, etc. To accomplish the goal, one typically "fits a model" to data. This entails finding the values of model parameters that are most consistent with data. Mathematically, the problem is equivalent to adjusting the values of the model parameters in order to minimize the difference between the model prediction and the actual measurement of tissue radioactivity ("data"). Consider what is called the weighted least-squares objective function, which measures this difference:

$$o(\mathbf{p}) = \sum_i [w_i(\text{model}_i(\mathbf{p}) - \text{data}_i)^2] \tag{12}$$

In this equation, data$_i$ represents data from frame "i" and model$_i(\mathbf{p})$ represents the model output, which is intended to predict data$_i$. Model output depends on the values of the parameter vector $\mathbf{p}$. The task is then to adjust values of the components of $\mathbf{p}$ to make model output most closely agree with data. Mathematically, we minimize $o()$ with respect to $\mathbf{p}$. We include weights, w_i, because we do not expect to achieve perfect agreement between model and data and because we do not have uniform confidence in the measurements. The value of w_i may be specified as the reciprocal of (an estimate of) the variance of data$_i$, in which case the value of **p** that minimizes $o()$ is a maximum likelihood estimate.[*]

While one could adjust the values of model parameters *manually* to find the ones that best explain data, such a search can often be done more efficiently and more objectively by a mathematical algorithm implemented on a computer. The Levenberg–Marquardt algorithm is popular for this application.[28–30] Because the value of $o(\mathbf{p})$ has a complex dependence on values of the parameters, a closed form solution, which minimizes $o(\mathbf{p})$ in the general case, is not available. Consequently, an iterative approach must be used. One starts with an initial guess of the parameter values and then adjusts the components of the parameter vector **p** in order to reduce the value of $o(\mathbf{p})$. The efficiency of this process depends on having a means to predict values of $o(\mathbf{p})$ as values of **p** are altered because this provides a basis for adjusting parameter values. For this purpose, algorithms often require an estimate of the derivative of the objective function $o()$ with respect to the parameter vector **p**. By differentiating Eq. (12) with respect to component j of the parameter vector, we obtain the expression

$$\frac{do}{dp_j} = 2 \sum_i \left[w_i(\text{model}_i - \text{data}_i) \frac{d\text{model}_i}{dp_j} \right] \tag{13}$$

Notably, this expression contains a term for the derivative of the model output with respect to component j of the parameter vector. These derivatives have particular significance and are given the name sensitivity functions.

One numerical approach to evaluating the sensitivity functions is to use finite differences. This approach is attractive because it is conceptually very simple: solve the model equations at one value of **p**, change the value of the jth component of **p** by a small amount denoted here as Δp_j, solve the model equations again, and then estimate the derivative as the difference in model output divided by Δp_j. Unfortunately, in practice, this approach is not very robust. It is not at all trivial to pick a value of Δp_j small enough so that the finite differences approximate the desired derivative but not so small that the differences are dominated by "noise" or numerical imprecision.

[*] Under certain assumptions about the data.
[28] K. Levenberg, *Q. Appl. Math.* **2,** 164 (1944).
[29] D. Marquardt, *S.I.A.M. J. Appl. Math.* **11,** 431 (1963).
[30] J. J. More, *in* "Numerical Analysis" (G. A. Watson, ed.), *Lecture Notes in Mathematics* **630,** p. 105. Springer-Verlag, New York, 1977.

A more robust—and recommended—approach to evaluating the sensitivity functions can be obtained by differentiating the state equations. To describe this approach, we have to take a step back and define notation for the composite set of differential equations for the state and output equations [Eqs. (5–7) and (10)]. Recall that the equations were all of the form $dx/dt = y$ with accompanying specified initial conditions. We can group these together by defining a vector $\mathbf{c}$ and a vector-valued function $f()$ that have components corresponding to the state and output equations and their variables. In the example given earlier, the vector $\mathbf{c}$ would have components

$$\mathbf{c} = [F \quad B \quad NS \quad h]^T \tag{14}$$

and the function $f()$ would be defined as

$$\frac{d\mathbf{c}}{dt} = f(\mathbf{c}, t, \mathbf{p}) = \left[\frac{dF}{dt} \quad \frac{dB}{dt} \quad \frac{dNS}{dt} \quad h \right]^T \tag{15}$$

with

$$\mathbf{p} = [K_1 \quad k_2 \quad k_{on} \quad k_{off} \quad B'_{max} \dots]^T \tag{16}$$

(The superscript T connotes the transpose; Eqs. (14–16) describe column vectors.) The initial condition for $\mathbf{c}$, called $\mathbf{c}_0$, is a column vector of the initial conditions of each of the state equations.

With the state and output equations expressed in this framework, we now obtain the equations needed for a robust approach to evaluating the sensitivity functions. Specifically, by differentiating Eq. (15) and its initial condition with respect to the parameter vector $\mathbf{p}$ we obtain

$$\frac{d\mathbf{S}}{dt} = \frac{\partial f}{\partial \mathbf{c}} \mathbf{S} + \frac{\partial f}{\partial \mathbf{p}} \quad \text{with} \quad \mathbf{S}_0 = \frac{\partial \mathbf{c}_0}{\partial \mathbf{p}}; \tag{17}$$

which is an initial value problem like Eq. (15) except that $\mathbf{S}$ is a matrix. The rows of $\mathbf{S}$ correspond to those of $\mathbf{c}$, whereas the columns of $\mathbf{S}$ correspond to different components of the derivatives. For example, the element in row 2, column 3 of $\mathbf{S}$ would be the derivative of B (element 2 of $\mathbf{c}$) with respect to k_{on} (element 3 of $\mathbf{p}$).

Numerical Solution of Differential Equations

Having presented a formalism of how we relate a model to experimental measurements, we next turn to the details of the numerical implementation of the solution of state, Eq. (15), and sensitivity equations, Eq. (17). These are both considered initial value problems. Numerically solving state equations entails programming Eq. (15) and selecting an appropriate ordinary differential equation solver. For example, in MATLAB one could use a solver from Shampine's ODEsuite,[31] whereas in C or FORTRAN one might use a member of the LSODE family of solvers.[32–34]

Conceptually, algorithms for solving these initial value problems begin with the initial value and use the Euler formula to approximate the solution at the next time step. For example,

$$\mathbf{c}(t + \Delta t) \cong \mathbf{c}(t) + f(\mathbf{c}, t, \mathbf{p}) \cdot \Delta t \tag{18}$$

Details of the implementation must include a strategy for selecting Δt to achieve a specified accuracy in the solution without requiring an excessive amount of computation. Fortunately, problems of this form are common and a number of algorithms are available. Generally, algorithms are classified as being designed for "stiff" or "nonstiff" equations. Details of these designations are beyond the scope of this Chapter, but suffice it to say that "stiff" equations are "hard" problems to solve in that the solver is forced to take very small steps in Δt. Special algorithms have been designed for stiff equations. In comparison to nonstiff solvers, stiff solvers trade-off more complex algorithms and evaluations in each step for the ability to take larger steps.

How does one determine if equations are stiff in any given case? The pragmatic approach is to try both stiff and nonstiff solvers. Well-written solvers have built-in methods to select step size (Δt) and still keep errors in the solution within a specified range. Under such conditions, using a nonstiff solver with stiff equations (and vice versa) would lead to computationally inefficient solutions.

We alert the reader to the availability of a MATLAB-based software package that implements methods for setting up and solving models such as those used to analyze dynamic PET data. The package includes implementations of state and sensitivity equations and functions for fitting models to data in order to estimate parameters. COMKAT[35] can be downloaded from www.nuclear.uhrad.com/comkat. It was written by one of the authors of this Chapter (R.F.M.) and is presently used by each of the authors in their research. COMKAT takes into account the details described

---

[31] L. F. Shampine and M. W. Reichelt, *S.I.A.M. J. Sci. Comput.* **18,** 1 (1997).

[32] A. C. Hindmarsh, *in* "Scientific Computing" (R. S. Stepleman, ed.), p. 55. North-Holland, Amsterdam, 1983.

[33] A. C. Hindmarsh and R. Serban, "User Documentation for CVODES: An ODE Solver with Sensitivity Analysis Capabilities. UCRL-MA-148813." Lawrence Livermore National Laboratory, Livermore, CA, 2002.

[34] J. R. Leis and M. A. Kramer, *ACM Trans. Math. Software* **14,** 61 (1988).

[35] R. F. Muzic, Jr. and S. Cornelius, *J. Nucl. Med.* **42,** 636 (2001).

in the preceding section so that its users do not have to be experts in numerical analysis.

Parameter Estimation Considerations

Selection of the Initial Guess. As mentioned in the previous section, algorithms estimate parameters by starting with an initial guess of the parameter values and adjusting them to minimize the value of the objective function. Care should be taken in selection of the initial guess. Algorithms often converge to the true parameter values only when the initial guess is "close enough" to the true values. How close is "close enough" is difficult to quantify in practice because it depends on the true parameter values, which are unknown, and also on the information content of data. In practice, as one is developing the fitting strategies for a particular application, one should try the estimation procedure with a range of initial guesses. Analysis of the resultant parameter estimates will provide insight into how close is "close enough."

Validating Parameter Estimates. When the optimization algorithm has converged to parameter values and the model output and data are in close agreement, one might assume that the parameter estimates are valid and even precise. This is not always the case. For example, there could be more than one set of parameter values that produce a model output that agrees well with data. One possibility is that there are multiple local minima in the objective function $o()$. Another possibility is that the parameters are correlated, meaning that different combinations of parameter values will lead to essentially the same model output. Consider plotting $o()$ as a function of values of two parameters with the height of the surface indicating the value of $o()$. If the surface is relatively flat, then a large change in the parameter values would give rise to a small change in $o()$. To achieve good precision in the parameter estimates, we would like to design the experiments to make the surface of $o()$ steep. A steep objective function means that data are very sensitive to the model parameters. Moreover, we want the surface to be steep in all directions. Consider an alternative case wherein the surface is shaped like a long narrow valley aligned with the parameter axes. Changes along the valley floor make hardly any difference in the value of $o()$. An experiment that yielded such an objective function would be insensitive to the parameter aligned with the valley, and it would not be possible to make reliable estimates for this parameter.

To investigate these possibilities, it is important to conduct simulation studies a priori. The basic steps of the study are as follows. (1) Create data; using representative parameter values, solve the model equations to create "perfect" data. (2) Add "noise" to perfect data to emulate the expected imprecision in the experimental data. (3) Fit simulated data with the proposed parameter estimation method. This process must be repeated numerous times with different noise realizations. Parameter estimates are then compared to the *known* values used to create data. In particular, one might calculate the error in the parameter estimates by subtracting the true values from each estimate and then summarize data in terms of the bias and precision of the estimates by calculating the mean and standard deviation of the error.

While the aforementioned techniques are important in establishing the validity of the parameter estimates, they are not necessarily complete. Simulation is but one component of validation. The next section describes another component: careful examination of the model fit to measured data.

Results and Interpretation

A fit to the TAC for an ROI drawn on the thalamus (middle curve on Fig. 4) is shown in Fig. 5. This fit results in estimates of the "best" parameter values that can explain data, but how do we know if these estimates are good? There are tests that must be done. One was suggested in the

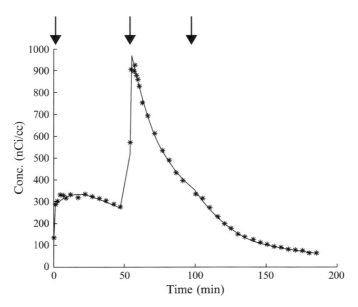

FIG. 5. Data from the time–activity curve from the thalamus (middle curve, Fig. 4). The solid curve indicates the fit of the model given in Eqs. (4–9) to data via with nonlinear parameter estimation described in the text. Injection times are indicated by vertical arrows.

previous section; namely, if fitting the model to simulated data sets reveals that multiple choices of parameters would result in equally good fits, there is little hope that fits to experimental data (which may not be strictly consistent with the model) will yield more identifiable parameters. However, assuming that the model appears to fit simulated data well and that minimizing the objective function yields unique parameters, what are the basic steps that must be followed to evaluate the quality of the results?

Examination of Residuals

Figure 6 shows a plot of normalized residuals derived from the fit to M-I data shown in Fig. 5. Normalized residuals, calculated as $[\text{model}_i - \text{data}_i]/\text{SD}(\text{data}_i)$, are a good way of determining the quality of the fit. Ideally, these residuals should be distributed normally with zero mean and unit standard deviation. Both of these conditions appear to be met in Fig. 6. The mean of the residuals will be obvious from their plot. Any order in the pattern of the residuals, however, may indicate that the model is deficient. To determine the nonrandomness of the residuals it is useful to perform a "runs" test. A run is defined as a series of adjacent residuals that are either all positive or all negative. The fewer the number of runs, the more

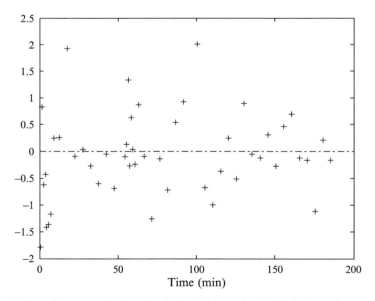

FIG. 6. Plot of the normalized residuals derived from fitted M-I thalamus data (shown in Fig. 5). Note the apparent zero-mean behavior of the residuals. See text for details.

likely that the fit is poor and that either the model or the fitting algorithm is suspect. (For examples using the runs test, see Bard.[36])

Parameter Precision

If fits to data are acceptable, then the investigator will want to report his/her findings in terms of the estimated parameter values and their approximate uncertainties (variance, standard deviation, confidence intervals, correlation, etc.). To report intra-trial variance, it is necessary to use an estimate of parameter variance because one fit to a data set yields only one estimate of each parameter. Many search algorithms (such as the Levenberg–Marquardt mentioned previously) will return a covariance matrix along with the optimal parameter set. The covariance matrix is usually approximated by the inverse of the weighted product of the sensitivity matrix (mentioned in the previous section) with its transpose ($Cov(\mathbf{p}) = [S^T W S]^{-1}$, where W is an $n \times n$ diagonal matrix whose elements are related inversely to the n data points in the TAC). The approximation is valid when the parameter values $\mathbf{p}$ are close to the optimal point. The diagonal elements of the resulting covariance matrix are the variances of the respective parameters. Parameters should be reported plus or minus a standard deviation (square root of the variance).

Normalization of the covariance matrix by its diagonal elements yields the correlation matrix. It is prudent to examine this matrix, whose diagonal elements are unity and whose off-diagonal elements are the covariances between parameters. Highly correlated parameters (e.g., correlation >0.95) are not separable. See Table II for an example correlation matrix produced from a six-parameter fit to [^{18}F]fallypride data shown in Fig. 5. If two parameters a and b are highly correlated, then, in practical terms, only their product (or their ratio) is identified. Neither of their values individually should be trusted because an increase in one could be completely offset by a comparable decrease (or increase) in the other with no decrement to the quality of the fit and no basis for choosing one combination of parameters over another with the same product. Consider the following practical scenario. If B'_{max} is highly correlated with k_{on} (as is often the case), then there will be multiple pairs of these parameters that will be equally plausible choices to explain the acquired data. Imagine further that we are trying to compare the on rate (k_{on}) of a tracer at the serotonin transporter site in two groups of subjects, who are known to express different genetic variants of the transporter, to test the hypothesis that the binding rate will be different. If one of the groups also tends to have fewer available

---

[36] Y. Bard, "Nonlinear Parameter Estimation," p. 213. Academic Press, New York, 1974.

TABLE II
CORRELATION MATRIX FOR THE DATA FIT DEPICTED IN FIG. 5[a]

	K_1	k_2	k_{on}	B'_{max}	k_5	k_{off}
K_1	1					
k_2	0.809	1				
k_{on}	−0.632	−0.181	1			
B'_{max}	0.612	0.751	**−0.197**	1		
k_5	0.13	0.496	0.429	0.604	1	
k_{off}	−0.374	−0.211	**0.416**	**0.162**	0.635	1

[a] Each element in the matrix is Corr(a,b). Diagonal elements are 1 because each parameter is completely correlated with itself. Thanks to the M-I experiment, there is very little correlation among any of the parameters B'_{max}, k_{on}, and k_{off} (see bold values). The correlation matrix is symmetric, so the top half of the matrix has not been shown; Corr(a,b) = Corr(b,a).

receptors at steady state (smaller B'_{max}) because of medication that blocks these sites (e.g., Prozac), then the medication will be a confound and the population on medication may be seen, artifactually, to have faster binding because the higher k_{on} merely balances a lower B'_{max} when data are fitted. The correlation matrix in Table II confirms that thanks to the M-I experiment, correlations among k_{on}, B'_{max}, and k_{off} have all been minimized. In contrast, in a single-injection experiment, the correlation between k_{on} and B'_{max} would be nearly 1.

Model Selection/Goodness of Fit

Often, even well-designed experiments produce data that do not justify the use of models of the desired complexity. That is, not all parameters of the model can be identified. In these cases, it may be necessary to opt for a simpler model by fixing some parameters and not estimating them. How do we know that the simpler model is appropriate? There are a number of popular criteria that gauge "goodness of fit." One such determinant of goodness of fit is the F statistic.[37] As in all statistical testing, it is conducted with reference to the question of whether to accept or reject the null hypothesis. In the case of model selection, the null hypothesis is that the simpler model is adequate to describe data. Another popular index is the Akaike critierion,[38] $AIC = \ln(SS) + 2P$, where SS is the weighted sum of

[37] E. M. Landaw and J. J. DiStefano, *Am. J. Physiol.* **246,** R665 (1984).
[38] H. Akaike, *in* "System Identification: Advances and Case Studies" (R. K. Mehra and D. G. Lainiotis, eds.), p. 27. Academic Press, New York, 1976.

squares that result from the fit to data and P is the number of parameters in the model. Thus, a "good fit" will correspond to a low AIC value, but AIC will be penalized if the fit is achieved through the use of extraneous parameters. In the case of the fit to [^{18}F]fallypride data shown for the thalamus in Fig. 5, data did not support use of both a k_5 and a k_6 parameter. It was found that setting k_6 identically to zero and estimating only k_5 was necessary and sufficient to fit data. To confirm that this was the appropriate model, the Akaike criterion was calculated for both six-parameter (k_6 set to 0) and seven-parameter fits, and the six-parameter fit was shown to be better.

Understanding and Designing M-I Experiments

Sensitivity Functions

The sensitivity functions described earlier are key to the procedure of minimizing the least-squares objective function. They are also central to understanding how data fitting in general and the analysis of M-I data in particular work. Once we understand what the sensitivity functions tell us, we can use them to improve the design of our experiments. For an example, consider the sensitivity functions plotted in Fig. 7. These curves correspond to the derivatives of the model with respect to the six parameters, **p**, that were estimated by fitting the model to data from thalamus (shown in Fig. 5). The sensitivity equations have been solved at the value of the parameters that minimized the objective function given in Eq. (12). First, we noted that the sensitivities are time-varying functions as we would expect from looking at Eq. (17). In other words, the sensitivity of the observed PET signal to any model parameter rises and falls throughout the course of the experiment. The PET signal may be most sensitive to one parameter at one moment and to another at the next. Early time data are usually the most sensitive to blood flow parameters (i.e., K_1, k_2), whereas late-time data are sensitive to receptor binding. In fact, the independent, time-varying status of each of the sensitivity functions is at the heart of parameter identifiability. In Figs. 7A and 7B, one can observe that the sensitivities to the K_1 and k_2 parameters are very nearly identical except that one is always positive and the other always negative. We explain this behavior by noting that these two parameters are both dependent on blood flow (K_1 = extraction fraction*flow; $k_2 = K_1$/volume of distribution). As blood flow increases, more tracer is delivered to the tissue and, hence, the effect on the measurable signal is a positive one. An increase in blood flow also means that k_2, the rate at which tracer leaves the tissue, will increase and we would expect the PET signal to be diminished. The fact that these time

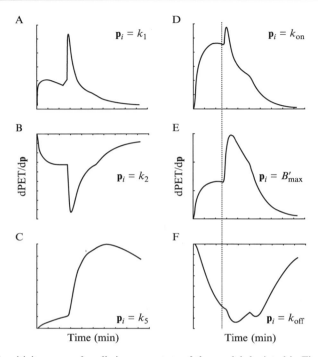

Fig. 7. Sensitivity curves for all six parameters of the model depicted in Fig. 2 evaluated over time and at the optimal parameter vector resulting from the fit to data shown in Fig. 5. Each curve is on its own scale. The vertical dotted line in D, E, and F corresponds to the time of the second injection. Everything on those plots to the left of the dotted line is the equivalent of sensitivity curves for a single injection study. See text for details.

courses mirror each other so closely means that they are not independent; in fact, they are nearly linearly dependent and so the parameters are highly correlated [Table II, $\text{Corr}(K_1, k_2) = 0.809$]. Thus, these parameters are not identified easily from the type of experiment that was performed with [^{18}F]-fallypride to estimate receptor binding in regions of moderate binding. Luckily, identification of the blood flow parameters is not the goal of the experiment. K_1 and k_2 are much better and more easily identified by an experiment involving a very sharp injection of tracer and rapid blood sampling to catch the fine detail of the input function.

More interesting from the standpoint of potential receptor–ligand characterization are the curves in Fig. 7D, E, and F. Figure 7D–F shows the time-varying derivatives of the PET with respect to k_{on}, B'_{max}, and k_{off}, respectively. If we consider just the first epoch in each curve (to the left of the dotted line), it is very hard to distinguish the role that is played by any of these parameters. Certainly, there would be no difference in effect between raising B'_{max} or raising k_{on} during the first epoch. Recall that the first epoch is merely a single-injection experiment (with a high SA tracer), and it is well known that k_{on} and B'_{max} are not identifiable from such a limited experiment. The effect of *lowering* k_{off} during this period would also be hard to differentiate from a concomitant rise in either of the other two parameters. As mentioned earlier in such cases, modelers must fall back to an identifiable parameter and not try to estimate both the "m" and the "n" as discussed in the introduction.

If we look over the entire study duration at the sensitivity curves for k_{on}, B'_{max}, and k_{off}, we can begin to appreciate that (1) they are each distinguishable from each other and (2) that it takes a sufficiently complicated experiment that manipulates occupancy to draw out differences in the processes represented by the three separate parameters. Recall that the second and third injections were termed "partial saturation" and "saturation" (see Table I). In fact, Christian *et al.*[2] observed that there is apparently a narrow range of partial saturations that, if achieved during the second phase of the M-I experiment, yield TACs for [^{18}F]fallypride experiments, which produce estimates of k_{on} and B'_{max} that are uncoupled. If the target level of occupancy is under- or overshot in these experiments, interestingly, the parameters remain correlated in the fitting. The potential success of M-I experiments has been explained previously in terms of the sensitivity coefficients.[39]

Using the Sensitivity Information for Design

How can we use this information that appears to be contained in the sensitivity curves objectively? This is the subject of what is known as sensitivity analysis and optimal experiment design. As learned in the section on parameter precision, the sensitivity matrix can be used to approximate the variances of each parameter estimate. Many scalar quantities can be derived from this matrix and used to compare different experimental designs. A classical index for optimization of an experiment is the D-optimal criterion. "D" refers to the determinant of the Hessian matrix ($H \approx S^T W S$) or, equivalently, to the determinant of the inverse of the covariance matrix. In either case, to achieve a D-optimal design, we seek to maximize the value of the determinant of the matrix. The matrix, in turn, contains information about the collective variances of the parameters. In a physical sense,

[39] E. D. Morris, A. A. Bonab, N. M. Alpert, A. J. Fischman, B. K. Madras, and B. T. Christian, *Synapse* **32**, 136 (1999).

the confidence region surrounding the optimal choice of parameters in parameter space is a n_p-dimensional ellipsoid (where n_p is number of parameters) whose axes are the eigenvectors of H. Maximizing the determinant of the Hessian matrix is equivalent to minimizing the volume of this confidence region and thus reducing the possible choices of the parameter vectors that yield an equally good fit to data. That is, maximizing det(H) is equivalent to minimizing overall variance of the parameters. Many other quantities can be derived from the Hessian matrix and used as design criteria to maximize or minimize some other aspect of a parameter or parameters (for examples related to PET experiments, see Muzic et al.[10,11,40]).

Because the Hessian matrix is a function of both the parameters and the experimental protocol, there are two points to consider. (1) Optimal design of experiments is iterative—it must be repeated as more becomes known about parameter values upon which Hessian-based criteria depend and (2) the variances of the parameter estimates can be improved by the best choice of protocols, that is, by optimizing over a set of design variables.

Design Variables. What are the design variables in the typical M-I PET experiment? There are two. First, the specific activities of the respective injections can be varied by mixing differing amounts of labeled and unlabeled ligand for each injection. Second, the time between injections can be varied. Because the specific activity (or equivalently the mass for a given radioactivity dose) will determine the occupancy level of receptors at a given time, we can appreciate that specific activity is the experimenter's tool for manipulating the receptor–ligand system to achieve decreased parameter correlation and increased parameter precision. In some circumstances, it may be necessary to put constraints on the design. For instance, if the total time of the experiment must be limited for reasons of convenience or safety, this will act as a constraint on the combination of times between injections. If the synthesis of the radiopharmaceutical is very difficult, it may be practically necessary to limit the design of the M-I study to a one synthesis. If so, then we constrain the choice of specific activities. In particular, the second and third injections will be limited to lower SA than the high SA material available for the first. To investigate more about this technique, the reader is directed elsewhere for uses of optimal design in PET and tracer kinetics.[1,2,5,10,11,36,41–45]

[40] R. F. Muzic, Jr., A. D. Nelson, G. M. Saidel, and F. Miraldi, *Ann. Biomed. Eng.* **22**, 43 (1994).

[41] J. Beck and K. Arnold, "Parameter Estimation in Engineering and Science," p. 419. Wiley, New York, 1977.

Conclusion

M-I PET studies are labor-intensive undertakings that demand not only experimental acumen, but also a synthesis of mathematical and numerical expertise as well. Despite the overhead associated with such experiments, they may be the only practical means of extracting certain kinetic information about tracer uptake and behavior *in vivo* from dynamic PET images. To the extent that it may be helpful and illuminating to determine precisely the values of all the *in vivo* kinetic parameters of a tracer, it is hoped that this Chapter served part as an introductory review, part as a tutorial, and part as an operating guide to a useful technique that merges functional imaging with tracer kinetics and optimal experiment design.

[42] E. R. Carson, C. Cobelli, and L. Finkelstein, "The Mathematical Modeling of Metabolic and Endocrine Systems," p. 129. Wiley, New York, 1983.
[43] D. Feng, D. Ho, K. K. Lau, and W. C. Siu, *Comput. Methods Progr. Biomed.* **59**, 31 (1999).
[44] J. A. Jacquez, "Compartmental Analysis in Biology and Medicine," 2nd Ed. The University of Michigan Press, Ann Arbor, 1988.
[45] E. D. Morris, G. M. Saidel, and G. M. Chisolm, 3rd, *Am. J. Physiol.* **261**, H929 (1991).

[12]   Imaging Dopamine Receptors in the Rat Striatum with the MicroPET R4: Kinetic Analysis of [11C]Raclopride Binding Using Graphical Methods

By DAVID L. ALEXOFF, PAUL VASKA, and JEAN LOGAN

Introduction

The promise of *in vivo* imaging in laboratory rodents with dedicated positron emission tomography (PET) cameras lies in bringing successful PET methodologies used in clinical and large animal studies to preclinical stages of new drug and radiopharmaceutical development.[1–3] For example, rapid *in vivo* screening of compounds labeled with positron emitters in a single rat or mouse is possible using small animal PET imaging. PET can describe more accurately the kinetics of these labeled drugs in a single

[1] S. P. Hume and T. Jones, *Nucl. Med. Biol.* **25**, 729 (1998).
[2] H. D. Burns, T. G. Hamill, W. Eng, B. Francis, C. Fioravanti, and R. E. Gibson, *Curr. Opin. Chem. Biol.* **3**, 388 (1999).
[3] S. P. Hume and R. Meyers, *Curr. Pharm. Des.* **8**, 1497 (2002).

rodent than traditional *ex vivo* methods that require sacrificing multiple animals. As well, measurements of saturation kinetics in rodents using PET, together with kinetic modeling or graphical analyses, can be used to determine pharmacokinetic constants such as B_{max} and K_d *in vivo* (see discussion in Hume and Meyers[3]). By comparison, *in vitro* measurements are often flawed, as specific values depend greatly on methodological details and do not include effects of bioavailability inherent to *in vivo* measures.[1] In addition, PET studies in rodents can be used to evaluate the relationship of receptor occupancy to drug efficacy *in vivo*, as demonstrated by Hirani *et al.*,[4] who characterized 5-HT$_{1A}$ receptor occupancy in the rat brain by the drug pindolol. Finally, high-resolution PET imaging can help bridge decades of biological research in laboratory rodents with human biology by allowing direct comparisons of animal research with clinical research.[5] For example, several rodent models of human diseases, including Parkinson's disease,[6] Huntington's disease,[6–8] and drug abuse,[9] have been studied using small animal PET.

Commercialization of small animal PET cameras by at least two manufacturers (microPET; Concorde Microsystems, Knoxville, TN, and HIDAC; Oxford Positron Systems, Oxford, UK) is expected to expand the use of PET methods in basic research and new drug development by making available cost-effective PET cameras designed specially for small animal imaging. For clinical PET research groups, acquisition of a dedicated animal PET permits simultaneous operation with clinical studies, making efficient use of expensive radiochemical syntheses. The first commercially available, dedicated animal cameras were designed for nonhuman primate scanning (for review, see Chatziioannou[5]). One of these early machines, the SHR-2000 (Hamamatsu, Japan), a ring-based system featuring a large diameter port (50.8 cm) permitting whole body viewing of rhesus monkeys, has also been used for rodent imaging.[10] A new large-aperture animal camera (SHR-7700) has become available from Hamamastu with increased sensitivity [three-dimensional (3D) mode] and better resolution.[5] However, few results from rodent studies using the SHR-7700 have been reported in the literature. A second commercially

[4] E. Hirani, J. Opacka-Juffry, R. Gunn *et al.*, *Synapse* **86**, 330 (2000).
[5] A. F. Chatziioannou, *Eur. J. Nucl. Med.* **29**, 98 (2002).
[6] S. P. Hume, A. A. Lammertsma, R. Myers *et al.*, *J. Neurosci. Methods* **67**, 103 (1996).
[7] K. Ishiwata, N. Ogi, N. Hayakawa *et al.*, *Nucl. Med. Biol.* **29**, 307 (2002).
[8] D. M. Araujo, S. R. Cherry, K. J. Tatsukawa, T. Toyokouni, and H. I. Kornblum, *Exp. Neurol.* **166**, 287 (2000).
[9] H. Tsukada, J. Kreuter, C. E. Maggos *et al.*, *J. Neurosci.* **16**, 7670 (1996).
[10] H. Umegaki, K. Ishiwata, O. Ogawa *et al.*, *Synapse* **43**, 195 (2002).

available animal camera (HIDAC; Oxford Positron Systems, UK) with a variable opening or 10–20 cm was designed specifically for small animal imaging.[11] More recent HIDAC designs have increased absolute sensitivity beyond 1% while still offering intrinsic resolution reported to be <1 mm. This animal camera does not function by detecting photons with arrays of scintillation detectors (like BGO and LSO), but relies on collecting and amplifying electrons created from interactions of photons with lead sheets drilled with an array of closely spaced (~0.5 mm) holes (0.4 mm diameter). HIDAC has been used successfully to image the GABA-benzodiazepine receptor subtype containing the $\alpha 5$ subunit in the rat brain.[12]

This Chapter presents detailed methods using a third commercially available small animal PET camera, the microPET R4 (Concorde Microsystems). The MicroPET R4 is a LSO scintillator ring-based tomograph borne from original research by Cherry and co-workers[13] at UCLA. A significant advance of the commercial microPET was the increase in absolute sensitivity achieved using four rings of detector blocks, resulting in an axial field of view of ~7.8 cm. This expanded axial field of view also allows for imaging of a whole mouse body in a single bed position. This work evaluates the quantitative capabilities of the microPET R4 by examining the effects of photon scatter, attenuation, and image reconstruction methods on determinations of dopamine receptor availability using graphical methods. Graphical analysis of kinetic data in neuroimaging in human and nonhuman primates has been a very successful, efficient, general-purpose methodology. Here we extend this method and its assumptions to the analysis of PET studies in the rat brain.

Animal Preparation

For experiments where dynamic measurements are required starting at the time the PET radioligand is injected, complete restraint or anesthesia is generally required to prevent motion artifact during imaging. In either case, target organ function may be affected by the specific immobilization technique employed. For example, studies of the conscious primate brain have shown that PET measures can be altered significantly by the administration of anxiety-provoking drugs, suggesting that stress must be minimized when imaging awake animals.[14] Although the use of anesthesia to

[11] A. P. Jeavons, R. A. Chandler, and C. A. R. Dettmar, *IEEE Trans. Nucl. Sci.* **46,** 468 (1999).
[12] A. Lingford-Hughes, S. P. Hume, A. Feeney *et al.*, *J. Cereb. Blood Flow Metab.* **22,** 878 (2002).
[13] S. R. Cherry, A. F. Chatziioannou, Y. Shao *et al.*, *J. Nucl. Med.* **40,** 1164 (1999).
[14] H. Takamatsu, A. Noda, Y. Murakami *et al.*, *J. Nucl. Med.* **44,** 1516 (2003).

immobilize an animal avoids these problems, it can have its own confounding effects on PET experiments. Studies in awake and anesthetized cats have shown that halothane anesthesia can increase measures of dopamine receptor-specific binding in the striatum compared to awake, restrained imaging.[15] This same work showed that ketamine anesthesia had no effect on PET measures of dopamine receptor binding in the striatum. Tsukada et al.[16] however, have shown that intravenous administration of ketamine can reduce [^{11}C]raclopride binding in the awake primate brain in a dose-dependent manner (3–10 mg/kg/h). Clearly, careful consideration must be given to the immobilization technique employed for PET imaging and its potential effect on PET outcome measures.

Rats are immobilized by anesthesia using a mixture of ketamine and xylazine. Approximately 40 min before scheduled injection of radioligand, the animal is knocked down with ketamine (100 mg/kg) and xylazine (10 mg/kg) by intraperitoneal (ip) injection. To allow for administration of radioligand, a lateral tail vein is catheterized using a 22-gauge catheter. The tail is first prepared by gently cleaning with soapy water and then an alcohol swab, followed by soaking in warm water. Anesthesia is maintained during the imaging session (2–4 h) by ip injection of ketamine/xylazine at a dose averaging 85 mg/kg/h. Depth of anesthesia is monitored using respiration rate and whisker twitch.

All animals used for this study are male Sprague–Dawley rats (Taconic, Germantown, NY) weighing 350–450 g. Rats are housed individually and food and water are provided *ad lib*. This work was approved by the IACUC of BNL, and all rats are housed and maintained in an AALAC-approved and accredited animal husbandry facility.

[^{11}C]Raclopride Preparation and Administration

Dopamine receptor imaging is carried out using carbon-11 labeled substituted benzamide [^{11}C]raclopride.[17] [^{11}C]Raclopride has been shown to bind specifically to dopamine D2 receptors (D2R) in humans[17] and has been used as an *in vivo* probe of dopaminergic function due to its sensitivity to changes in endogenous neurotransmitter.[18] [^{11}C]Raclopride is synthesized from ^{11}CH$_3$I according to a method described previously.[17] ^{11}CH$_3$I is prepared by a gas phase reaction of ^{11}CH$_4$ with I$_2$ after conversion of ^{11}CO$_2$

[15] W. Hassoun, M. Le Cavorsin, N. Ginovart *et al.*, *J. Nucl. Med. Mol. Imaging* **30,** 141 (2003).
[16] H. Tsukada, N. Harada, S. Nishiyama *et al.*, *Synapse* **37,** 95 (2000).
[17] L. Farde, H. Hall, E. Ehrin, and G. Sedvall, *Science* **231,** 258 (1986).
[18] N. D. Volkow, G.-J. Wang, J. S. Fowler *et al.*, *Synapse* **16,** 255 (1994).

to $^{11}CH_4$ as first described by Larsen *et al.*[19] and Link *et al.*[20] using an automated commercially available device (GE Medical Systems, Milwaukee, MN). [^{11}C]Raclopride doses are small fractions (<0.8 ml) of routine syntheses prepared for human subject studies on the same day. Specific activity determination (the ratio of radioactive atoms to total atoms) is made using mass measurements acquired during radiotracer purification by HPLC (Waters Novapak C18; 250 × 10 mm), and radioactivity measurements are obtained with a calibrated ion chamber (Capintec, Inc., Ramsey, NJ).

The range in dose of [^{11}C]raclopride injected for all studies is 165–551 μCi. Specific activity at time of injection varies from a minimum of 0.11 Ci/μmol to a maximum of 1.3 Ci/μmol. This corresponds to a range of injected raclopride mass of 0.5–18 nmol/kg of rat body weight.

MicroPET R4 Configuration

Imaging is carried out using the microPET R4 tomograph (Concorde Microsystems), which has a 12-cm animal port with an image field of view (FOV) of ~10 cm. Each animal is positioned prone on the microPET bed, centering the brain in the field of view. The full-width half-maximum resolution (FWHM) in the center of the FOV is approximately 1.85 mm and remains <2.5 mm FWHM across the rat brain.[21] The entire rat brain is easily imaged simultaneously using the large axial (rostral to caudal) FOV of the microPET R4. The rat head is supported and secured to the bed to approximate a flat-skull orientation in the camera field of view without the use of a head holder. Fully 3D list-mode data are collected for 60 min using an energy window of 250–750 keV and a time window of 10 ns (default settings recommended by the manufacturer). For maximum sensitivity (~3.4% absolute according to the manufacturer), coincidence data are binned into 3D sinograms using the full axial acceptance angle of the scanner. Obtaining maximum sensitivity is especially important in small animal PET to minimize the amount of injected radioactivity. Large injected doses of radioligand can confound measures of D2R availability using [^{11}C]raclopride due to competition at the receptor of the unlabeled compound associated with the radiotracer.[22] Based on these findings and results from simulations, it has been proposed that a minimum animal PET camera absolute efficiency of 3% is required to obtain quantitative images, whereas 1% is sufficient for region of interest analysis.[23]

[19] P. Larsen, J. Ulin, K. Dahlstrom *et al.*, *Appl. Radiat. Isot.* **48**, 153 (1997).

[20] J. M. Link, K. Krohn, and J. Courter, *Nucl. Med. Biol.* **24**, 93 (1997).

[21] C. Knoess, S. Siegel, A. Smith *et al.*, *Eur. J. Nucl. Med. Mol. Imag.* **30**, 737 (2003).

[22] S. P. Hume, J. Opacka-Juffry, R. Myers *et al.*, *Synapse* **21**, 45 (1995).

MicroPET data acquisition is started simultaneously with [^{11}C]raclo-pride injection. To preserve axial resolution, high sampling of the polar angle is used (21 segments). The binning produces 24 time frames (6 × 10 s, 3 × 20 s, 8 × 60 s, 4 × 300 s, 3 × 600 s) and includes subtraction of random coincidences collected in a delayed time window.

Data Corrections and Image Reconstruction

Accurate determination of radioactivity concentration using PET requires application of several corrections to coincidence lines of response recorded by the camera. These corrections include subtraction of random coincidences, correction for dead-time losses, subtraction of scattered coincidences, correction for photon attenuation, and normalization for varying individual detector efficiencies. Finally, the PET camera must be calibrated against a known radioactivity concentration. If all necessary corrections are not applied and/or validated, calibration should be done with a phantom approximating the geometry of each imaging experiment (e.g., primate brain vs mouse body). In fact, practical corrections for photon attenuation and scatter were not available from the microPET manufacturer at the time of these studies. Therefore, a scatter correction algorithm was implemented (described later) and its effect on measurements of dopamine receptor availability in the rat striatum was evaluated. Effects of photon attenuation were evaluated subsequently using software released by the manufacturer. Currently the microPET R4 is fully configured with scatter and attenuation correction capabilities. Scatter correction provided by the manufacturer is based on the work of Watson[24] and requires transmission data (i.e., attenuation map) of the object. This method was not evaluated in this work.

For all studies described here, the microPET R4 detector efficiencies were normalized using a point source normalization procedure provided by the manufacturer. This procedure produces a generally uniform image of a rat brain phantom (3 cm) across the central 45 planes of the camera.

Scatter Correction

Scatter is known to have an appreciable effect on 3D PET data in the human brain.[25] Given that the diameter of the adult rat head (~3 cm) is small compared to the mean free path of a 511-keV photon in tissue

[23] S. R. Meikle, S. Eberl, R. R. Fulton, M. Kassiou, and M. J. Fulham, *Nucl. Med. Biol.* **27,** 617 (2000).

[24] C. C. Watson, *IEEE Trans. Nucl. Sci.* **47,** 1587 (2000).

[25] S. R. Cherry and S.-C. Huang, *IEEE Trans. Nucl. Sci.* **42,** 1174 (1995).

(~10 cm), [^{11}C]raclopride imaging in the rat brain was expected to give microPET data with relatively small scatter fractions. However, the lack of septa in the microPET and its large axial FOV increase the probability that sinogram scatter fractions might be appreciable. Scattered events were subtracted from randoms-corrected sinograms based on a quadratic polynomial fit of projection data from outside the head in a manner similar to previously published "tail-fitting" methods[25,26] For improved statistics of the fit, radial projections from all azimuthal angles in the sinogram were summed. Thus, the same curve was subtracted from each projection within a given slice, although independent fits were made for each different slice. Due to the large transaxial field of view, it was possible to use conservative, fixed radial limits for the fit, which excluded all possible true coincidences. Note that this approach has the advantages of accounting for scatter from outside the field of view, as well as any other potential sources of uniform background in the data.

The effects of scatter correction using the "tail-fitting" algorithm produced a small but consistent effect on distribution volume ratio (DVR, see later), increasing the DVR an average of 3.5% (Table I; see also Alexoff et al.[27]). ROI data showed a more significant effect, especially when derived from brain regions of lower accumulation of tracer radioactivity, such as the cerebellum, where mean ROI pixel values from scatter-corrected images decreased as much as 20% compared to uncorrected image data.[27] For comparison, Yoa and colleagues[28] have reported a reduction in contrast of 3–5% in the rat brain due to scatter.

An interesting observation is the significant increase in sinogram background from the beginning of the study to the end. Scatter fractions determined from early time frames (<40 min) in this work to be ~40% in the rat head were similar to those reported by Knoess et al.[21] for a rat phantom using the same energy window. Scatter fractions at the end of study (60 min), however, were found to be significantly higher (~65%). If related to scatter, this higher scatter fraction is presumably caused by a redistribution of radioactivity from inside to outside the field of view. However, at the low activity levels near the end of a scan, other low-level sources of uniform background might also contribute, such as an imperfect randoms correction or the natural radioactivity of LSO. Further investigations using a noise equivalent counts analysis as a function of PET camera energy windowing support the hypothesis that a LSO coincidence background has a

[26] P. M. Bloomfield, R. Myers, S. P. Hume, T. J. Spinks, A. A. Lammertsma, and T. Jones, *Phys. Med. Biol.* **42**, 389 (1997).
[27] D. L. Alexoff, P. Vaska, D Marsteller et al., *J. Nucl. Med.* **44**, 815 (2003).
[28] R. Yoa, J. Seidel, C. A. Johnson et al., *IEEE Trans. Med. Imaging* **19**, 798 (2000).

TABLE I
EFFECT OF PET DATA CORRECTIONS ON DVR

Study	None	Scatter	%Change	Scatter + Atten.	%Change
1	1.876	1.935	3.14	1.909	−1.36
2	2.600	2.706	4.08	2.708	0.0738
3	2.023	2.028	0.247	2.014	−0.695
4	2.531	2.650	4.70	2.631	−0.722
5	2.620	2.764	5.50	2.684	−2.98
t test[a]			$p < 0.025$		$p < 0.125$

[a] Paired two-tailed t test.

measurable effect on microPET data.[29] Therefore, it is important to incorporate a background (scatter) subtraction technique into the recommended configuration of the microPET R4 in order to obtain accurate tracer kinetic information. Although the image-based scatter correction software included with the microPET R4 will not correct for this LSO background, LSO background was only significant at activity levels reached at the end of low-dose (<200 μCi) [^{11}C]raclopride studies.

Attenuation Correction

While the absolute attenuation of deep rat brain regions can be calculated to be significant (approaching 20–40%), the relative variation across the rat head is more likely to be closer to 10%, as all parts of the image are attenuated to some extent. In fact, Yoa et al.[28] (2000) detected nonuniformity due to attenuation in the rat brain to be ~12%. Image attenuation corrections determined previously for the striatum and cerebellum in adult Sprague–Dawley rats using a custom-built animal PET camera[30] were found to be the same (16%). Because the data analysis methods used here rely on the ratio of a D2R-rich region (striatum) with a region used to estimate nonspecific binding (cerebellum), a constant attenuation correction factor for the striatum and cerebellum could be ignored. To confirm this, the effect of attenuation correction on the graphical analysis of [^{11}C]raclopride binding in the rat striatum was evaluated as described here.

[29] P. Vaska and D. L. Alexoff, *J. Nucl. Med.* **44**, 138P (2003).
[30] R. Meyers, S. P. Hume, S. Ashworth, A. A. Lammerstma, P. M. Bloomfiled, S. Rajeswaran, and T. Jones, *in* "Quantification of Brain Function Using PET" (R. Meyers, V. Cunningham, D. Bailey, and T. Jones, eds.), p. 12. Academic Press, London, 1996.

Typically, attenuation correction is performed using transmission and blank data.[31] Alternatively, emission data can be used to create an attenuation map. MicroPET R4 software includes image segmentation, smoothing, forward projection, and inverse Fourier rebinning tools that allow the user to create 3D attenuation sinograms from emission image data. These attenuation files can then be used to correct PET sinogram data for photon attenuation according to established procedures. Using these tools, rat [^{11}C]raclopride images are converted to attenuation sinograms in the following manner. First, reconstructed dynamic data are summed in time to create a single image. This image is then smoothed with a five pixel (4.5 mm) FWHM Gaussian kernel and segmented into a uniform object ($\mu = 0.096$ cm^{-1}) describing primarily the head of the rat in the FOV of the microPET. The skull is not segmented separately. The thresholding set point for segmentation is set at approximately to 50% full scale on the log histogram plot. Proper thresholding for correct head size is validated using a profile tool to measure the dimensions of the segmented image. After segmentation, the image is forward projected into 3D attenuation sinograms. Note that this technique is only applicable where attenuating objects in the field of view also contain sufficient radioactivity that can be used to define their geometric boundaries through the segmentation process. The use of head holders (not used here) for positioning requires the use of transmission data to define the object boundaries in space. It is possible to obtain head-holder transmission data once accurately and then create an attenuation sinogram that can be added to segmented emission data for a complete attenuation map.

DVRs derived from images corrected for attenuation as described earlier are not statistically different (two-tailed paired t test; $p < 0.126$) than DVRs determined from data without attenuation correction (see Table I). This is consistent with previous findings of similar attenuation in the striatum and cerebellum of the adult Sprague–Dawley rat.[30]

Image Reconstruction

After the histogram process, microPET data are stored as a three-dimensional sinogram containing the parallel projections through the imaged object. Image reconstruction in PET is the process of creating a three-dimensional image volume from projection data. Optimizing reconstruction algorithms for PET is still an active field of research, especially using iterative techniques (for examples, see Riddel *et al.*,[32] Boellaard *et al.*,[33]

[31] E. J. Hoffman and M. E. Phelps, *in* "Positron Emission Tomography and Autoradiography. Principles and Applications for the Brain and Heart" (M. E. Phelps, J. C Mazziotta, and H. R. Schelbert, eds.), p. 237. Raven Press, New York, 1986.
[32] C. Riddell, R. E. Carson, J. A. Carrasquillo *et al.*, *J. Nucl. Med.* **42,** 1316 (2001).

and Mesina *et al.*[34]). The microPET R4 is configured with two reconstruction algorithms, both utilizing only two-dimensional data sets to reduce processing time. Because the R4 is a 3D machine, its 3D sinograms must be converted into 2D sinograms before an image can be reconstructed. This is done on the microPET R4 using the process of Fourier rebinning.[35] After Fourier rebinning, images are reconstructed by 2D-filtered back projection using a ramp filter with cutoff at one-half the Nyquist criteria (maximum sampling frequency) and by 2D-ordered subsets expectation maximization (16 subsets, 4 iterations) using software provided by the manufacturer. All images are reconstructed using a 128×128 matrix. Image pixel size is 0.85 mm transaxially with a 1.21-mm slice thickness.

Bias in structures surrounded by high background has been observed in images reconstructed using ordered subsets expectation maximization (OSEM).[33] This can especially be a problem in imaging the rat brain, as radioactivity uptake in extra cerebral structures such as the intraorbital lachrymal glands can be much higher than $[^{11}C]$raclopride uptake in brain tissue.[36] Studies evaluating OSEM reconstruction of rat brain images have shown that for a given resolution, the number of iterations required to reconstruct images of the rat head increases compared to the case of imaging the brain outside the head, resulting in increased noise in the image.[28] Other results from human brain FDG imaging suggest that OSEM data give unbiased results comparable with FBP.[37] Iterative methods have also been shown to improve signal detection characteristics using statistical parametric mapping.[34]

Results from both FBP and 2D OSEM reconstructions of $[^{11}C]$raclopride in the rat brain are compared in Table II. These two groups were significantly different at $p < 0.06$ using a two-tailed paired t test. Only the default OSEM setting (16 subsets, 4 iterations) provided by the software of the microPET was evaluated. It is clear from Table II that calculations of the DVR in the rat brain with $[^{11}C]$raclopride are sensitive to differences in reconstruction methods. Further experiments are required to determine optimal OSEM parameters for imaging dopamine receptors in the rat brain with $[^{11}C]$raclopride. MicroPET R4 software also allows for additional iterations using standard maximum likelihood methods after ordered subsets processing.

Reconstruction processing time (including Fourier rebinning) for the 24 frames was approximately 5 min for FBP compared to 5 h for OSEM (4×16) on a Pentium (1 GHz) PC.

[33] R. Boellaard, A. van Lingen, and A. A. Lammerstma, *J. Nucl. Med.* **42**, 808 (2001).

[34] C. T. Mesina, R. Boellaard, G. Jongbloed *et al.*, *Neuroimage* **19**, 1170 (2003).

[35] M. Defrise, P. E. Kinahan, and D. W. Townsend, *IEEE Trans. Med. Imaging* **16**, 145 (1997).

[36] S. P. Hume, R. Myers, P. M. Bloomfield *et al.*, *Synapse* **12**, 47 (1992).

[37] K. Oda, H. Toyama, K. Uemura *et al.*, *Ann. Nucl. Med.* **15**, 417 (2001).

TABLE II
EFFECT OF RECONSTRUCTION METHOD ON DVR

Study	2D FBP[a]	2D OSEM[b]	%Change
1	1.909	1.936	1.39
2	2.708	3.011	10.1
3	2.014	2.102	4.19
4	2.631	2.710	2.91
5	2.684	2.848	5.76

[a] Two-dimensional filtered back projection using ramp filter at Nyquist criteria after Fourier rebinning (128×128 image matrix).
[b] Two-dimensional ordered subset expectation maximization (16 subsets, 4 iterations) after Fourier rebinning (128×128 image matrix).

Graphical Analysis of Region of Interest Data

MicroPET images are analyzed using regions of interest (ROIs) created with software provided with the camera (ASIPro v.3.2) as described previously.[27] Briefly, elliptical regions of interest are drawn manually on a single plane of striatum and a single plane of the cerebellum (30 pixels). Separate ROIs are drawn on the left and right striatum (15 pixels each). The rat striatum is clearly visible as a bilateral structure on four to five coronal planes when imaged using high specific activity (<10 nmol/kg) [^{11}C]raclopride. Typical time–activity curves are depicted in Fig. 1 for the striatum and cerebellum.

Many physiological and molecular processes influence the uptake, distribution, and clearance of PET radiotracers. In order to compare specific measures such as apparent receptor concentration or affinity between subjects or scans, various mathematical techniques have been employed to isolate these parameters from kinetic data similar to those shown in Fig. 1. Many of these methods are based on a deconstruction of physiological processes into a small set of abstract, uniform compartments. These ideal compartments often represent a combination of real molecular processes. Under conditions of biological steady state during the imaging session, concentration in model compartments as a function of time can be described by a set a first-order linear equations (for examples, see Ichise et al.[38]). Solving this set of linear equations using standard methods yields values for model constants describing an exchange between compartments (e.g., receptor–ligand dissociation constant) or the concentration of a specific binding compartment (e.g., receptor binding). Often a combination of

[38] M. Ichise, J. H. Meyer, and Y. Yonekura, *J. Nucl. Med.* **42,** 755 (2001).

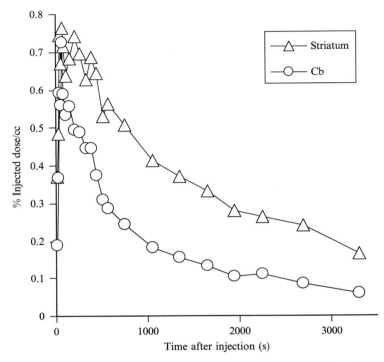

FIG. 1. MicroPET R4 time–activity data following injection of 170 μCi[^{11}C]raclopride (18 nmol/kg) in the lateral tail vein of an adult Sprague–Dawley rat.

model parameters is more useful to compare across subjects than individual parameters because of the variability in derived values (see Mintun et al.[39] for a discussion of binding potential). One of these combined parameters is the distribution volume ratio. The DVR can be calculated from compartmental model parameters or directly using graphical techniques. Software for either compartmental modeling or graphical analysis of ROI data was not included with the microPET R4. Software for kinetic analysis has been created in our laboratory over many years of research using clinical PET cameras. This software was applied directly to microPET R4 data for this work. The graphical analysis method used[40] is described briefly.

The simplicity and model independence of graphical methods make them attractive for PET data analysis. Although graphical methods are independent of any particular model structure, the slope can be interpreted

[39] M. A. Mintun, M. E. Racichle, M. R. Kilbourn et al., Ann. Neurol. 15, 217 (1984).
[40] J. Logan, J. S. Fowler, N. D. Volkow et al., J. Cereb. Blood Flow Metab. 16, 834 (1996).

in terms of a combination of model parameters for some model structure. The original graphical analysis method [Eq. (1)] requires plasma tracer measurements [Cp(t)] over the course of the experiment, as well as tissue uptake [ROI(t)] to calculate the distribution volume (DV) of the region of interest:

$$\frac{\int_0^t \text{ROI}(t')dt'}{\text{ROI}(t)} = \text{DV}\frac{\int_0^t \text{Cp}(t')dt'}{\text{ROI}(t)} + \text{int} \tag{1}$$

A plot of $\frac{\int_0^t \text{ROI}(t')dt'}{\text{ROI}(t)}$ vs $\frac{\int_0^t \text{Cp}(t')dt'}{\text{ROI}(t)}$ becomes linear for times $t > t^*$, and the slope is the total tissue distribution volume. By using a reference region in place of the plasma integral, graphical analysis can be extended to obtain DV ratios directly without blood sampling. This is particularly useful for microPET studies where it is difficult to obtain plasma sampling. By rearranging the graphical analysis equation, the plasma integral can be expressed in terms of the reference region radioactivity [REF(t)]. Rearranging Eq. (1) where REF is used in place of ROI gives

$$\int_0^t \text{Cp}(t')dt = \frac{1}{\text{DV}^{\text{REF}}}\left[\int_0^t \text{REF}(t')dt' - \text{int}^{\text{REF}}\,\text{REF}(t)\right] \tag{2}$$

Substituting for the integral of the plasma in the equation for the receptor ROI in Eq. (1) gives

$$\frac{\int_0^t \text{ROI}(t')dt'}{\text{ROI}(t)} = \frac{\text{DV}}{\text{DV}^{\text{REF}}}\left[\frac{\int_0^t \text{REF}(t')dt' - \text{int}^{\text{REF}}\text{REF}(t)}{\text{ROI}(t)}\right] + \text{int}^{\text{ROI}}$$

If the reference region is a one-tissue compartment model, then DV^{REF} is given by K_1/k_2 and int^{REF} is $-1/k_2$. This, however, is not a requirement for the method. For simplicity of expression, let $\text{int}^{\text{REF}} = -1/k_2'^{\text{REF}}$. In the case of a two-tissue compartment model for the reference region, this becomes

$$\frac{1}{k_2'^{\text{REF}}} = \frac{1}{k_2^{\text{REF}}}\left[\left(1 + \frac{k_3^{\text{REF}}}{k_4^{\text{REF}}}\right) + \frac{k_2^{\text{REF}}}{k_4^{\text{REF}}\left(1 + \frac{k_4^{\text{REF}}}{k_3^{\text{REF}}}\right)}\right]$$

where the error term is $\delta = \text{DVR}\left(\frac{1}{k_2^{\text{REF}}} - \frac{1}{k_2^{\text{REF}}}\right)\frac{\text{REF}(t)}{\text{ROI}(t)}$. The operational equation is

$$\frac{\int_0^t \text{ROI}(t')dt'}{\text{ROI}(t)} = \text{DVR}\left[\frac{\int_0^t \text{REF}(t')dt' + \frac{\text{REF}(t)}{k_2'^{\text{REF}}}}{\text{ROI}(t)}\right] + \text{int}' \tag{3}$$

In many cases (e.g., raclopride) the term $\text{REF}(t)/\bar{k}_2'^{\text{REF}}$ may be omitted.

The DVR or binding potential, which can be derived from the DVR, is used for comparing data sets. If the DV of the reference region has one tissue compartment, then the binding potential (BP) is related to the DVR by

$$DVR = \frac{\lambda^{ROI}}{\lambda^{REF}} (1 + BP) \text{ so that if } \lambda^{REF} = \lambda^{ROI}, \; BP = DVR - 1.$$

Summary of Results

Using the method described in this Chapter, the binding potential was computed for 15 $[^{11}C]$raclopride studies in the rat striatum using the micro-PET R4. In two of these studies, a 1-mg/kg blocking dose of raclopride was coadministered with the tracer to measure nonspecific binding of the tracer. The remaining studies were carried out with varying doses of $[^{11}C]$raclopride of varying specific activity. Data are summarized in Fig. 2 plotted as the binding potential as a function of raclopride mass injected. The curve is the best fit to a single-site ligand–receptor model proposed by Hume et al.[22] allowing for fitting of the modified equilibrium constants $appB_{max}$ and $appK_d$. Fitted values of $appB_{max}$ and $appK_d$ were 16.0 and 12.5 nmol/kg, respectively. Hume et al.[22] reported values of 8.7 and 17.1 nmol/kg in the rat striatum using a clinical PET camera with a reconstructed resolution of 5.5 mm in the center of the field of view. Image data from Hume et al.[22] were corrected for spillover from the intraorbital lachrymal glands but were not corrected for partial volume averaging. The microPET R4, with a FWHM resolution about half that of Hume et al.,[22] is expected to give a truer measure of striatal $[^{11}C]$raclopride concentration and therefore gives a larger binding potential (B_{max}/K_d). The increased resolution of a dedicated animal camera in this case should allow for increased sensitivity of in vivo determinations of $appB_{max}$ and $appK_d$ to pharmacological challenges and/or animal models of human disease.

Determinations of binding potential using the graphical method were compared with values obtained using a compartmental model without blood sampling.[41] Results are given in Table III. A paired two-tailed t test showed that values of DVR determined from the different methods were not significantly different ($p = 0.828$). These results suggest that assumptions used in both methods are not violated for $[^{11}C]$raclopride in the rat brain. Previous studies and analyses in humans have also confirmed that graphical methods and SRTM give nearly identical results.[42]

---

[41] A. Lammertsma and S. Hume, *Neuroimage* **4**, 153 (1996).

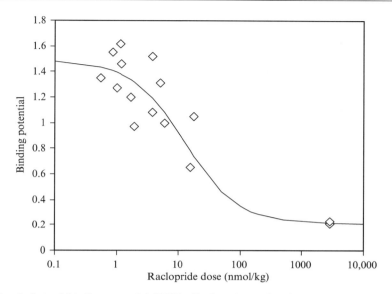

FIG. 2. Striatal binding potential (DVR – 1) plotted as a function of raclopride-injected mass calculated from specific activity measurements during tracer production. No additional raclopride mass was added except for the two high dose points where 1 mg/kg raclopride was coinjected with tracer. The solid line is a best fit to single-site binding model proposed by Hume et al.[22]

TABLE III
COMPARISON OF GRAPHICAL AND KINETIC MODEL DETERMINATIONS OF DVR

Study	Graphical DVR	SRTM^a DVR(BP + 1)
1	1.935	1.922
2	2.706	2.570
3	2.028	2.049
4	2.650	2.591
5	2.764	2.899

^{a} Simplified reference tissue method.[41]

Taken together, these results support the use of the noninvasive graphical analysis methodology with [^{11}C]raclopride and the microPET R4 to measure dopamine receptor availability in the rat striatum. Extension of the graphical method and its assumptions to rodent imaging, together with

[42] V. Sossi, J. E. Holden, G. Chan et al., J. Cereb. Blood Flow Metab. 20, 653 (2000).

the availability of commercial small animal PET cameras, should facilitate the application of small animal PET both in preclinical investigations of new drugs and in experimental studies of the relationship of receptor occupancy to animal behavior.

Acknowledgments

This work was carried out at Brookhaven National Laboratory under Contract DE-AC02-98CH10886 with the U.S. Department of Energy and supported by its Office of Biological and Environmental Research and the National Institutes of Health (EB002630).

[13]   Quantitative Analysis of Dopamine D_2 Receptor Kinetics with Small Animal Positron Emission Tomography

By Susanne Nikolaus, Markus Beu, Henning Vosberg, Hans-Wilhelm Müller, and Rolf Larisch

Introduction

During the last decade, small animal positron emission tomography (PET) has been employed increasingly for the investigation of dopamine D_2 receptor binding in animal models of reinforcement,[1–3] senescence,[4–6] and neurodegenerative disorders, including Parkinson's and Huntington's disease.[7–12]

[1] H. Tsukada, J. Kreuter, C. E. Maggos, E. M. Unterwald, T. Kakiuchi, S. Nishiyama, M. Futatsubashi, and M. J. Kreek, *J. Neurosci.* **16,** 7670 (1996).

[2] E. M. Unterwald, H. Tsukada, T. Kakiuchi, T. Kosugi, S. Nishiyama, and M. J. Kreek, *Brain Res.* **775,** 183 (1997).

[3] C. E. Maggos, H. Tsukada, T. Kakiuchi, S. Nishiyama, J. H. Myers, J. Kreuter, S. D. Schlussman, E. M. Unterwald, A. Ho, and M. J. Kreek, *Neuropsychopharmacology* **19,** 146 (1998).

[4] O. Ogawa, H. Umegaki, K. Ishiwata, Y. Asai, H. Ikari, K. Oda, H. Toyama, D. K. Ingram, G. S. Roth, A. Iguchi, and M. Senda, *Neuroreport* **11,** 743 (2000).

[5] M. Suzuki, K. Hatano, Y. Sakiyama, J. Kawasumi, T. Kato, and K. Ito, *Synapse* **41,** 285 (2001).

[6] H. Umegaki, K. Ishiwata, O. Ogawa, D. K. Ingram, G. S. Roth, J. Yoshimura, K. Oda, H. Matsui-Hirai, H. Ikari, A. Iguchi, and M. Senda, *Synapse* **43,** 195 (2002).

[7] J. Opacka-Juffry, S. Ashworth, R. G. Ahier, and S. P. Hume, *J. Neural Transm.* **105,** 349 (1998).

[8] E. M. Torres, R. A. Fricker, S. P. Hume, R. Myers, J. Opacka-Juffry, S. Ashworth, D. J. Brooks, and S. B. Dunnett, *Neuroreport* **6,** 2017 (1995).

A prerequisite for the accurate derivation of receptor-binding parameters with *in vivo* imaging methods is the measurement of the arterial free ligand concentration. Arterial input functions, however, are difficult to obtain in investigations performed on rat or mouse models, as sufficient amounts of blood may not be drawn repeatedly from animals of this small size. In the studies performed until now, the need for an arterial input function was avoided either by confinement to a semiquantitative approach expressing radioactive values as a percentage of the injected dose per tissue volume[4] or by applying dynamic,[1–3] noninvasive graphic,[5,11] and simplified reference tissue models.[6–10,12] For a thorough overview of the main concepts underlying model-based invasive as well as noninvasive quantification methods, the reader is referred to the excellent review article by Slifstein and Laruelle.[13]

As none of the described methods allow the separate assessment of B_{max} and K_D, we validated a method that applies saturation-binding analysis to *in vivo* imaging of small laboratory animals and thus allows the determination of K_D and B_{max} in analogy to *in vitro* experiments with nonlinear or linear (Scatchard) regression analysis.[14]

Model-based *in vivo* methods generally proceed from the notion of "compartment," which is a physiological or biochemical space characterized by homogeneous tracer concentrations, C(t). Thereby, the arterial vessel constitutes the first compartment, C_1, from which the radioligand enters the free tissue compartment, C_2. The free ligand in C_2 eventually binds to specific sites and, by this, enters the third compartment, C_3. Most radioligands also exhibit nonspecific binding. As equilibration between C_2 and the nonspecific compartment, C_2', is assumed to be rapid compared to the kinetics of specific binding, C_2 and C_2' are often pooled into one compartment, yielding a three-compartment configuration. In this model, the exchange of radioligand between compartments is governed by four rate constants, K_1 to k_4. When C_2 and C_2' are considered as separate compartments, a four-compartment configuration is obtained with six rate

[9] S. P. Hume, A. A. Lammertsma, R. Myers, S. Rajeswaran, P. M. Bloomfield, S. Ashworth, R. A. Fricker, E. M. Torres, I. Watson, and T. Jones, *J. Neurosci. Methods* **67,** 103 (1996).

[10] R. A. Fricker, E. M. Torres, S. P. Hume, R. Myers, J. Opacka-Juffry, S. Ashworth, D. J. Brooks, and S. B. Dunnett, *Neuroscience* **79,** 711 (1997).

[11] T. V. Nguyen, A. L. Brownell, Y. C. I. Chen, E. Livni, J. T. Coyle, B. R. Rosen, F. Cavagna, and B. G. Jenkins, *Synapse* **36,** 57 (2000).

[12] D. M. Araujo, S. R. Cherry, K. J. Tatsukawa, T. Toyokuni, and H. I. Kornblum, *Exp. Neurol.* **166,** 287 (2000).

[13] M. Slifstein and M. Laruelle, *Nucl. Med. Biol.* **28,** 595 (2001).

[14] S. Nikolaus, R. Larisch, M. Beu, F. Forutan, H. Vosberg, and H. W. Müller-Gärtner, *Eur. J. Nucl. Med.* **30,** 390 (2003).

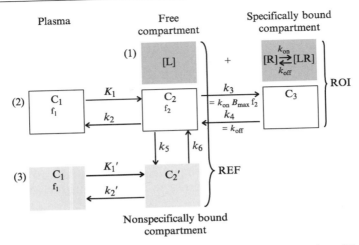

Fig. 1. Two (■, ▨)-, three (□)-, and four-compartment configurations (□ plus ▨), (1) *In vitro* two-compartment system consisting of the ligand-containing buffer solution and the receptor-containing tissue samples. (2) Three-compartment configuration consisting of arterial blood (C_1), free compartment (C_2), and receptor compartment (C_3). (3) Two-compartment configuration consisting of C_1 and the nonspecific compartment (C_2'). In the three-compartment model, C_2' and C_2 are pooled, whereas in the four-compartment model, C_2' and C_2 are assumed to exchange. B_{max}, concentration of receptors (mol/g); C_1, plasma activity of unmetabolized radioligand (Bq/ml); C_2, tissue radioligand activity in the free compartment (Bq/ml); C_2', tissue radioligand activity in the nonspecifically bound compartment /Bq/ml); C_3, tissue radioligand activity in the specifically bound compartment (Bq/ml); f_1, free fraction of unmetabolized radioligand in plasma; f_2, free fraction of radioligand in C_2; K_D, equilibrium dissociation constant with $K_D = k_{off}/k_{on}$, k_{on}, k_{off}, association [(g/ml × min)$^{-1}$] and dissociation (min^{-1}) rate constants, respectively; K_1, transfer constant from plasma to free compartment (min^{-1}); K_1, transfer constant from plasma to nonspecific compartment (min^{-1}); k_2, transfer constant from free compartment to plasma (min^{-1}); k_2, transfer constant from nonspecific compartment to plasma (min^{-1}); k_3, k_4, transfer constants between free and specifically bound compartment (min^{-1}); k_5, k_6, transfer constants between free and nonspecific compartment in the four-compartment configuration (min^{-1}); [L], concentration of unbound radioligand (Bq/ml); [LR], concentration of bound radioligand (Bq/ml); [R], concentration of unbound receptors (mol/g); REF, reference region; and ROI, region of interest.

constants, K_1 to k_6 (Fig. 1). *In vitro* radioligand-binding assays can be conceived as two-compartment systems with both compartments localized in the same test tube and representing the ligand-containing buffer solution and the receptor-containing tissue samples, respectively (Fig. 1). In transferring the principle of autoradiographic evaluation to *in vivo* imaging methods, the free ligand concentration, [L], can be assigned to the pooled compartments $C_2 + C_2'$, whereas the concentrations of unbound receptors, [R], and receptor–radioligand complexes, [LR], adhere to C_3. Specific binding is determined by subtracting the nonspecific and free radioactivity concentrations, [L], as measured in the reference region, REF, from total radioactivity concentrations, [LR], in the region of interest, ROI.

The interaction of receptor and radioligand can be described by

$$[L] + [R] \leftrightarrow [LR] \tag{1}$$

As this reaction is governed by the association and dissociation constants k_{on} and k_{off}, at equilibrium

$$k_{on}[L][R] = k_{off}[LR] \tag{2}$$

according to the laws of mass reaction. Because the equilibrium dissociation constant, K_D, is defined as k_{off}/k_{on} and the receptor density, B_{max}, is $[R] + [LR]$, substitution in Eq. (2) and rearrangement yields

$$[L][R] = \frac{B_{max}[L]}{K_D + [L]} \tag{3}$$

At tracer dose, with $[L] \ll K_D$, the ratio of bound to free radioligand is obtained by rearrangement of Eq. (3) with

$$\frac{[LR]}{[L]} = \frac{B_{max}}{K_D} \tag{4}$$

This last expression is referred to as the binding potential, BP. When [LR] is plotted against [L], B_{max} and K_D as in *in vitro* saturation-binding analysis may be derived from the resulting hyperbolic curve by nonlinear regression analysis.

The saturation-binding approach per definition requires the saturation of receptor binding and, thus, the application of increasing radioligand concentrations. As a consequence, varying and low specific activities, which may confound the performance of PET investigations according to the tracer principle, may become advantageous when the method of saturation-binding analysis can be applied. This approach is described for the investigation of striatal D₂ receptor binding in the rat with the D₂ receptor antagonist N-[¹⁸F]-methlybenperidol (FMB) and a dedicated small animal tomograph.

Material and Methods

Instrumentation

The small animal tomograph was developed at the Central Laboratory for Electronics, Research Center Jülich, Jülich, Germany ("TierPET"[15]). The camera consists of two orthogonal pairs of detectors mounted on an aluminium wheel. Each detector block contains a matrix of 400 yttrium aluminium perovskite scintillator crystals ($2 \times 2 \times 15$ mm^3; Preciosa Crytur Ltd., Turnov, Czech Republic), which are coupled to a position-sensitive photomultiplier (R2487; Hamamatsu Photonics, Herrsching, Germany). Axial and transaxial fields of view (FOV) have a diameter of 40 mm. The sensitivity is 3.24 cps/kBq for a center-detector distance of 80 mm. The spatial resolution is 2.1 mm [full width at half-maximum (FWHM)]. One cps/mm^3, as registered with the PET camera, corresponds to 444 Bq/mm^2.[16]

Radiochemistry

FMB is synthesized as described previously.[17] Radiochemical purity is assessed with high-performance liquid chromatography and, routinely, exceeds 98%. For the performance of saturation-binding analyses, the specific activity of the radioligand should cover one order of magnitude. In our validation study, specific activity ranged from >11 to >100 TBq/mmol.

Protocol

Anesthesia. For short-time inhalation anesthesia, the rat is placed into a lockable acrylic box containing a pad moistened with 1-chloro-2,2,2-trifluoro-ethyldifluoromethylether (Forene; Abbott GmbH, Wiesbaden, Germany). The anesthetized animal is administered a mixture of ketaminehydrochloride (Ketavet; Pharmacia GmbH, Erlangen, Germany; concentration, 100 mg/ml; dose, 0.9 ml/kg) and xylazinehydrochloride (Rompun; Bayer Vital GmbH, Leverkusen, Germany; concentration, 0.02 mg/ml; dose, 0.4 ml/kg) into the gluteal muscle.

Radioligand Application. Upon shaving and skin disinfection, the thorax of the rat is opened on the right side at the level of the third rib. FMB is diluted in 0.9% saline containing 10% ethanol and is injected into the right jugular vein. Thereby, application is facilitated if the needle (Microlance 25G5/8; Becton Dickinson, Drogheda, Ireland) is fixed in the pectoral muscle. Injection into the tail vein is also feasible. In this case, we recommend usage of a winged needle infusion set (Butterfly-23 with

[15] S. Weber, A. Terstegge, H. Herzog, R. Reinartz, P. Reinhart, F. Rongen, H. W. Muller-Gartner, and H. Halling, *IEEE Trans. Med. Imaging* **16**, 684 (1997).
[16] S. Nikolaus, R. Larisch, M. Beu, H. Vosberg, and H. W. Müller-Gärtner, *J. Nucl. Med.* **42**, 1691 (2001).
[17] S. M. Moerlein, J. S. Perlmutter, J. Markham, and M. J. Welch, *J. Cereb. Blood Flow Metab.* **17**, 833 (1997).

an outer needle diameter of 0.6 mm; Venisystems, Abbott, Sligo, Ireland). The injected amount of radioactivity should be adjusted to the performance characteristics of the employed small animal tomograph. With the employed small animal scanner, a rat has to be administered at least 74 MBq in order to achieve a sufficient count rate. In our study, the injected radioactivity dose amounted to 158 ± 24 MBq/kg (mean $\pm$ SD).

Data Acquisition. After surgical dressing, the rat is positioned on the object tablet with an associated acrylic head holder comprising ear and tooth bars. The head holder (Institute of Medicine, Research Center Jülich, Jülich, Germany) allows both the fixation and the accurate and reproducible positioning of the rat head. The motor-controlled object tablet is moved along the x, y, and z axes such that the striata are localized centrally within the FOV. During scanning of the brain, the body of the animal is kept within a lead tube (wall thickness, 20 mm) in order to reduce perturbations due to scattered photons. The tube is perfused consistently by warm water to maintain a body temperature of 37°. Images are acquired for 36 min (six time frames of 6 min each) with angular steps of 7.5° (30 s per angular step).

Evaluation

Images are analyzed with the multipurpose imaging tool (version 2.57; Advanced Tomo Vision, Erfstadt, Germany). After summation of the six time frames, striata are localized individually for each animal on coronal sections. Two standard circular ROIs with diameters of 2.5 mm are drawn around the centers of both striata. The positions of the ROIs are adjusted individually for each animal. The obtained radioactivity values are corrected for radioactive decay. As time–activity curves have shown that the equilibrium between radioactivity concentrations of specifically bound and both nonspecifically bound and free FMB is reached at 20 min postinjection, left and right radioactivity concentrations in time frames four to six are averaged.[18]

The investigation of small targets is biased by the partial-volume effect. This nonlinear artifact leads to an underestimation of the radioactivity concentration in objects smaller than twice the resolution of the employed scanning device. On the coronal sections, where the ROIs have been delineated, the mesiolateral striatal diameters amount to ~2.5 mm, which is in the range of the spatial resolution of the system. Phantom studies with

[18] H. Ito, J. Hietala, G. Blomqvist, C. Halldin, and L. Farde, *J. Cereb. Blood Flow Metab.* **18,** 941 (1998).

water-filled cylinders of various radii have shown that in a target with a diameter of 2.5 mm, radioactivity is underestimated by approximately 60%. On the basis of these measurements, the mean radioactivity values of time frames four to six are fixed to 60% of the true striatal radioactivity, and the respective 100% values are calculated (see discussion later).

Radioligand accumulation in the retroorbital Harderian glands may lead to an overestimation of radioactivity concentrations in adjacent tissues due to spillover effects. To estimate the various amounts of spillover affecting striatal and cortical radioactivity concentrations, Gaussian model functions (Fig. 2; $y = A_{HG}\exp(-0.5[(x - m_{HG})/s_{HG}]^2 + A_{STR}\exp(-0.5[(x - m_{STR})/s_{STR}]^2 + A_{COR}\exp(-0.5[(x - m_{COR})/s_{COR}]^2$ are fitted to line-activity profiles through the Harderian gland, striatum, and adjacent occipital cortex with A_{HG}, A_{STR}, and A_{COR} as normalization factors; m_{HG}, m_{STR}, and m_{COR} as the x coordinates of the peaks; and s_{HG}, s_{STR}, and s_{COR} as the FWHM values of glandular, striatal, and cortical curves. After decomposition of the sum function into a glandular, striatal, and cortical portion, the amount of spillover is determined by relating the integrals of the overlaps to the integrals of the curves. The percentual contributions of one curve to another are subtracted for each side from the mean striatal radioactivity concentrations.[19]

Cortical radioactivity as obtained from Gaussian fitting provides an estimate of the free and nonspecifically bound radioligand in brain tissue (see discussion later). From striatal and cortical radioactivity concentrations (MBq/mm^3) and the known specific activity at the beginning of the fourth time frame, both striatal and cortical molar radioligand concentrations ($fmol/mg$) are calculated. For each animal, cortical molar concentrations are subtracted from left and right striatal radioligand concentrations in order to obtain specific binding. Specific binding obtained for left and right striatum is averaged.

For nonlinear regression analysis (Fig. 3; GraphPad Prism, GraphPad Software, San Diego, CA), saturation-binding curves are generated by fitting the hyperbolic function $y = \frac{B_{max}x}{K_D+x}$ to the data set with x and y corresponding to the concentrations of free, [L], and bound radioligand, [LR], respectively. For linear regression analysis, the straight line function $y' = -\frac{1}{K_D}x' + \frac{B_{max}}{K_D}$ is fitted to Scatchard transformed data with x' and y' corresponding to [LR] and [LR]/[L], respectively. Thereby, the x intercept of the Scatchard line represents B_{max}, whereas the slope equals $-1/K_D$.

[19] M. Beu, S. Nikolaus, R. Larisch, S. Weber, H. Vosberg, and H. W. Müller-Gärtner, *Nuklearmedizin* **39**, 162 (2000).

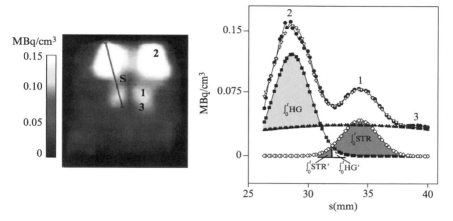

FIG. 2. Characteristic transversal slice of a Sprague–Dawley rat head obtained with a PET camera (left). In order to estimate the influence of the retroorbital radioactivity on the determination of striatal radioactivity concentrations, Gaussian model functions (∇) are fitted to the line activity profiles (s; ●) of both striata (1) and Harderian glands (2) of each side (right). After decomposition of the sum function into the three components, "striatal" ($\diamond$), "retroorbital" (■), and "background" radioactivity" (▲), the overlap ($\int_0^t STR'/\int_0^t HG$ and $\int_0^t HG/\int_0^t STR$) between striatal and orbital curves is taken as a measure for spillover. Cortical radioactivity (3) is used to estimate free and nonspecific binding. For each animal, mean cortical values are subtracted from striatal radioactivity concentrations as averaged over time frames four to six.

Results and Discussion

In our validation study, K_D and B_{max} values of FMB as determined with nonlinear regression analysis amount to 6.2 nM and 16 fmol/mg, respectively (Fig. 3). With linear regression analysis, K_D and B_{max} values of 5 nM and 15.3 fmol/mg, respectively, are obtained (Fig. 3, inset). *In vitro* assessment of K_D and B_{max} with storage phosphor autoradiography has yielded values within the same order of magnitude (K_D: 4.4 nM, B_{max}: 84 fmol/mg[20]). For the sake of comparability, the *in vivo* estimation of specific binding has also been applied to autoradiographic data. Thereby, the determination of specific binding in the striatum by subtraction of cortical binding instead of striatal binding in the presence of a competitor has led to similar results ($K_D = 7.9$ nM, $B_{max} = 70$ fmol/mg[20]). The agreement of

---

[20] S. Nikolaus, R. Larisch, M. Beu, K. Hamacher, F. Forutan, H. Vosberg, and H. W. Müller, *J. Nucl. Med.* **44**, 618 (2003).

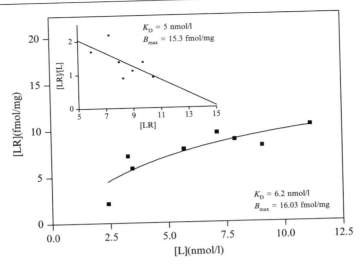

FIG. 3. Nonlinear regression analysis of FMB binding to striatal dopamine D_2 receptors. Binding data were acquired from eight Sprague–Dawley rats with an animal PET device. The free radioligand concentration in the occipital cortex was used as the concentration parameter. (Inset) Scatchard analysis of binding data contingent on linear transformation.

values shows that K_D and B_{max} can be measured in the living brain in analogy to *in vitro* saturation-binding studies.

For small animal scanners with the characteristics of the TierPET as to sensitivity and spatial resolution, we propose using cortical rather than cerebellar radioactivity concentrations as an estimate of free and nonspecific radioligand binding. Due to the low D_2 receptor concentration and the lack of anatomical landmarks it is difficult to delineate cerebellar ROIs in a reproducible manner. Cortical radioactivity concentrations are negligible as well[21] and have the advantage that they may be determined on the same slices used for striatal ROI definition (Fig. 2). Thereby, we recommend usage of the occipitocortical region, as previous investigations have shown that occipital D_2 receptor concentrations fall short of frontal ones by approximately one-third.[22] Because occipital D_2 receptor density is that low, radioactivity concentrations in this region can be assumed to provide a reasonable estimate of free and nonspecific radioligand concentrations in brain tissue.

[21] R. M. Kessler, N. S. Mason, J. R. Votaw, T. De Paulis, J. A. Clanton, M. S. Ansari, D. E. Schmidt, R. G. Manning, and R. L. Bell, *Eur. J. Pharmacol.* **223,** 105 (1992).

[22] M. S. Lidow, P. S. Goldman-Rakic, P. Rakic, and R. B. Innis, *Proc. Natl. Acad. Sci. USA* **86,** 6412 (1989).

In our *in vivo* saturation-binding analysis, K_D values correspond to *in vitro* results, whereas B_{max} values are lower compared to *in vitro* results, although still lying within the same order of magnitude. In comparing *in vivo* and *in vitro* measurements, various issues should be taken into account. First, the exact determination of the anterioposterior stereotaxic coordinates of the PET slices is hampered when no concurrent morphological imaging may be performed. Noncorresponding sectional planes between PET and autoradiography thus offer one explanation for the difference between receptor densities obtained *in vivo* and *in vitro*.

Second, the applied partial-volume correction factor is estimated on the basis of phantom studies and enters as a constant factor into the quantification of radioactivity values irrespective of the actual striatal diameters on the slices selected for ROI definition. Thus, if striatal diameters fall short of 2.5 mm, the used correction factor leads to underestimations of radioactivity concentrations compared to autoradiography.

Third, the accuracy of our method depends on the correctness of the assumption that C_2 and C_2' may be pooled into one compartment. Actually, we derive [L], i.e., both the free and the nonspecifically bound radioligand concentration, from the cortical reference region. Thus, in our approach, BP is given by $\frac{[LR]}{[L]} - 1$. As [L] corresponds to $C_2 + C_2'$, the nonspecifically bound radioligand enters into the denominator in addition to the free radioligand concentration. Thus, the correctness of parameter estimation suffers from the fact that the free fraction of radioligand in C_2, f_2, may not be determined accurately. Actually, with our method, we derive B'_{max} $(=f_2 B_{max})$ and thus underestimate B_{max} consistently.

Because a reference region is used for the estimation of $C_2 + C_2'$ in the simplified reference tissue model,[23] the portion of nonspecifically bound radioligand in this approach may also lead to an underestimation of BP. Actually, instead of B_{max}/K_D, $f_2 B_{max}/K_D$ is obtained as an outcome measure. With the dynamic approach,[24–26] the receptor parameter $k_3 = k_{on}(B_{max} - \frac{C_3(t)}{SA} f_2$ is estimated with SA denoting the specific activity of the radioligand. If SA is high and the radioligand is administered in tracer amounts, then $\frac{C_3(t)}{SA} \ll B_{max}$ and an estimate of k_3 is provided by the product of k_{on}, B_{max}, and f_2. Thereby, f_2 usually is inferred from *in vitro*

[23] A. A. Lammertsma and S. P. Hume, *Neuroimage* **4**, 153 (1996).

[24] M. A. Mintun, M. E. Raichle, M. R. Kilbourn, G. F. Wooten, and M. J. Welch, *Ann. Neurol.* **15**, 217 (1984).

[25] J. S. Perlmutter, K. B. Larson, M. E. Raichle, J. Markham, M. A. Mintun, M. R. Kilbourn, and M. J. Welch, *J. Cereb. Blood Flow Metab.* **6**, 154 (1986).

[26] D. F. Wong, A. Gjedde, H. N. Wagner, Jr., R. F. Dannals, K. H. Douglass, J. M. Links, and M. J. Kuhar, *J. Cereb. Blood Flow Metab.* **6**, 147 (1986).

studies or is estimated from the kinetics in tissue regions lacking specific binding. If the latter is the case, the product of B_{max} and k_{on} provides an estimate of BP, the goodness of which, as is the case for *in vivo* saturation binding and the simplified reference tissue approach, depends on the accuracy, with which f_2 may be determined. A similar problem occurs with the noninvasive graphic model,[27] which yields the distribution volume ratio, DVR, as an outcome measure. The DVR is the ratio of the equilibrium distribution volume in the ROI to that in the REF. Thereby, DVR = BP + 1 if the distribution volumes of REF and C_2 are assumed to be equal.[13] Thus, also in this approach, existing nonspecific radioligand binding leads to inaccurate BP estimates. As the tissue-to-plasma efflux constant, k_2, is required for this kind of analysis, the goodness of BP estimates furthermore may depend on the feasibility of inferring k_2 from an independent sample.

In discussing possible errors of parameter estimation, their degree, however, should be viewed in relation to other error sources inherent to the method of *in vivo* PET imaging, such as spillover and partial-volume effect. For the nonspecific binding moreover holds that it is characterized by both tissue and radioligand properties, which are not assumed to vary over time or between subjects. Thus, f_2B_{max} and K_D, f_2B_{max}/K_D, $f_2k_{on}B_{max}$, and DVR as obtained with *in vivo* saturation-binding analysis, the simplified reference tissue model, the dynamic approach, and the noninvasive graphic model, respectively, can be considered to provide reasonable estimates of the receptor parameters in question. Thereby, the advantage of *in vivo* saturation-binding analysis is the provision of $B_{max}(')$ and K_D as separate values and direct measures of receptor density and affinity.

In vivo saturation-binding analysis represents an equilibrium method, which for the first time was described by Phelps and collaborators.[28] So far, *in vivo* saturation-binding analysis with small animal tomographs has been applied in two studies: Nikolaus and collaborators[14] investigated time-dependent changes of B_{max} in the rat 6-hydroxydopamine model, whereas Tsukada and collaborators[1] utilized the saturation-binding approach as a complement to PET in order to analyze whether the observed effect on D_2 receptor binding was due to alterations in K_D or B_{max}. These investigations are examples for scientific questions requiring methods for the separate determination of K_D and B_{max} in small laboratory animals.

[27] J. Logan, J. S. Fowler, N. D. Volkow, G. J. Wang, Y. S. Ding, and D. L. Alexoff, *J. Cereb. Blood Flow Metab.* **16**, 834 (1996).

[28] M. E. Phelps, J. C. Mazziotta, and H. R. Schelbert, "Positron Emission Tomography and Autoradiography: Principles and Applications for the Brain and Heart." Raven Press, New York, 1986.

With the *in vivo* saturation-binding approach, the benefits of two concepts are brought together. By applying the *in vitro* method to *in vivo* imaging, K_D and B_{max} may be determined in analogy to *in vitro* experiments preserving all advantages of the *in vivo* method, such as the possibility to re-use an animal for several investigations and to assess receptor kinetics in the living subject instead of utilizing cryosections or membrane homogenates as mere models of cerebral tissue and receptor surface. Finally, an enormous disadvantage of *in vivo* methods, the struggle with varying and low specific activities, is changed to the opposite, as with the method of saturation binding, increasing concentrations of radioligand have to be applied in order to reach the state of receptor saturation.

It is important to note, however, that there is one fundamental difference between *in vivo* saturation-binding analysis yielding B_{max} and K_D and other approaches quantifying BP or related outcome measures: with the saturation-binding method, various radioligand concentrations must be applied to several animals. In contrast, in the other approaches, outcome measures may be obtained for *one* animal with *one* radioligand concentration. Thus, basically, with *in vivo* saturation-binding analysis, more animals are required, leading to a greater influence of interindividual variance. Theoretically, application of more than one concentration to a single animal is also feasible; in this case, individual values of both B_{max} and K_D may be determined for each animal. PET investigations, however, should lie a sufficient time apart to account for radioactive decay and to permit recovery from aftereffects of anesthesia and surgery.

Acknowledgments

This work was supported by a grant from the "Forschungskommission" of the Faculty of Medicine, Heinrich-Heine-University, Düsseldorf, Germany. The authors acknowledge Dr. Simone Weber from the Central Laboratory for Electronics and Dr. Karl Hamacher from the Institute of Nuclear Chemistry (Forschungszentrum Jülich GmbH, Jülich, Germany) for their contributions to the experiments.

[14]   Magnetic Resonance Imaging in Biomedical Research: Imaging of Drugs and Drug Effects

By Markus Rudin, Nicolau Beckmann, and Martin Rausch

Introduction

Magnetic resonance imaging (MRI) and spectroscopy (MRS) have become established technologies in modern biomedical research, providing relevant information at various stages of the drug discovery and development process.[1–3] Strengths of MRI are (1) *high soft tissue contrast*, which is governed by a multitude of parameters; (2) *noninvasiveness*, which is of relevance when studying chronic diseases and allows for translation of study protocols from animals to humans; and (3) the high *chemical specificity* of MRS, allowing the identification of individual analytes based on compound-specific resonance frequencies, an important prerequisite for the *in vivo* study of tissue metabolism. The principal disadvantage of MRI/MRS is the *inherently low sensitivity* due to the low quantum energy involved as compared to optical spectroscopy, leading to a low degree of spin polarization. Only nuclei with high natural abundance and high intrinsic magnetic moment, such as protons, provide sufficient signal intensity for imaging applications. In addition, exogenous contrast agents (CA, paramagnetic or superparamagnetic compounds) with different levels of specificity are often used to modulate local signal intensities. Whereas CAs are the only source of image signal in nuclear and optical imaging methods, MRI CAs only alter the intrinsic MR signal originating from tissue water. Hence, contrast-enhanced MRI methods always suffer from a high background signal, reducing the sensitivity of the approach.

These properties of MRI/MRS have determined the applications in pharmaceutical research: The focus of applications in the last decade(s) has been the quantitative assessment of drug effects on tissue morphology, physiology, and metabolism in animal models of human disease or in patients.[1–3] It is not the scope of this Chapter to provide yet another review; instead, it discusses some methodological aspects of applying MRI/MRS to biomedical research.

[1] M. Rudin, N. Beckmann, R. Porszasz, T. Reese, T. Bochelen, and A. Sauter, *NMR Biomed.* **12,** 69 (1999).

[2] N. Beckmann, R. P. Hof, and M. Rudin, *NMR Biomed.* **13,** 329 (2000).

[3] N. Beckmann, T. Mueggler, P. R. Allegrini, D. Laurent, and M. Rudin, *Anat. Rec.* **265,** 85 (2001).

Preclinical *in vivo* characterization of a drug candidate comprises the study of its pharmacokinetic (PK) and pharmacodynamic properties. MRI/MRS was applied predominantly during late phases of the discovery process, such as the optimization of a lead compound, the profiling of a potential development candidate, and during early clinical development. More recently, molecular imaging approaches based on MRI involving target-specific contrast principles have been described that provide information relevant for target validation or the elucidation of molecular pathways.[4–6] These novel MRI approaches are of high potential value; nevertheless, they are severely limited by the low sensitivity of MRI, requiring high amplification of the molecular signal, and by the fact that MRI contrast agents are, in general, bulky and not delivered readily to the molecular target.

This Chapter focuses on conventional MRI/MRS applications, discussing their role in pharmacokinetic and -dynamic studies, and addresses some issues related to the development of MRI-based biomarkers.

Drug Imaging and Pharmacokinetic Studies

The inherently low sensitivity of nuclear magnetic resonance has prevented any widespread attempts of using the technique for mapping the biodistribution of drug candidates *in vivo*. Let the detection limit for a given acquisition time be $p_l = 10^{12}$ spins. At a magnetic field strength of 9.4 T (proton resonance frequency $\nu_0 = 400$ MHz) the polarization of proton spins at room temperature is $\Delta p = \exp(-h\nu_0/kT) \approx 10^{-5}$ (i.e., out of 10^5 spins there is an excess of only one oriented parallel to the field and, hence, detectable). Conventional MRI maps the distribution of water protons in tissue, with the average tissue proton concentration being approximately $c_t = 80$ M. For the detection limit of 10^{12} spins, this then corresponds to a volume of

$$V = \frac{10^6 [\mu l] \cdot p_l}{c_t \cdot N_A \Delta p} = 2 \times 10^{-4} \mu l \tag{1}$$

with N_A being Avogadro's number. This volume element (voxel) translates into linear pixel dimensions of 120 μm, a typical value for animal MRI studies. For endogenous metabolites studied commonly by MRS or

[4] M. Rudin and R. Weissleder, *Nat. Rev. Drug Disc.* **2,** 123 (2003).
[5] A. Y. Louie, M. M. Huber, E. T. Ahrens, U. Rothbacher, R. Moats, R. E. Jacobs, S. E. Fraser, and T. J. Meade, *Nat. Biotechnol.* **18,** 321 (2000).
[6] R. Weissleder, A. Moore, U. Mahmood, R. Bhorade, H. Benveniste, E. A. Chiocca, and J. P. Basilion, *Nat. Med.* **6,** 351 (2000).

spectroscopic imaging methods, the tissue concentration is of the order of $c_t = 1$ mM and, assuming the same data acquisition protocol, the volume would increase to 170 μl, corresponding to a voxel of (5.5 mm)^3.

We now apply this simple consideration to estimate drug levels that can be detected in a volume corresponding to the whole rat brain (2 ml) or rat liver (15 ml). For a compound with a molecular mass of 500 Da, doses of $c_t = 40$ and 6 mg/kg should be detectable, corresponding to tissue concentrations of 80 and 12 μM for brain and liver, respectively, assuming homogeneous distribution throughout the body and neglecting any concurrent drug clearance. However, this would be only feasible when drug signals are well separated from the signals of the endogenous metabolites at millimolar concentration, which is not the case for proton spectroscopy due to a limited chemical shift dispersion.

This is the reason why essentially all NMR drug biodistribution studies reported to date focused on magnetic nuclei such as ^{13}C, ^{19}F, and ^{31}P. Fluorine is especially promising, as its magnetic properties are comparable to those of protons, whereas the degree of polarization and hence the sensitivity is 2.5-fold and four times weaker for ^{31}P and ^{13}C for the same number of magnetic nuclei, respectively.

Given these perspectives, the attempt to image drug biodistribution using MRI/MRS approaches is not very promising. In fact, the few spectroscopic studies reported to date have sampled large detection volumes. In these cases the spatial resolution is provided by the selectivity of the radiofrequency excitation/detection scheme using surface coils. Most of the examples relate to drugs used in oncology, and some of these pharmacokinetics studies using MRS have even been translated into the clinics. A compound studied with ^{19}F MRS is 5-fluorouracyl (5-FU).[7–10] Such studies allowed the prediction of patient response to 5-FU based on individual PK data.[8,9] ^{31}P MRS has been used for PK studies of ifofsamide,[11] whereas other cancer drugs have been investigated using ^{13}C[12] and ^{1}H MRS.[13] In all these studies, three favorable factors apply: (1) doses of the anticancer drugs administered were relatively high (100–800 mg/kg), (2) the volume

[7] M. J. W. Prior, R. J. Maxwell, and J. R. Griffiths, *Biochem. Pharmacol.* **39,** 857 (1990).

[8] M. P. N. Findlay, M. O. Leach, D. Cunningham, D. J. Collins, G. S. Payne, J. Glaholm, J. L. Mansi, and V. R. McCready, *Ann. Oncol.* **4,** 497 (1993).

[9] H. P. Schlemmer, P. Bachert, W. Semmler, P. Hohenberger, P. Schlag, W. J. Lorenz, and G. vanKaick, *Magn. Reson. Imaging* **12,** 497 (1994).

[10] P. M. McSheehy, S. P. Robinson, A. S. E. Ojugo, E. O. Aboagye, M. B. Cannell, M. O. Leach, I. R. Judson, and J. R. Griffiths, *Cancer Res.* **58,** 1185 (1998).

[11] G. S. Payne, C. R. Pinkerton, and M. O. Leach, *Int. Soc. Magn. Reson. Med.* **7,** 1588 (1999).

[12] D. Artemov, Z. M. Bhujwalla, R. J. Maxwell, J. R. Griffiths, I. R. Judson, M. O. Leach, and J. D. Glickson, *Magn. Reson. Med.* **34,** 338 (1985).

of interest sampled in the study was large (>20 ml), and (3) high specificity was achieved by studying nonendogenous nuclei (^{19}F), by exploiting unique chemical shift properties (^{31}P), or by using spectral editing techniques to distinguish drug signal from metabolite resonances (^{13}C, ^{1}H).

The percentage of drugs containing fluorine or phosphorus, however, is low (5%), and the doses commonly used are well below the values discussed. Hence, *in vivo* MRS is of limited value for PK studies and, even in the most favorable cases, the spatial and temporal resolutions achieved are rather poor. It is therefore not surprising that the focus of using magnetic resonance techniques has been on the pharmacodynamic readouts (i.e., on the analysis of drug effects on tissue morphology, physiology/function, and endogenous metabolism).

Noninvasive Assessment of Drug Efficacy/Pharmacodynamic Studies

The vast majority of MRI/MRS applications in pharmacological research address pharmacodynamic effects of drugs in animal models of human diseases or in patients. The first step is the morphological, physiological/functional, and/or metabolic characterization of a disease phenotype on both a *qualitative* and a *quantitative* basis. The MRI signal behavior is governed by a variety of independent parameters, which are determined by the microstructural environment of the tissue water in a voxel. Those are the proton density, the various relaxation times [spin–lattice relaxation time (T_1), spin–spin relaxation or phase-memory time (T_2), free induction decay time (T_2^*)] the diffusion properties of the tissue water as characterized by the apparent diffusion coefficient (ADC) or when accounting for anisotropy by the diffusion tensor, incoherent motion within a voxel due to perfusion, and coherent motion due to macroscopic blood flow. A detailed description of the various MRI contrast mechanisms is beyond the scope of this Chapter and the reader is referred to the literature.[14] Pathology leads to morphological and physiological alterations and hence to concomitant changes of MRI contrast parameters. These changes are dynamic and will evolve as pathology evolves (Fig. 1) and may be used as a diagnostic tissue signature.[15] By suitable choice of the experimental parameters, the contrast-to-noise ratio (CNR) between a structure of

[13] Q. He, Z. M. Bhujwalla, R. J. Maxwell, J. R. Griffiths, and J. D. Glickson, *Magn. Reson. Med.* **33**, 414 (1995).

[14] F. W. Wehrli, J. R. MacFall, and T. H. Newton, *in* "Advanced Imaging Techniques" (T. H. Newton and D. G. Potts, eds.), p. 81, 117. Clavadel Press, San Anselmo, 1983.

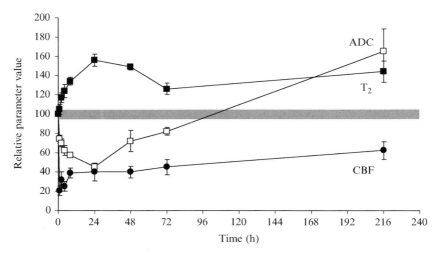

FIG. 1. Time dependence of MRI parameters in rat cerebral cortex in a model of human embolic stroke. Cerebral ischemia was induced by permanent occlusion of the left middle cerebral artery, leading to a reduction of cortical cerebral blood flow (CBF) to 20% of its contralateral value. Within minutes, the apparent water diffusion coefficient (ADC) in the affected area started to decrease, reaching a minimum of 50% of its baseline value at 24 h and pseudo-normalizing between 3 and 5 days. The most commonly used parameter for the quantitative morphometric analysis of infarct volume is the transverse relaxation time T_2, with contrast in T_2-weighted images being maximal at 24 and 48 h postinfarction. The relative profile defined by the parameters (ADC, CBF, T_2) changes as a function of time and constitutes a signature reflecting the tissue state. All values are given relative to the respective value of the contralateral, normal hemisphere (indicated by the gray line). Values represent mean ± SEM (see Rudin et al.[16]).

interest and its environment can be optimized, explaining the high value of MRI as a diagnostic tool.

Qualitative Characterization of Disease Phenotype and Assessment of Drug Efficacy

Qualitative Characterization. Qualitative characterization of a disease phenotype is the basis for clinical diagnosis. The multivariate nature of the MRI signal often allows for staging of the disease, which is of relevance

[15] K. M. A. Welch, J. Windham, R. A. Knight, V. Nagesh, J. W. Hugg, M. Jacobs, D. Peck, P. Booker, M. O. Dereski, and S. R. Levine, *Stroke* **26,** 1983 (1995).
[16] M. Rudin, D. Baumann, D. Ekatodramis, R. Stirnimann, K. H. McAllister, and A. Sauter, *Exp. Neurol.* **169,** 56 (2001).

for the stratification of a patient/animal population to be included in drug evaluation studies. Qualitative or semiquantitative tissue characterization can also be applied to assess therapy response. As an example, the efficacy of immunosuppressive treatment in a rat kidney allograft transplantation model has been evaluated using a score accounting for morphological tissue appearance on MR images.[17] As in qualitative clinical diagnostics, the quality of such results critically depends on the skills of the interpreter. Analyses have to be carried out blindly, and even then it cannot be excluded that different operators will arrive at different conclusions.

Quantitative Analysis: Morphometric and Physiological Imaging

In Vivo *Morphometry.* Quantitative analysis of biomedical imaging data is based on morphometric or densitometric measures. Morphometric readouts are, for instance, the volume of a pathology, such as the infarct volume in stroke or the tumor volume in oncology studies, or distance measures, such as the thickness of articular cartilage in models of arthritis. Preferentially, such measures are carried out in an automated fashion, i.e., with minimal operator interaction. The critical step in morphometric image analysis is image segmentation, which is relatively simple for high CNR, allowing for segmentation based on intensity thresholds (Fig. 2A). However, due to limited CNR, this approach is not generally applicable, and there is still a need to develop automated three-dimensional segmentation algorithms. For current practical applications, operator interaction is still required. Nevertheless, reproducibility is, in general, good so that the uncertainty introduced by the methodology is significantly smaller than the biological variability (Fig. 2B).

Quantitative structural information may also be derived by determining the relative or absolute values of MRI contrast parameters. Alteration of T_2 and ADC values in ischemic brain tissue following cerebral infarction indicates the severity of the ischemic insult[16]; regional normalization of these values following cytoprotective therapy reflects drug efficacy.[18]

Physiological Parameters. Quantitative densitometric analysis is applied in order to derive physiological or functional information from dynamic MRI data sets. In such experiments the MRI signal intensity is monitored in response to a pharmacological or physiological challenge or to the passage of an exogenous contrast agent. Derivation of physiological information from dynamic MRI data sets involves modeling within the

[17] N. Beckmann, J. Joergensen, K. Bruttel, M. Rudin, and H. J. Schuurman, *Transpl. Int.* **9**, 175 (1996).
[18] A. Sauter and M. Rudin, *Stroke* **17**, 1228 (1986).

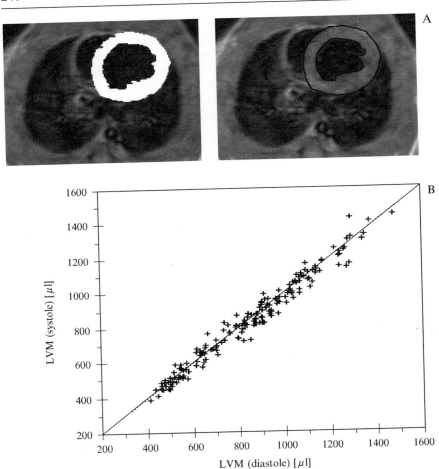

Fig. 2. *In vivo* morphometric analysis of MR images of the rat heart. One section of a transverse multislice data set covering the heart is shown. The myocardial left ventricular mass (LVM) is derived by determining the cross-sectional area in each slice either by intensity-based segmentation (A, left) or by operator-interactive tracing of the structure boundary (A, right). By adding the selected areas in all slices and multiplying with the interslice separation and the tissue density (1.05 g/cm³), the LVM can be estimated. The reproducibility of the estimate is illustrated (B) comparing two subsequent analyses of the LVM mass, one during diastole and one during systole. The average standard deviation is 8% (see Rudin *et al.*[19]).

[19] M. Rudin, B. Pedersen, K. Umemura, and W. Zierhut, *Basic Res. Cardiol.* **86,** 615 (1991).

framework of physiological models. Three examples illustrate the approach.

TISSUE PERFUSION. Assessment of tissue perfusion is based on the tracer dilution method.[20] Administration of an intravascular contrast agent leads to a transient change in signal intensity in the tissue caused by a transient increase in the relaxation rate R_2 (and R_2^*), the amount of which is proportional to the local tissue concentration of the tracer $c_t(t)$,

$$\Delta R_2(t) = R_2(t) - R_{20} = \alpha_2 c_t(t) V \tag{2}$$

with R_{20} and $R_2(t)$ being the relaxation rates prior and at time t after administration of the contrast agent, α_2 is the molar relaxivity (in $M^{-1}s^{-1}$) of the contrast agent (accounting for the susceptibility difference $\Delta\chi$ between intra- and extravascular space induced by the contrast agent), and V is the local tissue blood volume. Analysis of the tracer profile $c_t(t)$ versus t allows deriving relative hemodynamic parameters such as tissue blood flow and tissue blood volume.[21] Determination of absolute hemodynamic parameters requires calibration of the perfusion maps by the arterial input function.[22]

Tissue perfusion is a critical parameter for tissue survival and function, and both relative and absolute perfusion assessments are highly relevant for both diagnosis and evaluation of the therapy response. Typical applications in drug discovery are studies of focal cerebral ischemia, (e.g., induced by occlusion of the middle cerebral artery in rats).[23,24]

TISSUE OXYGENATION. The delivery of oxygen and nutrients to tissue is critical for functional integrity or even survival; hence, the oxygenation level is a relevant parameter for the characterization of the tissue state. Ogawa et al.[25] observed significant increases in proton R_2^* relaxivity in rat brain during severe hypoxia, which were attributed to increased intravascular levels of deoxygenated hemoglobin. While oxygenated hemoglobin (Hb-O_2) is diamagnetic, deoxygenated hemoglobin (Hb) is paramagnetic and acts as an endogenous CA. The proton relaxation rate $\Delta R_2(=1/\Delta T_2)$ depends on the fraction of intravascular deoxyhemoglobin $(1-Y)$ according to[26]

[20] P. Meier and K. L. Zierler, J. Appl. Physiol. 6, 731 (1954).
[21] M. Rudin, N. Beckmann, and A. Sauter, Magn. Reson. Imaging 15, 551 (1997).
[22] M. Rausch, K. Scheffler, M. Rudin, and E. Radü, Magn. Reson. Imaging 18, 1235 (2000).
[23] A. Sauter, M. Rudin, and K. H. Wiederhold, Neurochem Pathol. 9, 211 (1988).
[24] A. Sauter, T. Reese, R. Pórszász, D. Baumann, M. Rausch, and M. Rudin, Magn. Reson. Med. 47, 759 (2002).
[25] S. Ogawa, T. M. Lee, A. R. Kay, and D. W. Tank, Proc. Natl. Acad. Sci. USA 87, 9868 (1990).
[26] K. Scheffler, E. Seifritz, R. Haselhorst, and D. Bilecen, Magn. Reson. Med. 42, 829 (1999).

$$\Delta R_2(f, Y) = \alpha \gamma f (1 - Y) \Delta \chi B_0 \qquad (3)$$

where α is a proportionality factor, γ is the proton gyromagnetic ratio, f is the blood volume fraction, $\Delta \chi$ is the susceptibility difference between fully deoxygenated and fully oxygenated blood, and B_0 is the static magnetic field. For human hemoglobin, Y is derived from the blood oxygenation curve according to[27]

$$\frac{Y}{1 - Y} = \left(\frac{pO_2}{P_{50}} \right)^{2.8} \qquad (4)$$

with pO_2 being the partial pressure of oxygen in blood and P_{50} the partial pressure at which 50% of the hemoglobin is oxygenated. The proton relaxation rate is highly sensitive to changes in pO_2 around normoxic levels. Determination of absolute tissue pO_2 values from measurements of MRI relaxation rates, however, is hardly feasible, as the signal intensity depends on geometrical factors describing the vascular system (vessel diameter, spacing, and orientation), the rate of water diffusion, and the static magnetic field strength.[27] For the majority of practical applications, tissue oxygenation is analyzed qualitatively or semiquantitatively (relative values). This is illustrated by some examples.

Blood oxygenation level-dependent (BOLD) contrast forms the basis of brain activity studies using functional MRI (fMRI).[28,29] Local changes in blood oxygenation are caused by neuronal activation through metabolic and hemodynamic coupling. fMRI is widely applied to study the functional architecture of the brain and for the characterization of pathologies of the central nervous system (CNS). fMRI is also used to study the CNS response to the administration of neuro-active compounds,[30,31] providing readouts comparable to glucose utilization measurements using positron emission tomography (PET). More recently, fMRI has been applied to study functional recovery following cytoprotective treatment in a rat model of human embolic stroke.[24]

[27] R. P. Kennan, J. Zhong, and J. C. Gore, *Magn. Reson. Med.* **31,** 9 (1994).
[28] K. K. Kwong, J. W. Belliveau, D. A. Chesler, I. E. Goldberg, R. M. Weisskoff, B. P. Poncelet, D. N. Kennedy, B. E. Hoppel, M. S. Cohen, and R. Turner, *Proc. Natl. Acad. Sci. USA* **89,** 5675 (1992).
[29] S. Ogawa, D. W. Tank, R. Menon, J. M. Ellermann, S. G. Kim, H. Merkle, and K. Ugurbil, *Proc. Natl. Acad. Sci. USA* **89,** 5951 (1992).
[30] Y. C. I. Chen, W. R. Galpern, A. L. Brownell, R. T. Matthews, M. Bogdanov, O. Isacson, J. R. Keltner, M. F. Beal, B. R. Rosen, and B. G. Jenkins, *Magn. Reson. Med.* **38,** 389 (1997).
[31] T. Reese, B. Bjelke, R. Porszasz, D. Baumann, D. Bochelen, A. Sauter, and M. Rudin, *NMR Biomed.* **13,** 43 (2000).

Oxygenation (and perfusion) in neoplastic tissue is critical for therapy response, and knowledge of this physiological parameter would allow one to optimize treatment regimens.[32] BOLD MRI has indeed been applied successfully to assess tumor blood flow and oxygenation under a variety of conditions relevant to tumor therapy.[33] Knowledge of the tumor oxygenation state is also relevant when studying angiogenesis: The transcription factor hypoxia-inducible factor-1 (HIF-1) is a key regulator of oxygen homeostasis, with high HIF-1 levels stimulating the formation of neovasculature.[34]

Gas exchange is the major function of the lungs, and ventilation can be probed by analyzing changes in the relaxation rates of lung parenchyma induced by oxygen. Molecular oxygen has a triplet ground state, i.e., two unpaired electrons, is weakly paramagnetic, and thus constitutes a source of contrast in MRI. This property has been explored successfully to derive regional ventilation-related information from the human lung.[35,36] While qualitative and semiquantitative information is obtained readily, absolute quantification is again difficult; for example, the dissolution of O_2 in blood suggests that the tissue relaxation rates are also dependent on perfusion rates or blood volume, in addition to the degree of tissue oxygenation.[37]

VASCULAR PERMEABILITY. Interstitial and vascular spaces are separated by a tight barrier, which, under normal conditions, limits the free passage of molecules. Certain pathologies can be accompanied by a reduction of barrier integrity: inflammatory events, for instance, often lead to a transient increase of the vascular permeability. Angiogenesis, as often found in tumors, is characterized by the formation of new vessels with increased permeability compared to normal tissue, which can be assessed by the injection of CA into the blood circulation, followed by monitoring their accumulation in the lesions.[38,39] Because the CA cannot enter the intracellular space, its distribution is restricted to two compartments: blood plasma and

[32] R. J. Gillies, N. Raghunand, G. S. Karczmar, and Z. M. Bhujwalla, *J. Magn. Reson. Imaging* **16,** 430 (2002).

[33] F. A. Howe and S. P. Robinson, *NMR Biomed.* **14,** 497 (2001).

[34] C. W. Pugh and P. J. Ratcliffe, *Nat. Med.* **9,** 677 (2003).

[35] R. R. Edelman, H. Hatabu, E. Tadamura, W. Li, and P. V. Prasad, *Nat. Med.* **2,** 1236 (1996).

[36] K. W. Stock, Q. Chen, M. Morrin, H. Hatabu, and R. R. Edelman, *J. Magn. Reson. Imaging* **9,** 838 (1999).

[37] V. M. Mai, B. Liu, J. A. Polzin, W. Li, S. Kurucay, A. A. Bankier, J. Knight-Scott, P. Madhav, R. R. Edelman, and Q. Chen, *Magn. Reson. Med.* **48,** 341 (2002).

[38] H. Larsson, M. Stubgaard, J. L. Frederiksen, M. Jensen, O. Henriksen, and O. Paulson, *Magn. Reson. Med.* **16,** 117 (1990).

[39] P. S. Tofts and A. G. Kermode, *Magn. Reson. Med.* **17,** 357 (1991).

interstitial space. Following injection, the CA will diffuse along the concentration gradient from the circulation into the interstitial space at a rate determined by vascular permeability. Provided that the concentration of the CA remains nearly constant in the circulation over time, its concentration in tissue will reach a steady state, whose amplitude is determined by the extravascular extracellular space (i.e., the leakage space).

Quantitative assessment of vascular permeability is based on dynamic contrast-enhanced T_1 mapping (DCE-MRI), allowing one to calculate leakage rate and leakage space from the temporal profile of CA uptake. The spin–lattice relaxation rate R_1 is proportional to the concentration c_t of CA in a voxel,

$$\Delta R_1 = \alpha_1 c_t v_e \tag{5}$$

with α_1 being the molar longitudinal relaxivity of the CA and v_e its leakage space (interstitial space).

Several analytical approaches have been used for data analysis, with the two-compartment model being one of the most relevant.[40] In such a model, the concentration of CA in tissue c_t is determined by the leakage space v_e, the permeability surface product k, the blood volume fraction v_p, and the concentration of CA in the circulation c_p:

$$c_t(t) = k \int_0^t \exp(-k(t - t')/v_e) c_p dt' + v_p c_p \tag{6}$$

Because the elimination half-life time of MR contrast agents is usually larger than 10 min, the temporal profile of c_p can be described by a step function. Furthermore, the concentration of the tracer in circulation is often negligible in tumors compared to tissue. It is therefore possible to simplify Eq. (6) and model c_t by

$$c_t(t) = v_e c_{p0}[1 - \exp(-kt/v_e)] \tag{7}$$

where c_{p0} describes the initial concentration of the CA. The permeability of the vasculature is hence proportional to the initial slope of the enhancement curve (Fig. 3).

Several types of gadolinium-based CA are available for DCE-MRI. Only low molecular weight compounds such as Gd-DOTA or Gd-DTPA have been approved for clinical applications. Due to their small size, their diffusion rate is very high, rendering accurate determination of the initial enhancement curve often difficult. Therefore, there is a strong focus on

[40] P. S. Tofts, *J. Magn. Reson. Imaging* **7**, 91 (1997).

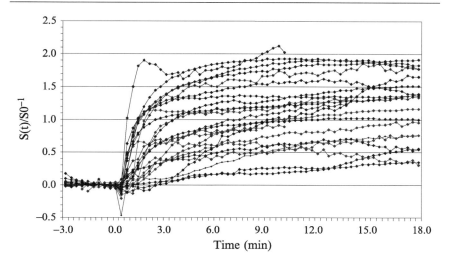

FIG. 3. Interanimal variability in a tumor xenograft model. Change in relative tumor signal intensity following administration of GdDTPA in B16 melanoma primary tumors implanted in a mouse ear. The initial uptake reflects the vascular (permeability)*(surface) product, whereas the final signal reflects the total tracer concentration in the tissue and, hence, is a measure for the extracellular leakage space. The large variability in this tightly controlled model illustrates the value of noninvasive readouts using each animal as its own control.

the development of macromolecular CA providing lower diffusion rates. Macromolecular CA can be based on low molecular Gd-chelating molecules bound to larger structures such as macrocyclic arms,[41] dendrimers,[42] or albumin.[43] Because evaluation for clinical use is still ongoing, they are today used only in preclinical applications.

These three examples illustrate that while the assessment of absolute physiological parameters from MRI data is feasible, the analysis is generally not straightforward; the models used are complex and include many assumptions/approximations. Hence, the majority of physiological MRI applications use semiquantitative analysis (i.e., parameter values in the region of interest in relation to a reference tissue).

[41] K. Turetschek, E. Floyd, D. M. Shames, T. P. Roberts, A. Preda, V. Novikov, C. Corot, W. O. Carter, and R. C. Brasch, *Magn. Reson. Med.* **45,** 880 (2001).
[42] H. E. Daldrup-Link, D. M. Shames, M. Wendland, A. Muhler, A. Gossmann, W. Rosenau, and R. C. Brasch, *Acad. Radiol.* **7,** 934 (2000).
[43] K. Turetschek, S. Huber, T. Helbich, E. Floyd, K. S. Tarlo, T. P. Roberts, D. M. Shames, M. F. Wendland, and R. C. Brasch, *Acad. Radiol.* **9,** S112 (2002).

In addition to being highly desirable with regard to animal care, noninvasive imaging, provides the advantage of allowing *paired study designs*, which improve statistical power. Longitudinal studies largely eliminate effects of interindividual variation (e.g., by relating the tissue state assessed following a therapeutic intervention to baseline values prior to treatment). An example illustrates this. Pituitary hyperplasia in rats induced by chronic stimulation (4 weeks) with estradiol is considered a model for prolactinoma.[44] Variability in this model is large, with the coefficient of variation (COV) for the pituitary volume following stimulation being of the order of 100%. This requires group sizes of 30 to 40 animals in order to detect a 50% therapy effect at a statistical level of $p \leq 0.05$. By relating the measurements following a 4-week drug treatment with the long-lasting somatostatin analog octreotide to the baseline values, this interindividual variation could be largely eliminated. The COV of the relative volume measures was found to be 5 to 10% only, and significant results demonstrating that octreotide treatment dose dependently reduced established hyperplasia could be demonstrated with $n = 4$ rats per group. Paired study design is attractive in case of high interindividual variability of pretreatment values as encountered frequently in cancer studies due to the inherent heterogeneity of neoplastic tissue. Even under controlled conditions (e.g., after the subcutaneous implantation of tumor xenografts) considerable data scattering is observed (Fig. 3).

The study of pharmacodynamic drug properties forms the backbone of MRI applications in drug research today. They rely on absolute or relative quantitative analysis of MRI data. While initially MRI was used predominantly to derive morphometric information *in vivo*, today physiological readouts have attracted much interest as sensitive indicators of tissue state and potentially early readouts of pharmacological efficacy.

Disease and Efficacy Biomarkers as Bridge between Preclinical and Clinical Drug Evaluation

The clinical evaluation of drug candidates that target chronic indications such as degenerative diseases involves large patient populations, is time-consuming and expensive. It is important to minimize the risk of such trials by optimizing the therapy regimen and by proper selection of the patient population. Thus, biomarkers that are indicative of the disease or provide mechanistic information on the efficacy or potential safety issues of the drug are of high value. The noninvasive character of the imaging

---

[44] M. Rudin, U. Briner, and W. Doepfner, *Magn. Reson. Med.* **7,** 285 (1988).

approaches, particularly MRI and MRS, makes them attractive tools for the development of such biomarkers. A number of MRI/MRS biomarkers based on structural and functional readouts have been proposed for clinical drug evaluation in indications such as neurodegeneration, multiple sclerosis, oncology, and osteoarthritis.

For *neurodegenerative* disorders, morphological (total brain atrophy, hippocampal atrophy) and functional biomarkers (altered task-related brain activity) have been proposed. Alternatively, MRS has been applied to analyze differences in endogenous tissue metabolism in patients with mild cognitive impairment (MCI) and diagnosed Alzheimer's disease (AD) as compared to normal aging.[45,46] Characteristic spectral changes were increased *myo*-inositol signals in MCI and both increased *myo*-inositol and decreased *N*-acetylaspartate signals (as marker of neuronal loss) in AD. However, in all these cases, both the sensitivity and the specificity of the proposed markers are questionable, limiting their applicability for clinical therapy evaluation. A more promising approach is based on plaque imaging using a plaque-specific PET ligand.[47]

PET has been used extensively for characterizing pharmacokinetic and pharmacodynamic properties of drug candidates by analyzing their receptor interaction, their effects on neurotransmitter systems, or general metabolic readouts such as glucose utilization. fMRI can provide similar information on drug-induced functional responses of CNS structures. This has been demonstrated in animal studies for a number of compounds interacting with various neurotransmitter systems[30,31] A distinctive advantage of the fMRI method is the high spatial (100 μm) and temporal (seconds to minutes) resolution. The fMRI method does not map the drug–receptor interaction per se, but rather the functional consequence thereof.

Clinical end points in *oncology* trials are tumor shrinkage and ultimately patient survival. Various potential biomarkers for the early assessment of drug efficacy have been proposed comprising both markers associated with a specific mechanism (e.g., angiogenesis or apoptosis) and general disease markers (e.g., tumor metabolism or tumor proliferation). Neovascularization is essential for the growth of primary tumors and metastases; hence, the assessment of angiogenesis might predict tumor malignancy, as well as its responsiveness to therapy. A clinically established method

[45] M. Catani, A. Cherubini, R. Howard, R. Tarducci, G. P. Pelliccioli, M. Piccirilli, G. Gobbi, U. Senin, and P. Mecocci, *Neuroreport* **12,** 2315 (2001).
[46] K. Kantarci, G. E. Smith, J. Ivnik, R. C. Petersen, B. F. Boeve, D. S. Knopman, E. G. Tangalos, and C. R. Jack, Jr., *J. Int. Neuropsych. Soc.* **8,** 934 (2002).
[47] E. D. Agdeppa, V. Kepe, A. Petri, N. Satyamurthy, J. Liu, S. C. Huang, G. W. Small, G. M. Cole, and J. R. Barrio, *Neuroscience* **117,** 723 (2003).

applied to evaluate antiangiogenetic drugs is DCE-MRI using contrast agents such as GdDTPA, which leaks into the extracellular space.[39,40,48] Alteration in tracer uptake by the tumor reflects changes in vascular permeability.[49] Such vascular permeability measurements have indeed been used extensively to assess the effects of antiangiogenic VEGF inhibitors at both preclinical and clinical levels.[50]

Efficacy biomarkers target general tumor properties such as metabolism and proliferation or microstructural changes of neoplastic tissue. MRS revealed significant alterations in tumor phospholipids and energy metabolism in response to drug treatment prior to detectable changes in tumor volume.[51] Finally, the apparent diffusion coefficient (ADC) of tissue water seems to predict successful tumor therapy: Significant increases of ADC values were observed within a few days of treatment with a cytostatic drug indicative of cell shrinkage, which was at least in part due to apoptosis.[52]

Disease progression in *degenerative joint disease* is commonly assessed by x-radiographical analysis of the joint gap. A reduction of the distance between the bone structures reflects the degeneration of articular cartilage and is a readout of advanced disease. Significant efforts have been devoted to the development of more sensitive biomarkers that could indicate subtle changes in the articular matrix that precede net structural loss. Cartilage stability is provided by a macromolecular network, with the main constituents being collagen and proteoglycans. Approaches proposed aim, therefore, at measuring the total content of macromolecules or of proteoglycans (PGs) in particular. The former is based on magnetization transfer (i.e., measuring the exchange rate between bulk and macromolecular-bound water). Reduction of the macromolecular pool, particularly collagen, or loss of its structural integrity leads to a decreased exchange rate.[53,54]

[48] O. Tyninnen, H. J. Aronen, M. Ruhala, A. Paetau, K. von Boguslawski, O. Salonen, J. Jaaskelainen, and T. Paavonen, *Invest. Radiol.* **34,** 427 (1999).

[49] C. van Dijke, R. C. Brasch, T. P. Roberts, N. Weidner, A. Mathur, D. M. Shames, F. Desmar, P. Lang, and H. C. Schwickert, *Radiology* **198,** 813 (1996).

[50] J. Drevs, R. Muller-Driver, C. Wittig, S. Fuxius, N. Esser, H. Hugenschmidt, M. A. Konerding, P. R. Allegrini, J. Wood, J. Hennig, C. Unger, and D. Marme, *Cancer Res.* **62,** 4015 (2002).

[51] J. L. Evelhoch, R. J. Gillies, G. S. Karczmar, J. A. Koutcher, R. J. Maxwell, O. Nalcioglu, N. Raghunand, S. M. Ronen, B. D. Ross, and H. M. Swartz, *Neoplasia* **2,** 152 (2000).

[52] T. L. Chevenert, L. D. Stegman, J. M. G. Taylor, P. L. Robertson, H. S. Greenberg, A. Rehemtulla, and B. D. Ross, *J. Natl. Cancer Inst.* **92,** 2029 (2000).

[53] M. L. Gray, D. Burstein, L. M. Lesperance, and L. Gehrke, *Magn. Reson. Med.* **34,** 319 (1995).

[54] D. Laurent, J. Wasvary, J. Yin, M. Rudin, T. C. Pellas, and E. O'Byrne, *Magn. Reson. Imaging* **19,** 1279 (2001).

Proteoglycans are one of the major constituents of the cartilage matrix, contributing to cartilage resiliency through a negative electrostatic force. The early phase of osteoarthritis is associated with a loss of PGs, leading to an impaired biomechanical support function of cartilage, which in turn contributes to its further degradation.[55] Noninvasive assessment of the PG content (e.g., using delayed gadolinium-enhanced MRI) might constitute a sensitive biomarker of osteoarthritis.[56,57] This MRI technique yields an estimate of the fixed charged density (FCD) of cartilage, reflecting negatively charged side chains of PGs. The negatively charged CA $Gd(DTPA)^{2-}$ penetrates the interstitial fluid of cartilage to reach an equilibrium concentration that is governed by (1) the $Gd(DTPA)^{2-}$ concentration gradient and (2) electrostatic interactions (inversely proportional to the FCD). Quantitative analysis of the Gd-induced changes in proton relaxation rate (R_1) allows one to estimate the FCD and hence the PG concentration. In model systems, a good correlation between changes in relaxation rates and biochemically determined PG levels has been obtained,[57] emphasizing the potential of MRI readouts as disease-relevant biomarkers.

The development of noninvasive biomarkers is highly relevant for clinical drug evaluation. They would help to stratify patient populations, optimize the therapy regimen (dosing, timing), or might even be used as surrogates for a clinical end point. Prerequisite is a careful validation of the biomarker.

Conclusion and Outlook

Today, *in vivo* imaging techniques, particularly MRI, have become indispensable in biomedical research. MRI is being used at various steps in drug discovery and development with focus on disease phenotyping and drug evaluation during lead optimization and compound profiling. Strengths of the MRI approach are: (1) high information content (i.e., the MR signal is governed by multiple independent parameters providing high soft tissue contrast, a key characteristic for comprehensive tissue characterization) and (2) noninvasiveness, which offers advantages in study design (paired design with increased statistical power) and is a prerequisite for translational studies. The principal disadvantage of MRI is low sensitivity

[55] G. Grushko, R. Schneiderman, and A. Maroudas, *Connect. Tissue Res.* **19,** 149 (1989).
[56] A. Bashir, M. L. Gray, J. Hartke, and D. Burstein, *Magn. Reson. Med.* **41,** 857 (1999).
[57] D. Laurent, J. Wasvary, M. Rudin, E. O'Byrne, and T. Pellas, *Magn. Reson. Med.* **49,** 1037 (2003).

largely precluding pharmacokinetic studies. Low sensitivity is also a significant drawback in the study of target-specific (molecular imaging) applications, which have raised considerable interest recently as tools for the validation of potential drug targets, for the analysis of molecular pathways involved in drug action, and for the identification of molecular biomarkers.[4] Nevertheless, MR approaches to visualize gene expression have been described[5,6] demonstrating that target–specific information can be derived in favorable situations. MRI approaches to monitor the migration of magnetically labeled cells with high spatial resolution seem more promising as demonstrated for tracking of macrophages,[58–63] stem cells,[64] and progenitor cells.[65]

Despite the recent developments in molecular and cellular imaging, MRI in biomedical research will be used primarily for pharmacodynamic studies (i.e., for the evaluation of drug efficacy models of human disease using morphological and physiological readouts). In addition, the development of MRI-based biomarkers for translational applications such as the rapid evaluation of a therapeutic concept in the clinics will increase rapidly in importance. Molecular and cellular imaging techniques, particularly optical and nuclear methods, will complement the conventional structural and functional imaging approaches.

[58] V. Dousset, C. Delalande, L. Ballarino, B. Quesson, D. Seilhan, M. Coussemacq, E. Thiaudiere, B. Brochet, P. Canioni, and J. M. Caille, *Magn. Reson. Med.* **41**, 329 (1999).
[59] M. Rausch, A. Sauter, J. Frohlich, U. Neubacher, E. W. Radü, and M. Rudin, *Magn. Reson. Med.* **46**, 1018 (2001).
[60] M. Rausch, D. Baumann, U. Neubacher, and M. Rudin, *NMR Biomed.* **15**, 278 (2002).
[61] M. Rausch, P. Hiestand, D. Baumann, C. Cannet, and M. Rudin, *Magn. Reson. Med.* **50**, 309 (2003).
[62] N. Beckmann, C. Cannet, M. Fringeli-Tanner, D. Baumann, C. Pally, C. Bruns, H. G. Zerwes, E. Andriambeloson, and M. Bigaud, *Magn. Reson. Med.* **49**, 459 (2003).
[63] N. Beckmann, R. Falk, S. Zurbrügg, J. Dawson, and P. Engelhardt, *Magn. Reson. Med.* **49**, 1047 (2003).
[64] M. Hoehn, E. Kustermann, J. Blunk, D. Wiedermann, T. Trapp, S. Wecker, M. Focking, H. Arnold, J. Hescheler, B. K. Fleischmann, W. Schwindt, and C. Buhrle, *Proc. Natl. Acad. Sci. USA* **99**, 16267 (2002).
[65] J. W. Bulte, S. Zhang, P. van Gelderen, V. Herynek, E. K. Jordan, I. D. Duncan, and J. A. Frank, *Proc. Natl. Acad. Sci. USA* **96**, 15256 (1999).

[15]   Enzyme-Dependent Fluorescence Recovery after Photobleaching of NADH: *In Vivo* and *In Vitro* Applications to the Study of Enzyme Kinetics

By Christian A. Combs and Robert S. Balaban

Introduction

Enzyme-dependent fluorescence recovery after photobleaching (ED-FRAP) has been described as a method of monitoring enzymatic activities in microsamples *in vitro* or cells using confocal microscopy,[1] as well as in larger samples of enzyme solutions or organelle suspensions.[2] This approach basically relies on the transient photobleaching or photolysis of a fluorophore product of an enzymatic reaction and observing its enzyme-dependent replenishment. Because the degree of photobleaching or photolysis can be controlled through the power of irradiation, concentration-dependent kinetics can be determined in a system with minimal perturbations.[2] The initial description of this technique used NADH as the fluorophore in *in vitro* samples, as well as in imaging experiments in intact cells. Due to the key role that NADH plays in cellular energy metabolism, as well as redox signaling and other cellular regulatory processes, this has remained one of the most active uses of this approach.

This Chapter provides a brief overview of the use of NADH fluorescence in biochemistry and biology and the photophysics of NADH. The use of NADH ED-FRAP to evaluate the turnover of NADH *in vitro* and *in vivo* is described with a focus on experimental details. Examples of performing NADH ED-FRAP experiments on isolated enzymes, mitochondria, and cells are outlined. Approaches in the analysis of NADH ED-FRAP data, along with an evaluation of the known limitations, are discussed.

Use of NADH Fluorescence in Biochemistry and Biology

NADH fluorescence has a long history of use in biochemistry. The first reported use of NADH fluorescence in the analysis of enzymatic reactions was Greengard.[3] The major advantage of NADH fluorescence assays was the improved specificity and sensitivity of the fluorescence measurement when compared to absorption spectroscopy studies. This high sensitivity

---

[1] C. A. Combs and R. S. Balaban, *Biophys. J.* **80,** 2018 (2001).

[2] F. Joubert, H. M. Fales, H. Wen, C. A. Combs, and R. S. Balaban, *Biophys. J.* **86,** 629 (2004).

[3] P. Greengard, *Nature* **178,** 632 (1956).

METHODS IN ENZYMOLOGY, VOL. 385   0076-6879/04 $35.00

resulted in the miniaturization of enzymatic assays to very small samples or very dilute reaction mixtures (for reviews and some methods, see Passonneau and Lowry[4]). Typically, NADH fluorescence assays are conducted under conditions far from equilibrium with regards to reactants and products to simplify the interpretation of the rate of change [NADH] with enzyme activity. The fluorescence of NADH is also very sensitive to its microenvironment, providing both spectral shifts (usually blue shifts with binding in hydrophobic environments)[5] and changes in emission lifetimes.[6] The high sensitivity of the NADH fluorescence detection and the relatively high concentration of NADH in intact tissues led to the ultimate application of monitoring NADH levels in intact cells and tissues as pioneered by Chance and colleagues in the late 1950s and early 1960s.[5,7,8] These studies demonstrated that NADH could be followed dynamically in intact tissues, providing a unique view into the steady-state [NADH] in cells as well as the dynamics of changes to new steady states with workload or various metabolic challenges. One of the major limitations of these types of studies was the inability to determine whether a change in [NADH] was due to a change in production or consumption of NADH. This is especially the case in single cell or *in vivo* studies where the net consumption of NADH is difficult to determine from oxygen consumption measurements. It was also appreciated very early on that many of the assumptions used for *in vitro* studies did not apply to the complex milieu in intact cells. First, other fluorophores potentially interfere with the NADH signal, most notably NADPH and FAD, whereas the high ratio of enzymes/proteins to NADH results in a significant fraction of bound NADH resulting in spectral shifts and changes in emission lifetimes. The change in fluorescence lifetimes resulted in different pools of NADH, potentially contributing to the fluorescence signal out of proportion to a given pool [NADH]. Finally, in intact tissues, the optical properties of the tissue can interfere with both excitation and emission light, requiring compensating techniques to correct for these potentially serious modifying factors.[9–11]

[4] J. V. Passonneau and O. H. Lowry, "Enzymatic Analysis: A Practical Guide." Humana Press, Totowa, NJ, 1993.
[5] B. Chance, *Science* **137,** 499 (1962).
[6] M. Wakita, G. Nishimura, and M. Tamura, *Biochem. (Tokyo)* **118,** 1151 (1995).
[7] B. Chance and B. Thorell, *J. Biol. Chem.* **234,** 3044 (1959).
[8] B. Chance, J. R. Williamson, D. Jamieson, and B. Schoener, *Biochem. Zeit.* **341,** 357 (1965).
[9] R. Brandes, V. M. Figueredo, S. A. Camacho, and M. W. Weiner, *Am. J. Physiol.* **266,** H2554 (1994).
[10] S. A. French, P. R. Territo, and R. S. Balaban, *Am. J. Physiol.* **275,** C900 (1998).
[11] G. J. Puppels, J. M. C. C. Coremans, and H. A. Bruining, *in* "Fluorescent and Luminescent Probes for Biological Activity" (W. T. Mason, ed.). Academic Press, London, 1999.

Photophysics of NADH

NADH is a naturally fluorescent molecule, whereas its oxidized product, NAD, is not. This situation provides a useful fluorescence assay for the interconversion of NADH and NAD. NADH is typically excited by ~350-nm light with a Stokes shift of typically 100 nm, placing the emission maximum in the 440- to 460-nm range. The fluorescence efficiency of free NADH in solution is low and has only a 0.4-ns mean fluorescence lifetime.[6] The NADH fluorescence lifetime can be enhanced or quenched by binding, depending on the molecular environment. In general, hydrophobic domains result in a blue shift of the NADH emission and a greater than 10-fold enhancement of fluorescence lifetimes. In general, enzymes that have been demonstrated to enhance NADH fluorescence emission include alcohol dehydrogenase, glutamate dehydrogenase, malate dehydrogenase, and lactate dehydrogenase, whereas other enzymes, for example, glyceraldehydes 3-phosphate dehydrogenase, can quench the emission signal significantly [for reviews and more information, see Estabrook (1962)[12] and Lowry and Passonneau (1993)[4]]. Thus, the emission frequency and efficiency of NADH fluorescence cannot be considered a constant in a complex biological system.

The photoexcitation of NADH to an excited state can lead to several biochemical consequences: the simple emission of a photon and relaxation to the ground state, the transmission of energy to the surrounding lattice, again returning the NADH to the ground state (quenching), or the degeneration of the NADH to the NADH$^+$-free radical ($\bullet$NADH$^+$) together with a hydrated electron (e_H). Both of these products have numerous metabolic paths, depending on the environment, with many leading to the stoichiometric production of NAD$^+$ from NADH. These reaction pathways have been studied extensively in the literature and are presented in Fig. 1 as adapted from Joubert *et al.*[2] If the free radicals $\bullet$NADH$^+$ or $\bullet$NAD do not form covalent reactions with other molecules, including themselves forming dimers,[13] the photolysis reaction results in the stoichiometric generation of NAD$^+$ from NADH with the production of water or reduced electron acceptors to consume the released electrons. The resulting production or consumption of a proton can provide some insight into the mechanism of the free radical metabolism in these reactions.[2] The net photolysis reaction of generating NAD from NADH in the steady state, such as in imaging NADH fluorescence in a microscope,[1] can be seen as a competitive oxidizing reaction in cells where the oxidation of NADH is the major source of energy in intermediary metabolism. One of the major concerns with this photolysis reaction is the generation of effectively two moles of

[12] R. W. Estabrook, *Anal. Biochem.* **4**, 231 (1962).
[13] A. V. Umrikhina, A. N. Luganskaya, and A. A. Krasnovsky, *FEBS Lett.* **260**, 294 (1990).

UV Light

$$\text{NADH} \longrightarrow \text{NADH}^* \longrightarrow \cdot\text{NADH}^+ + e_{aq}^- \longrightarrow \cdot\text{NAD} + e_{aq}^- + H^+$$

Free radical scavenging pathways in mitochondria

$$2\text{NADH}^* + 2H^+ + O_2 \longrightarrow 2\text{NAD}^+ + 2H_2O \qquad \text{With oxygen}$$

$$\text{NADH}^* + 2\text{REC}_{oxi} \longrightarrow \text{NAD}^+ + H^+ + 2\text{REC}_{red} \quad \text{With any electron acceptor}$$

FIG. 1. Two putative reaction pathways for NADH photolysis products and associated reaction intermediates in mitochondria. NADH*: NADH in excited state, •NAD and •NADH: metabolite free radical, e_{aq}^-: hydrated electron, REC_{oxi} and REC_{red}: oxidized and reduced forms of any electron acceptor (examples in the mitochondria include the cytochromes). Adapted from Joubert et al. (2003).

free radicals per mole of NADH converted to NAD^+. If the environment does not eliminate these free radicals effectively, then it is likely that irreversible damage to the cell or enzyme system could result, along with a potentially poor efficiency of converting NADH to NAD^+.

Cellular NADH Fluorescence

In cells and tissue the 450-nm autofluorescence is dominated by NADH and NADPH, depending on the relative concentration of these metabolites. Values for [NADH], [NAD], [NADP], and [NADH] can vary to a large degree in different types of tissue (Table I). Table I also shows that absolute quantification of the total pyridine nucleotide pool from tissue is dependent on the assay method, as is shown by the variability of this measurement on the same type of tissue but reported by different groups. In addition, the specific tissue compartment is important regarding the binding and fluorescence efficiency. For example, binding in mitochondria has been shown to increase the mean lifetime of NADH fluorescence by over six-fold,[6] consistent with the much larger levels of fluorescence usually observed from within the mitochondrial compartment.[7,13a–c] The specific binding sites in mitochondria have been ill defined, but likely include malate dehydrogenase, isocitrate

[13a] J. Eng, R. M. Lynch, and R. S. Balaban, Biophys. J. 55, 621 (1989).
[13b] F. F. Jobsis and J. C. Duffield, J. Gen. Physiol. 50, 1009 (1967).
[13c] E. M. Nuutinen, Basic Res. Cardiol. 79, 49 (1984).

TABLE I
REPORTED RATIOS OF CONCENTRATIONS OF NADH, NAD, NADP, AND NADPH FROM VARIOUS TISSUES AND SPECIES

Tissue	Species	NADH/NAD	NADPH/NAD	$\sum$NAD(H)/$\sum$NADP(H)	Reference
Liver	Rat (tissue section)	0.15	4.4	2.0	Chance et al. (1965)[14a]
		0.12	4.7	1.4	Matschinsky et al. (1978)[13d]
				0.6	Klingenberg et al. (1959)[13e]
		0.2	2.9	1.7	Klingenberg (1985)[14]
				7.9	Sies et al. (1974)[18]
Heart	Pig (isolated mitochondria)	0.4	0.4	7.1	Livingston et al. (1996)[15]
	Pig (isolated myocytes)	0.4	5.14		Livingston et al. (1996)[15]
	Rat (tissue section)			4.2	Klingenberg et al. (1959)[13e]
	Rat (tissue section)	0.04	1.0	8.2	Chance et al. (1965)[8]
	Rat (tissue section)	0.2	2.1	6.6	Klingenberg (1985)[14]
Brain	Rat (tissue section)	0.1	2.0	32.0	Klingenberg (1985)[14]
		0.1			Schulman et al. (1974)[16]
				6.5	Klingenberg et al. (1959)[13e]
Kidney	Rat (tissue section)			5.3	Klingenberg et al. (1959)[13e]

[13d] F. M. Matschinsky, C. S. Hintz, K. Reichlmeier, B. Quistorff, and B. Chance, *in* "Microenvironments and Metabolic Compartmentation" (P. Srere and R. W. Estabrook, eds.), p. 149. Academic Press, New York, 1978.

[13e] M. Klingenberg, *Biochem. Zeit.* **332,** 47 (1959).

[14] M. Klingenberg, *in* "Methods of Enzymatic Analysis" (H. U. Bergmeyer, ed.), Vol. 7, p. 251. VCH, Weinheim, FDR, 1985.

[14a] B. Chance, B. Schoener, K. Krejci, W. Russmann, H. Russmann, H. Schnitger, and T. Bucher, *Biochem. Zeit.* **341,** 325 (1965).

[15] E. Livingston, R. A. Altschuld, and C. M. Hohl, *Pediatr. Res.* **40,** 59 (1996).

[16] M. P. Schulman, N. K. Gupta, A. Omachi, G. Hoffman, and W. E. Marshall, *Anal. Biochem.* **60,** 302 (1974).

dehydrogenase, 3-hydroxyacyl-coenzyme A dehydrogenase, and NADH dehydrogenase (assuming activity and binding are related; for a table of activities, see Combs and Balaban[1]). In the cytosol, NADH binding can also enhance emission with alcohol dehydrogenase, whereas the prominate glycolytic enzyme glyceraldehyde 3-phosphate dehyrdrogenase quenches NADH fluorescence.[4,12] In the heart, the autofluorescence in the 450-nm region has been shown to be dominated by mitochondrial NADH levels, whereas in the liver, containing ADH and significant levels of NADPH, the interpretation is much more complex.[17,18] In the kidney, the 450-nm signal is primarily NADH but, the significant concentration of mitochondrial NADPH also likely contributes to the signal.[19] This factor complicates the conversion of fluorescence to absolute concentration values for NAD(P)H species.

Although NADH fluorometry is an effective means of measuring relative [NADH] in intact cells, it remains a challenge to image this molecule at high resolution in intact cells with minimal damage for the following reasons. (1) The relatively low quantum yield and small absorption cross section of NADH generally require a relatively high intensity excitation source compared to most extrinsically applied probes. This leads to direct photodynamic effects of the high-energy UV light used as well as just generalized heating from nonspecific absorption. (2) NADH photolysis leads to the significant generation of free radicals that can damage tissue independent of the direct photodynamic effects. (3) The overall concentration of NADH is low in most cells, especially those with low numbers of mitochondria (i.e., many cell culture systems). All of these effects are significant issues when just imaging NADH, to rapidly photolyze NADH for ED-FRAP experiments will naturally enhance these complications. Two-photon excitation has been proposed as a method of reducing the total impact of the excitation scheme by effectively generating UV light only at the focal plane of the cell. Indeed, the power dependence of cellular toxicity does seem to follow the square of the power rather than be linear with power, suggesting that generation of the two-photon condition dominates tissue damage.[20,21] This of course assumes that the absorption of light in the infrared is approximately 6 OD less than in the UV or much less destructive when absorbed. In many cultured cell systems this is likely the case; however, in tissues containing high concentrations of mitochondria, myoglobin, and hemoglobin, this might not be the case and the investigator should proceed with caution. Furthermore, many biological effects are

[17] R. S. Balaban and J. J. Blum, *Am. J. Physiol.* **242**, C172 (1982).
[18] H. Sies, D. Haussinger, and M. Grosskopf, *Hoppe-Seyler's Z. Physiol. Chem.* **355**, 305 (1974).
[19] R. S. Balaban, V. W. Dennis, and L. J. Mandel, *Am. J. Physiol.* **240**, F337 (1981).
[20] A. Hopt and E. Neher, *Biophys. J.* **80**, 2029 (2001).
[21] K. Konig, T. W. Becker, P. Fischer, I. Riemann, and K. J. Halbhuber, *Opt. Lett.* **24**, 113 (1999).

non-linear with temperature resulting in a non-linear dependence on power. This is discussed further later.

NADH ED-FRAP Measurements of Dehydrogenase Activity in Enzyme Systems and Isolated Mitochondria *In Vitro*

NADH ED-FRAP has been shown to be effective in measuring dehydrogenase activity in several enzyme systems *in vitro* as well as suspensions of isolated mitochondria. This section includes an example of an optical system used for acquiring data, examples of data collected, and a discussion of the interpretation of data collected.

Optical System for Microcuvette ED-FRAP Experiments

The design constrains for a ED-FRAP system in microcuvettes have several important elements: (1) the reaction chamber, (2) the high-intensity photolysis light source, and (3) the monitoring system. Our current system for this experiment is outlined in Fig. 2. The reaction chamber is a simple 3×3-mm quartz cuvette transparent on all four sides (Spectrocell Inc., RF-303/45). This chamber has excellent UV transmission characteristics as well as a reasonable cross-sectional area for laser illumination. This cuvette is mounted in a modified commercially available temperature-regulated cuvette holder with four optical windows (Flash 200, Quantum Northwest Corp, Spokane, WA). This system was modified by the manufacturer to hold a 3×3-mm cuvette in the center of the system with good thermal contact. The photolysis pulse originates from a pulsed laser, providing a ~6 ns 355-nm light pulse. The peak power generated in this system is 64 mW over ~0.2 cm^2. The laser beam is split equally to impinge on opposite sites of the cuvette simultaneously to avoid inner filter/scattering effects generating unequal irradiation distribution within the reaction chamber. The beam diameter is adequate to homogeneously irradiate a 50-ml sample placed in the bottom of the cuvette after careful alignment. This is demonstrated by the lack of change in NADH, due to diffusion within the cuvette, many minutes after a photolysis reaction in a simple buffered NADH solution (see Fig. 2). In subsequent studies on isolated mitochondria and simple NADH solutions, it was found that the dual beams were not required for even illumination, but this approach is still in use to assure even illumination. Thus, subsequent designs may not require this feature. In this system the power and the number of pulses impinging on the sample are altered to provide different amounts of photolysis.

Finally, a detection system is required to monitor either the excited molecule, in this case NADH, or other chromophores or reaction elements. NADH fluorescence is monitored using a bifurcated fiber-optic impinged

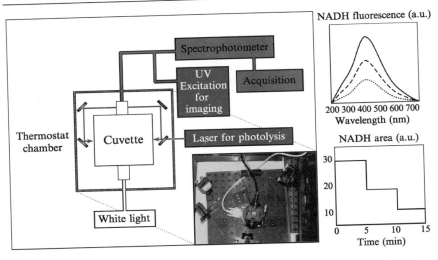

FIG. 2. Schematic diagram of hardware for ED-FRAP experiments involving enzyme solutions or organelle suspensions in a cuvette as described in the text. *Left:* Schematic diagram of system. Photolysis light is from a laser that is split and impinged on a temperature-controlled cuvette. The fluorescence or absorbance within the cuvette can then be followed using the remaining two ports on the holder. (Inset) An actual picture of the cuvette-holding system. *Top right:* Fluorescence spectra of NADH during control and after two photolysis pulses from the laser. *Bottom right:* Time course of NADH fluorescence after two sequential photolysis pulses. Note that the fluorescence signal is constant after a photolysis reaction, attesting to the homogeneity of the irradiation field within the entire cuvette. (See color insert.)

on a surface of the cuvette normal to the path of the photolysis beam. A constant wave excitation is provided by a 500-W Hg/Xe arc lamp (Oriel, Inc.) and a 360-nm band-pass excitation filter (Edmund Scientific, Inc.) impinged on one of the fibers. Fluorescence detection is accomplished using a linear array installed in a standard PC (Ocean Optics Inc., Model PC2000) to collect a broad spectral bandwidth (335–1064 nm) with a minimal scan time of ∼5 ms (see spectra in Fig. 2). Faster and more sensitive detection can be accomplished using an appropriately filtered photomultiplier tube. However, the scanning spectrophotometer is useful for analyzing the spectral characteristics or the probe understudy, as well as multiple optical fluorescent probes (i.e., pH- or Ca^{2+}-sensitive probes). In addition, by impinging a white light source on the other cuvette face normal to the photolysis beam, absorption spectroscopy of the sample can also be conducted, permitting a wide variety of probes to be used, including mitochondria cytochromes, hemoglobin, or light scattering.[2] The open top of the cuvette permitts the insertion of microelectrodes (pH) or temperature probes to measure other aspects of the photolysis reaction.

Efficiency and Products of the NADH Photolysis

Using the system outlined in Fig. 2, the products and yield (percentage conversion of NADH to NAD) were accessed using NMR, mass spectrometry, and classical enzymatic techniques by Joubert *et al.*[2] These investigators demonstrated that most, if not all, of the photolyzed NADH in simple NADH solutions ended up as NAD^+ independent of the photolysis level. These authors also confirmed the multiple pathways for NADH* oxidation to NAD^+. In more complex systems, including some isolated enzyme systems, the yield was lower as the level of hydrolysis increased. The yield could be as low as 50% in isolated mitochondria (see later) and 70% in some enzyme systems (alcohol dehydrogenase). The other products or reasons for this lower yield in complex systems are not yet fully resolved and remain an area of investigation. Most likely the interactions of the free radicals generated in this process contribute to this reduction in yield or direct effect of the intense illumination.

ED-FRAP Measurements of Isolated Enzyme Systems

In contrast to simple NADH solutions, the photolysis of NADH to NAD^+ in a cuvette that had previously reached a reaction steady state for a given dehydrogenase system results in a recovery of NADH that is catalyzed by the enzymatic reaction. An example of this type of study is shown in Fig. 3, which is adapted from Joubert *et al.*[2] for the 3-hydroxybutyrate dehydrogenase (HDH) system. The recovery rate clearly increases with increasing NADH photolysis due to two factors: the increase in the substrate for NADH regeneration, NAD, as well as the removal of the product inhibition by NADH (see later). The rate of recovery is also directly dependent on the amount of enzyme in the solution, confirming that the recovery rate is a measure of the enzymatic activity in the chamber (data not shown).[1,2] Similar results have been obtained in other enzyme systems, including the alcohol dehydrogenase (ADH) and glutamate dehydrogenase (GDH) systems. These data demonstrate that the NADH ED-FRAP methodology is feasible in isolated enzyme systems and permits the evaluation of rapid kinetics of enzyme systems near equilibrium or established steady states. However, the kinetic description of these reactions is complicated by the fact that both the reactant (NAD^+) and the product (NADH) are exchanged stoichiometrically by the nanosecond photolysis pulse, resulting in a complicated kinetic scheme.

Kinetic Model Description of NADH ED-FRAP

To extract the most information from ED-FRAP recovery data and to understand the underlying assumptions and limitations of this technique more fully, it is necessary to examine the kinetic models of the recovery

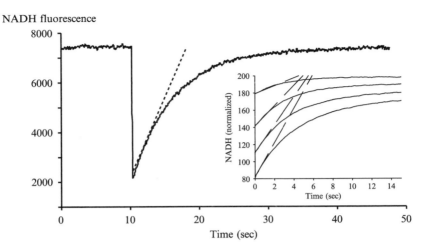

NADH fluorescence

Fig. 3. Time course of NADH fluorescence in an ED-FRAP experiment involving a solution of hydroxybutyrate dehydrogenase (HDH) and reaction substrates as measured from the cuvette system outlined in the text and shown schematically in Fig. 2. (Inset) The effect of increasing NADH photolysis (decreasing [NADH] and increasing [NAD]) on the initial rate of recovery (from Joubert et al.[2]).

behavior. For simplicity, this will first be presented for *in vitro* isolated dehydrogenase enzyme data. This description is a brief overview of the complete equations provided by Joubert et al.[2]

In the simplest case, where no net flux occurs and the reaction is in steady state, the initial rate of the recovery can be analyzed as the B_i–B_i kinetic mechanism presented here:

$$A + NAD + E \leftrightarrow E\text{-}NAD + A \leftrightarrow (E\text{-}NAD\text{-}A \leftrightarrow E\text{-}NADH\text{-}B)$$
$$\leftrightarrow E\text{-}NADH + B \leftrightarrow B + NADH + E$$

Following rearrangement of this expression to express it as a function of NADH and NAD, assuming that changes in [A] and [B] are negligible, the equation for initial velocity of the NADH following the photolysis pulse becomes

$$V_i = \frac{C1\,[NADH_p]}{1 + C2\,[NADH_p]} \tag{1}$$

where $NADH_p$ is defined as the amount of photolyzed NADH according to

$$[NADH_o] = [NADH_i] - [NAD_p] \qquad (2)$$

$$[NAD_o] = [NAD_i] + \alpha[NADH_p] \qquad (3)$$

where $[NAD_i]$ and $[NADH_i]$ are the initial concentrations of NAD and NADH before the photoconversion pulse, $[NAD_o]$ and $[NADH_o]$ are the concentrations of NAD and NADH after the photoconversion pulse, and α is the amount of photolyzed NADH that was converted to NAD. In addition, C1 and C2 are complex expressions of the form

$$C1 = \frac{\left(\frac{V_{maxf}}{K_{NAD}}\right)\left(\alpha + \frac{[NAD_i]}{[NADH_i]}\right)}{\left(1 + \frac{[NAD_i]}{K_{NAD}} + \frac{[NADH_i]}{K_{NADH}}\right)} \quad \text{and} \quad C2 = \frac{\left(\frac{\alpha}{K_{NAD}} - \frac{1}{K_{NADH}}\right)}{\left(1 + \frac{[NAD_i]}{K_{NAD}} + \frac{[NADH_i]}{K_{NADH}}\right)}$$

where V_{maxf} is the maximum rate of the forward reaction and K_{NADH} and K_{NAD} are the inhibitory and affinity constants for NADH and NAD, respectively. This is a useful form of this equation, as all of the V_{maxf} influence is in C1, permitting the detection of changes in enzyme activity that are simply linked to changes in enzyme activity.

These relationships have been shown to be effective in extracting kinetic information *in vitro*.[2] In particular, kinetic information is extracted most easily in two simple cases. Where $[NAD_i]/[NADH_i] \gg 1$ and $[NAD_i] \gg K_{NAD}$, Equation (1) simplifies to

$$V_i = V_{maxf} - \frac{V_{maxf}}{[NADH_i]}[NADH_o] \qquad (4)$$

In this example, the initial rate of recovery allows the determination of V_{maxf} directly from the plot of initial recovery rate vs $[NADH_o]$ and extrapolating to $[NADH_o] = 0$. Also, the full recovery is monoexponential and a rate constant $k = V_{maxf}/[NADH_i]$ can be measured. Figure 4 shows examples of these relationships. As predicted, the initial rate of recovery is linearly related to the amount of NADH photolysed (Fig. 4A and B) and the estimated V_{maxf} is proportional to the amount of enzyme added to the cuvette (Fig. 4C).

In the second case, where $[NAD_i]/[NADH_i] \gg 1$ and $[NAD_i] \ll K_{NAD}$ (NAD_i is small), C1 and C2 from Eq. (1) are simplified to become

$$C1 = \frac{\left(\frac{V_{maxf}}{K_{NAD}}\right)\alpha}{\left(1 + \frac{[NAD_i]}{K_{NAD}} + \frac{[NADH_i]}{K_{NADH}}\right)} \qquad (5)$$

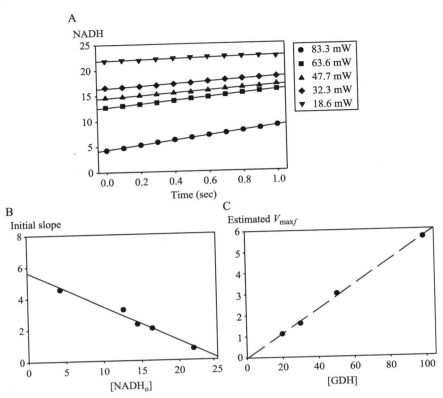

FIG. 4. Initial slope analysis of NADH ED-FRAP experiments involving glutamate dehydrogenase and reaction substrates. (A) Time courses of recovery over the first 1 s following varying levels of laser power (and concomitant levels of NADH photolysis) used for photolysis. (B) The relationship of increasing initial recovery rate to the level of [NADH$_o$] (and therefore increasing levels of NADH photolysis). An estimation of V_{maxf} can be obtained by extrapolation to [NADH$_o$] = 0, as outlined in the text. (C) Estimated values of V_{maxf} as a function of [GDH]. Modified from Joubert et al.[2]

$$C2 = \frac{\left(\frac{\alpha}{K_{NAD}} - \frac{1}{K_{NADH}}\right)}{\left(1 + \frac{[NADH_i]}{K_{NADH}}\right)} \tag{6}$$

In this second case, by varying [NADH$_p$] and [NADH$_i$], it is possible to extract all kinetic parameters from C1 and C2. Figure 5 shows an example of case 2 where V_i increases with [NADH$_p$] with a Michaelis–Menton dependence and 1/C1 is related linearly to [NADH$_i$]. In this example,

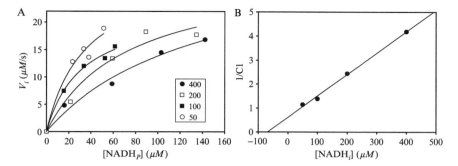

FIG. 5. Initial slope analysis of NADH ED-FRAP experiments involving alcohol dehydrogenase and reaction substrates. (A) The relationship between the initial rate and the amount of photolyzed NADH, which is consistent with a Michaelis and Menten relationship. (B) The linear relationship observed between the parameters $1/C1$ and [NADH] consistent with Eq. (1). Modified from Joubert *et al.*[2]

Joubert *et al.*[2] were able to extract most of the kinetic parameters for an enzyme system from Eq. (2) and the $1/C_1$ and V_i relationship.

NADH ED-FRAP Measurement of Kinetic Parameters in Isolated Mitochondria

To interpret NADH ED-FRAP images in intact cells, the experiment must be characterized under more controlled conditions. As discussed earlier, most of the NADH fluorescence signal in heart tissue comes from the mitochondrial NADH pool, thus characterizing that the kinetic behavior of NADH resynthesis in this system will help in the interpretation of cellular or tissue data. The nature of recovery data from NADH ED-FRAP experiments conducted on isolated mitochondria is much more complicated than *in vitro* enzyme systems for many reasons. First, all matrix dehydrogenases will contribute to the resynthesis of NADH. This means that the recovery rate in NADH ED-FRAP *in vivo* is proportional to the NADH-generating capacity of the entire system, integrating all of the enzymes weighted by their relative activity. Second, the concentration of the total NADH–NAD pool is fixed and it is impossible to modify [NADH] or [NAD] independently. Third, the kinetic parameters for NAD and NADH in the mitochondrial matrix are not known and therefore modeling and extracting quantitative kinetic data from recovery rates are more difficult. Fourth, there is usually a net flux through the NADH pool in intact systems. Fifth, given the more complex environment, the possibility for nonspecific photodynamic or free radical interactions is much higher. Sixth, the high ratio of [protein] to [NADH] results in a significant pool of NADH bound to macromolecules, complicating the interpretation of the fluorescence signal. All of these complications are also present in the intact cell, but can be better controlled and evaluated in isolated mitochondria preparations. Given the complex photochemistry of NADH and the mitochondria matrix, it was necessary to examine the effect of NADH ED-FRAP experiments on overall mitochondrial function, including the maximum rate of ATP production and inner membrane leak rates. Joubert *et al.*[2] demonstrated no measurable deleterious effects of the UV light on ATP synthetic rates or permeability of the mitochondria membranes.

An example of typical NADH ED-FRAP data taken from isolated mitochondria actively producing ATP under state 3 conditions[22] with carbon substrates (glutamate and malate), ADP and P_i, in a cuvette in the apparatus outlined earlier is shown in Fig. 6. Under these conditions, a net flux through NADH occurs as outlined:

$$\text{NAD} \underset{F_b}{\overset{F_f}{\rightleftarrows}} \text{NADH} \xrightarrow{F_{net}} \text{NAD}$$

where F_f and F_b represent forward and reverse flux through the dehydrogenases, including site 1, whereas F_{net} represents the net flux of reducing equivalents to oxygen. Thus, the recovery curve is complicated by the oxidation of NADH for use by oxidative phosphorylation. Even though this curve is similar to the recovery curves of isolated dehydrogenase systems and increased with decreasing [NADH] and increasing [NAD], the interpretation of these data is much more complex (see Joubert *et al.*[2]).

To eliminate the net flux, F_{net}, as well as any backflux of reducing equivalents from FADH or the cytochromes through coenzyme Q,[1] the inhibitor rotenone was used to simplify the analysis for evaluating the effects of different putative signaling molecules on site 1 NADH-generating capacity. This approach permitted a much more simplified interpretation of the recovery rate in terms of the net generation of NADH, without the complication of a significant net flux. Two molecules that have been proposed to increase the production of NADH are Ca^{2+} and inorganic phosphate (P_i). These molecules were studied in two separate studies using NADH ED-FRAP in the presence of rotenone and are shown in Fig. 7. Ca^{2+} [2] and P_i[23] both increased the ED-FRAP in an additive fashion consistent with the notion that both of these molecules can activate NADH generation in intact mitochondria.

One of the most interesting results from these studies was the observation that the quantitative analysis of the rate of resynthesis of NADH revealed that the calculated unidirectional rate of NADH formation was

[22] B. Chance and C. M. Williams, *Adv. Enzymol.* **17**, 65 (1956).

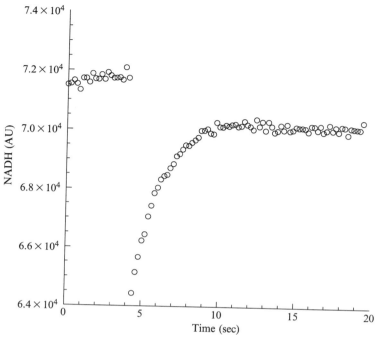

FIG. 6. Time course of NADH fluorescence in an ED-FRAP experiment involving isolated mitochondria from porcine heart as measured from the cuvette system outlined in Fig. 1. Mitochondria were preincubated in a large 1.5-ml suspension to determine respiratory rates and to establish steady-state conditions. At a given time, a 50-μl sample was placed in the cuvette and the NADH ED-FRAP experiment was conducted. Mitochondria were respiring on 5 mM glutamate and malate with 5 mM P_i and 500 nM Ca^{2+} with 400 μM ADP. The concentration of mitochondria was 2 nmol cytochrome A/ml. The experiment was conducted at room temperature.

only slightly faster than the measured net rate of utilization by oxidative phosphorylation.[2] These data imply that the reactions forming NADH are far from equilibrium, complicating the use of enzyme equilibria for calculating NADH/NAD values, which is a common practice.[4] In addition, data suggest that the generation of NADH may be a very significant limiting process near maximum ATP production rates.

These isolated mitochondria studies demonstrate that NADH ED-FRAP studies can be conducted on intact mitochondria with no observable consequence regarding the function of mitochondria. Analysis reveals that simplifying the study by eliminating the net flux associated with oxidative phosphorylation is useful for probing the characteristics of NADH generation through site 1. However, a more complete model permitting reverse

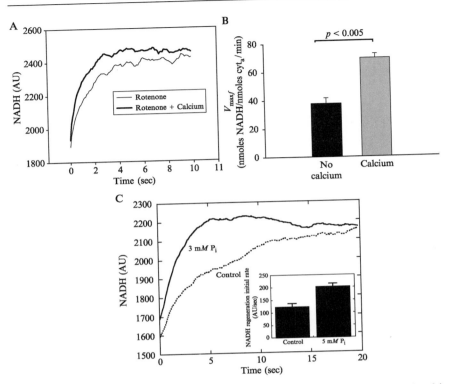

Fig. 7. Comparison of NADH ED-FRAP initial recovery rates from isolated mitochondria suspensions with known stimulators of mitochondrial dehydrogenases. (A and B) The effect of calcium on a typical recovery time course and the average of replicates (adapted from Joubert et al.[2]). In this experiment, mitochondria were pretreated with the site 1 inhibitor rotenone to avoid NADH oxidation by the electron transport system. (C) The effect of inorganic phosphate on the same process. From Bose et al.[23]

electron flow and a net flux will be required to use this approach under more physiological conditions. These initial studies suggest that the NADH generation process is not even near equilibrium and may contribute to the rate limitation of oxidative phosphorylation. These results are consistent with the notion that the rate of recovery of NADH might be closely related to the net rate of NADH utilization in oxidative phosphorylation. If these observations are applicable to the intact cell, then the NADH ED-FRAP rate might be used to measure the local metabolic rate within cells when performed in an imaging experiment.

[23] S. Bose, S. French, F. J. Evans, F. Joubert, and R. S. Balaban, J. Biol. Chem. 278, 39155 (2003).

Cellular and Tissue NADH ED-FRAP Imaging

Overview

NADH ED-FRAP imaging of cells and tissues can potentially be conducted using wide-field epifluorescence, confocal, or two-photon imaging methods. Each method has potential advantages, and all have been used to successfully image steady-state NADH fluorescence in cells and tissues (for reviews, see Balaban and Mandel,[24] Masters and Chance,[25] Piston and Knobel,[26] and Puppels *et al.*[11]). Just as in the *in vitro* methods presented earlier, NADH ED-FRAP imaging requires a powerful UV light source to quickly photooxidize NADH and excite NADH fluorescence. The available imaging methods differ greatly in terms of optical sectioning ability, depth of penetration into tissue, cost, and speed of acquisition. Wide-field NADH ED-FRAP imaging is potentially faster, less expensive, and easier to conduct than confocal or two-photon methods, but lacks optical sectioning capabilities and is limited to the very surface of the tissue. Confocal ED-FRAP is most appropriate for the imaging of isolated cells, where local NADH regeneration rates may be of interest. Two-photon ED-FRAP potentially provides the ability to image deep into tissue or to do repeated ED-FRAP experiments in a single cell or group of cells to generate a three-dimensional map of NADH flux rates. To our knowledge, wide-field and two-photon NADH ED-FRAP imaging experiments have not been conducted. This section reviews data from confocal studies (from cardiac myocytes) and discusses the potential for ED-FRAP imaging using wide-field and two-photon microscopy.

Confocal NADH ED-FRAP Imaging

Figure 8 shows the results of a typical ED-FRAP imaging experiment on a intact living cardiac myocyte using a point scanning confocal microscope. In this experiment the photolysis was accomplished by raster scanning a high-level irradiation over the entire cell several times before reinitiating the low-power imaging sequence. The high-power photolysis irradiation was generated by increasing the transmission properties of the acousto-optical tunable filter (AOTF) from values used in normal imaging excitation. Although an AOTF makes this process easier in that the user can generate customized ROIs for photobleaching, these results could have

[24] R. S. Balaban and L. J. Mandel, *in* "Noninvasive Techniques in Cell Biology" (S. Grinstein and J. K. Foskett, eds.), p. 213. Wiley, New York, 1990.

[25] B. R. Masters and B. Chance, *in* "Fluorescent and Luminescent Probes for Biological Activity" (W. T. Mason, ed.), p. 361. Academic Press, London, 1999.

[26] D. W. Piston and S. M. Knobel, *Methods Enzymol.* **307,** 351 (1999).

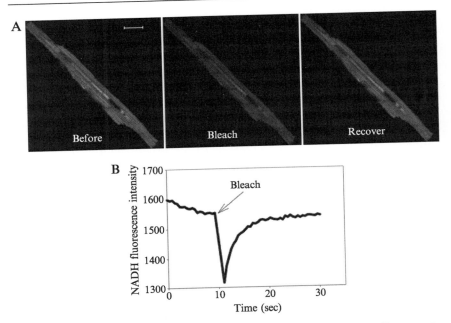

FIG. 8. NADH ED-FRAP images and time course from isolated perfused cardiac myocytes. (A) NADH fluorescence images (2 μm slice thickness) from a myocyte before photobleaching, immediately after, and following recovery. Scale bar: 10 μm. (B) Representative plot of NADH fluorescence intensity recovery after a bleach pulse. From Combs and Balaban.[1]

been accomplished using the fast exchange of neutral density filters, polarizing optics in the light path, or a separate irradiation source to illuminate the scanned field of view. The speed of the irradiation and subsequent imaging is important with this type of dynamic measurement, as is discussed further later. As was seen in *in vitro* data, the time course of the NADH signal intensity shows a decrease in fluorescence and an exponential recovery (Fig. 8). Also, as in *in vitro* data, the recovery rate is proportional to the level of NADH photolysis (Fig. 9). These data are typical of what can be achieved with a point-scanning confocal microscope and the appropriate laser. It is important to note that the NADH fluorescence from intact cardiac myocytes is overwhelmingly dominated by the mitochondrial NADH signal.[13a,c] This may not be the case in other cell types or tissues where cytosolic contributions or high NADPH levels may complicate the interpretation (Table I). Despite the similarities between isolated mitochondria and cellular data, there are many complicating factors, both optical and physiological, that make confocal NADH ED-FRAP imaging more challenging than acquiring *in vitro* data.

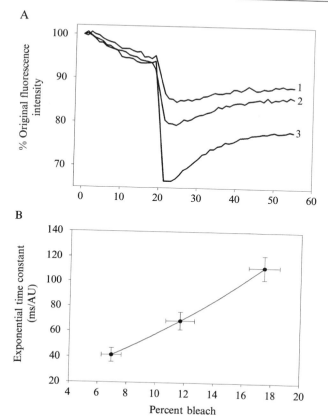

FIG. 9. The rate of recovery of the autofluorescence of NADH is proportional to the bleach level. (A) Time courses of NADH ED-FRAP at three different levels of photolysis of NADH measured in repeated experiments on a single cardiac myocyte. (B) Relationship between exponential recovery rates and degree of photolysis from experiments conducted on isolated cardiac myocytes. From Combs and Balaban.[1]

Optically, the most important factors in NADH ED-FRAP imaging are related to obtaining a high enough signal-to-noise ratio from all regions or subregions to be measured and providing enough light to photolyze NADH quickly. Data presented in Figs. 8 and 9 were collected using a Zeiss LSM 510 scanning confocal microscope, a C-apochromat (NA 1.2) lens, an 80-mW argon laser that provided 351-nm excitation light, an AOTF for fast changes in illumination power, and a long-pass 360-nm emission filter. Images were collected from 1.5-μm slices through the center of the cell. To photolyze NADH effectively, it is necessary to have a powerful UV laser and high numerical aperture (NA) objective to obtain high irradiance at the illuminated spot. Maximum irradiance of the illuminating spot can be calculated according to the following equation[27]:

$$Ip = \left(\frac{\pi}{2}\right) PT \left(\frac{\eta}{\lambda}\right)^2 (1 - \cos\alpha)(3 + \cos\alpha) \tag{7}$$

where P is the input power in watts, T is the transmission factor of the objective, η is the refractive index of the medium, λ is the wavelength of the illuminating light in meters, $\alpha(\alpha = \sin^{-1}(NA/\eta))$ is the semiaperture angle of the objective, and NA is the numerical aperture of the objective. The final units are in W/mm^2. Figure 10 shows the relationship between light levels at the back aperture of the microscope and peak irradiance at the illuminated spot as a function of the NA of the objective. Figure 10 illustrates the importance of using high NA objectives to photolyze NADH.

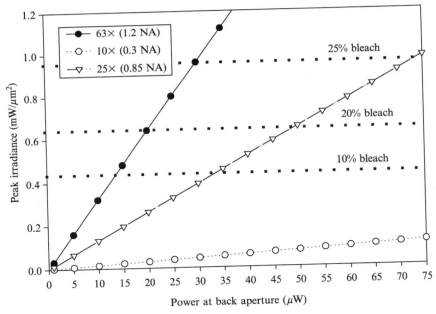

FIG. 10. Relationship between light power at the back aperture of the objective and the maximum irradiance of the irradiated spot as a function of the numerical aperture of the objective. Dotted lines represent power levels necessary to photolyze various percentages of NADH in isolated cardiac myocytes in a point-scanning confocal microscope as discussed in the text.

[27] C. J. Cogswell and K. G. Larkin, *in* "Handbook of Confocal Microscopy" (J. B. Pawley, ed.), p. 127. Plenum Press, New York, 1995.

Using a C-apochromat (NA 1.2) lens, we have found that it is necessary to use ~10 μW (1.1 mW/μm^2, ~15× the levels necessary to image NADH) at 351 nm to induce 10% photolysis of NADH with a pixel dwell time of 1.6 μs over a 512 × 200 FOV for 1 s. This corresponds to a duty cycle of only 0.001% of irradiation on each pixel over the 1-s time period (i.e., each pixel was only irradiated for 1 ms during the 1-s irradiation). Effective quantification of recovery rates usually requires at least a 10% initial photolysis of NADH as well as several higher photolysis levels to estimate the dependencies on [NADH] and [NAD] as outlined in the previous sections. Levels higher than 75 μW (or 30% NADH photolysis) (approximately 1.75 mW/μm^2) cause visible damage to cardiac myocytes (cell hypercontraction, blebbing, etc.). Thus, irradiation power in this range should be avoided to prevent nonspecific effects.

The temporal aspects of the intermittent irradiation used in confocal and two-photon microscopy scanning have significant effects in this dynamic system. Indeed, it can be shown that the 1-s intermittent irradiation used in our studies outlined earlier likely underestimates the true perturbation to the system, as NADH recovery occurs between the irradiation pulses. This makes the photolysis level not only a function of the irradiation power, but of duty cycle and the biochemical recovery kinetics. This is simulated in Fig. 11, where the recovery rate of NADH is assumed to be a single exponential with a time constant of 0.25, 0.5, 1, and 2 s^{-1}. As seen in this simulation, the NADH recovers between each equally powered irradiation pulse, resulting in an underestimate of the NADH perturbation that is a function of the recovery rate constant. Using this approach, just the effect of different trains of irradiation pulses on the net [NADH] could be used to extract the rate constant of recovery in a well-defined system. At equilibrium, it can be shown that a simple exponential recovery process will respond to a periodic irradiation with the following characteristics:

$$NADH_{ob} = NADH_o - \left[NADH_o \, MF_p \, \exp(-F_r \, k) \right] \qquad (8)$$

where $NADH_{ob}$ is the observed NADH signal, $NADH_o$ is initial [NADH], MF_p is the mole fraction of NADH photolyzed per pulse, F_r is the framing rate (sec), and k is the apparent rate constant of the biochemical resynthesis of NADH. If MF_p can be determined accurately, this may provide a steady-state estimate of k overcoming the temporal resolution problems inherent in these raster-scanning techniques.

This limitation of the scanning techniques suggests that the initial rate methods used with the 6-ns laser excitation *in vitro* are not applicable unless severely reduced FOVs are used to minimize the framing rates or a brightfield illumination is used for initiating the photolysis. Based on these concerns, using a simple raster-scanning approach, only the relative changes

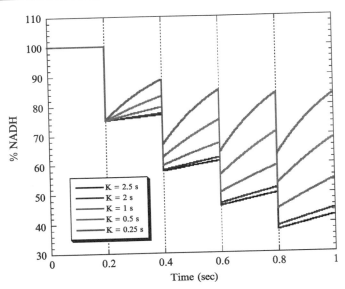

FIG. 11. Simulation of NADH recovery during a confocal microscopy imaging experiment. The plot is NADH concentration as a function of time. Calculations were based on Eq. (8), which assumes that the rate of recovery of NADH is constantly occurring over the imaging process. The rate constant for the metabolic recovery of NADH was varied as shown in the figure key. The framing rate was 200 ms, whereas each photolysis reaction consumed 25% of the NADH. Simulation was run numerically using IDL (RSI) running on a laboratory PC. This model likely underestimates the effect of metabolic recovery on NADH, as the recovery rate is assumed constant and it is known that the recovery rate increases with decreasing NADH (see Fig. 9). (See color insert.)

in the rate constants of recovery can be evaluated assuming a simple exponential recovery. Steady-state approaches outlined in Eq. (8) might prove useful in overcoming this limitation for these scanning approaches.

Another concern in these types of experiments is the fact that the imaging experiment itself also induces some NADH photolysis. This is due primarily to the power requirements needed to achieve adequate NADH fluorescence signal to noise ratios (SNR) as discussed earlier. Thus, the imaging experiment also results in a photolysis of NADH, but at a much lower level than attained by the photolysis pulses. We found that a new steady state is established under the imaging conditions where the NADH/NAD ratio is decreased, which increases the net biochemical synthetic rate to match the additional oxidation induced by the photolysis process.[1] This is illustrated in Fig. 12, where the initiation of imaging reduces $NADH_{ob}$ but $NADH_{ob}$ recovers quickly to near- $NADH_o$ levels after interrupting

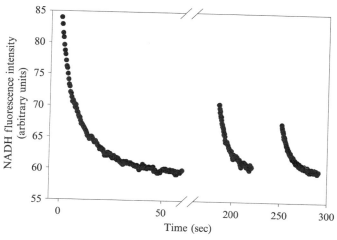

Fig. 12. Balance between photolysis and metabolic generation of NADH over time in an isolated perfused cardiac myocyte. This study was conducted on a perfused single cardiac myocyte at room temperature using a Zeiss 510 confocal microscope. The signal was integrated from an ROI consuming most of the myocyte to enhance the SNR of the measurement. The initiation of imaging results in a monotonic decrease in NADH to a new steady state. The imaging experiment was paused twice where the NADH signal recovers to almost the initial conditions. These results are consistent with a dynamic steady state between NADH photolysis and NADH resynthesis in the fluorescence monitoring of NADH using confocal microscopy. From Combs and Balaban.[1]

the imaging series. Clearly the steady-state level is dependent on the power used and the ability of the biochemical processes to compensate, and the attempted "noninvasive" measurement of NADH has resulted in a rather significant perturbation to the cellular redox state. Thus, this approach might not be as noninvasive as we would like. The kinetics of the approach to steady state and the steady-state level alone may also be useful in evaluating aspects of the metabolic capacity of the tissue as discussed earlier. Efforts are currently underway to explore this option. As seen in Fig. 12, care must also be taken to assure that the steady state is reached before an ED-FRAP experiment to make sure that the approach to steady state does not obscure the recovery rate.

Physiologically, some complicating factors in the NADH ED-FRAP measurement include motion and diffusion artifacts. Motion is problematical, as it is necessary to sample the same tissue region during the recovery process. This will likely limit the use of dynamic NADH ED-FRAP techniques to quiescent cells or well-behaved paced cells. The steady state approaches with homogenous excitation schemes may be more tolerant of tissue motion. FRAP techniques were designed originally to look for the diffusion of fluorescent-labeled molecules.[28] In heart and many other cells the majority of the NADH signal originates from mitochondria, where the NADH is "trapped" with the micrometer-scaled organelle. Thus, diffusion is not a serious concern unless significant signal is obtained from the free cytosolic fraction.[1] In some cell types, mitochondria are very mobile (in our hands, kidney cells) and the bulk motion of the mitochondria could interfere with the measurement. Under these conditions the voxel sizes or regions of interest should be increased to minimize these effects, naturally compromising the spatial resolution of the measurement.

Even with the limitations of this approach, the detection of physiological modifications of NADH-generating capacity can be shown in intact cells. Figure 13, from Combs and Balaban,[1] shows a comparison of whole cell recovery rates between various treatments known to increase dehydrogenase activities *in vivo*. In these experiments, the NADH recovery rate was measured as a rising exponential. Recovery rates were expressed as an exponential rate constant. Rate constant measurements for *in vivo* measurements have been expressed in terms of fluorescence units per unit time.[1] Conversion to absolute concentration values has not been

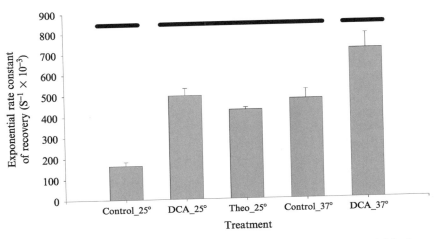

FIG. 13. The NADH recovery exponential rate constant is temperature- and dehydrogenase activity level-sensitive in cardiac myocytes. DCA at the two temperatures stands for superfusion with 5 mM DCA and 5 mM pyruvate added to the nutrient medium ($n = 5$ at each temperature). Theo_25° stands for superfusion with theophyline (17 mg/ml). The combination of DCA and pyruvate is a potent stimulator of dehydrogenase activity. Lines at the top indicate significant statistical differences. From Combs and Balaban.[1]

[28] M. Edidin, Y. Zagyansky, and T. J. Lardner, *Science* **191**, 466 (1976).

attempted *in vivo* due to the difficulty of converting fluorescence units to absolute concentration values for the reasons outlined earlier. In a well-characterized system where the total [NAD(H)] is known from chemical analysis and tissue or cellular fluorescence values have been calibrated to levels of relating to the fully oxidized and reduced NAD(P)H concentrations, it may be possible to express rate constant measurements in terms of concentration values per unit time.[2] All in all, comparing rates of recovery (or resistance to photolysis) between treatments is valid where the level of NADH photolysis and sampling conditions are identical.

The dynamic range of the rate constant measures of NADH flux in cardiac myocytes is demonstrated by varying temperature, as well as chemical stimulation of dehydrogenase activity. In these studies, using room temperature and 37° together with DCA and theophylline to activate dehydrogenase activity directly, a dynamic range of over five-fold is demonstrated in intact cardiac myocytes. Furthermore, all of the agents, including temperature, activated the NADH regeneration rate in the intact cardiac myocyte. The rate constant of recovery in a cell can also be made into an image of the cell. Figure 14 shows a pix map of exponential recovery rates in individual cardiac myocytes. In most cases, pix maps show a homogeneous recovery rate when cells are unstimulated; however, differences are apparent after stimulation (same cell imaged before and after the addition of DCA). This approach may be used to examine differences among mitochondrial metabolism in subregions of cells.

In summary, confocal imaging of NADH ED-FRAP recovery rates is a useful tool for making a relative measure of overall NADH flux in conditions where the amount of photolysis and recovery are similar and where starting [NADH] is not qualitatively different. Absolute quantification of metabolic parameters such as mitochondrial NADH affinities and maximum rates remains problematic due to the poor temporal resolution of these raster-scanning techniques and the inability to generate large dynamic range changes in [NADH] and [NAD].

Potential for Multiphoton and Wide-Field NADH ED-FRAP Imaging

Imaging metabolic rates within cells under true *in vivo* conditions (i.e., intact animal or human) is a very attractive idea. In their normal environment, cells would be expected to have the full complement of factors (known and unknown) that influence metabolism and function. In addition, cells in intact tissues would be expected to be more viable and perform normal differentiated functions when compared to those isolated by enzymatic digestion or cells adapted to the culture environment. We have found these later issues to be very significant in studies on muscle cells.

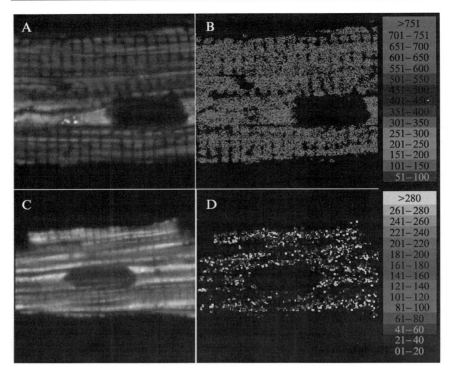

FIG. 14. Pix maps of NADH recovery rate show local NADH flux rates within isolated cardiac myocytes. (A and C) NADH fluorescence intensity images (1.5 μm slice thickness). (B) A false color pix map of exponential recovery rates. (D) A false color pix map of the difference in exponential recovery rate between superfusion with nutrient medium and 5 mM DCA of the cell depicted in C. From Combs and Balaban.[1] (See color insert.)

Of the available techniques for optical imaging, multiphoton approaches have the most potential for examining deep (hundreds of micrometers) into living tissue, although significant challenges to performing NADH ED-FRAP experiments exist. Perhaps the most significant challenge is tissue movement during the photolysis pulse or during imaging. For instance, it has been shown that 2-photon microscopy produces beautiful NADH images from isolated quiescent cardiac myocytes,[29a] but to date, this has not been replicated in an intact beating heart. Two-photon microscopy will be even more sensitive than confocal microscopy due to the spatially localized excitation beam in the axial dimension. Indeed, fluorophore excitation probability decreases as the fourth power of the excitation cone diameter.[29]

[29] W. Denk, J. H. Strickler, and W. W. Webb, *Science* **248,** 73 (1990).
[29a] S. Huang, A. A. Heikal, and W. W. Webb, *Biophys. J.* **82,** 2811 (2002).

This is beneficial in that it provides inherent sectioning ability on the excitation side, but movements on the order of less than a micrometer in the z dimension can prevent the ability to complete an NADH ED-FRAP experiment. To some extent this can be ameliorated with a non two-photon UV light source for the photolysis pulse; however, depth penetration of the UV pulse would still be a problem. The other major problem is associated with the large amounts of laser power needed for two-photon microscopy. Two-photon excitation is dependent on a very intense photon irradiation many orders of magnitude higher than single-photon techniques. The need for excessive power presents two major problems for two-photon NADH ED-FRAP experiments. First, it typically requires at least 5–10 mW of laser power to image NADH in isolated cells or at the surface of tissues. As depth increases and scattering and absorption occur, much more power is needed. If the ratio of the photolysis pulse power to the imaging pulse power is similar to the one-photon case (at least 15-fold), then at the surface it would require approximately 75 mW to photolyze NADH by 10%. The photolysis pulse would need to be much higher as depth increases. This logically leads to questions of tissue damage at higher laser powers. This is a particular concern at 700–740 nm, where NADH is most often excited for two-photon microscopy. Both myoglobin and hemoglobin absorb light in this range, and therefore the potential exists for significant tissue heating both in and out of the focal plane. It has been estimated that red blood cells may heat up by as much as 60 K at 100 mW under some conditions.[30] Even at somewhat longer wavelengths (780–800 nm), which should be less damaging, there have been reports of two-photon light pulses causing oxidative stress and deformations of the organization of mitochondrial inner membranes even at imaging powers less than 10 mW.[31,32] Although the amount of power used for imaging is a large concern in regard to potential damage, it should be pointed out that different types of cells have different sensitivities to light levels and that the exact imaging parameters (e.g., pixel dwell time) must be taken into account when considering potential photodamage. All in all, photodamage may be the most important issue in determining the usefulness of two-photon microscopy in NADH ED-FRAP.

Wide-field NADH ED-FRAP also has the potential for examining *in vivo* metabolism. Wide-field techniques have been used effectively to examine steady-state NADH fluorescence in many different preparations

[30] K. Konig and U. K. Tirlapur, *in* "Confocal and Two-Photon Microscopy: Foundations, Applications, and Advances" (A. Diaspro, ed.), p. 191. Wiley-Liss, New York, 2002.
[31] H. Oehring, I. Riemann, P. Fischer, K. J. Halbhuber, and K. Konig, *Scanning* **22**, 263 (2000).
[32] U. K. Tirlapur, K. Konig, C. Peuckert, R. Krieg, and K. J. Halbhuber, *Exp. Cell Res.* **263**, 88 (2001).

and tissues, including Langendorf-perfused whole hearts,[13c,33–36] brain,[37] and kidney.[38] An example of a wide-field NADH fluorescence image of a canine heart *in vivo* is shown in Fig. 15. This was collected using a flash xenon light source filtered to 340–380 nm gated to the cardiac cycle. The

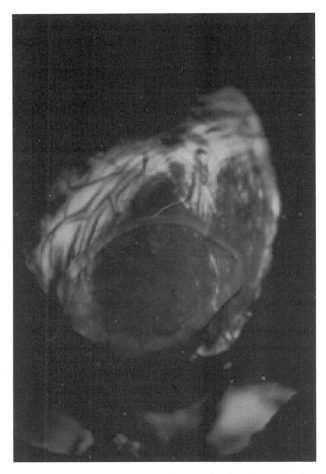

FIG. 15. Whole-heart NADH fluorescence image of a beating canine heart *in vivo*, as described in the text. Basically, this image was collected by gating a 360-nm filtered flash xenon light source to the EKG. The image was collected by simply leaving the shutter of a camera open for the four heartbeats used to create this image. The image was collected with 100 ASA color film. The yellow fluorescence is from epicardial fat, especially around the vessels. The arterial and venous blood vessels can be observed directly, with arterial vessels being brighter than venous vessels, presumably due to the hemoglobin oxygenation state. The transparent ring on the surface of the heart is made of a silicon-based material to make a well on the surface of the heart to hold drugs and extrinsic probes on the surface of the heart. (See color insert.)

image was collected using a color film camera with a high-pass >400-nm filter. The main challenges for NADH ED-FRAP in this context are movement, specificity of the NADH fluorescence signal, depth penetration, and the ability to provide enough light to photolyze NADH. Calculations show that light power will not be a problem if a focusing lens of high NA is used in conjunction with a commercially available arc or laser and the appropriate excitation filters using photon density requirements from our experience in confocal microscopy. The major limitation will be the depth of penetration, only tens of micrometers,[39] and the influence of primary and secondary inner filters on the emission signal.[9,10,35] Although a NADH ED-FRAP wide-field excitation scheme has not been attempted *in vivo* to our knowledge, it is likely that this approach will provide first data from *in vivo* tissues due to the limitation in the multiphoton approaches listed previously.

Summary

In summary, ED-FRAP is an effective technique for both qualitative measurement of NADH or NADPH turnover, depending on tissues and conditions, in intact cells and quantitative measurement of dehydrogenase activity *in vitro*. In theory, either the recovery from NAD(P)H photolysis or the approach to steady state can be used to estimate NAD(P)H turnover. In isolated enzyme systems, this technique has been shown to measure dehydrogenase activity directly. In isolated organelles, cells, and tissues, the interpretation of NADH ED-FRAP becomes more difficult due to complex reaction pathways and may only provide relative measures of NADH turnover, although some data suggest that NADH ED-FRAP may provide information on the net utilization of NADH under conditions where the back flux is minimal. Each of the imaging methods that could be used in conjunction with NADH ED-FRAP has advantages and limitations. Imaging methods provide localized turnover measurements to the subcellular level, but trade-offs must be considered concerning optical sectioning ability, imaging and photolysis power, temporal resolution, imaging depth, and size of field of view. In addition, the biological implications of

[33] C. H. Barlow and B. Chance, *Science* **193**, 909 (1976).
[34] C. Ince, J. F. Ashruf, J. A. Avontuur, P. A. Wieringa, J. A. Spaan, and H. A. Bruining, *Am. J. Physiol.* **264**, H294 (1993).
[35] A. P. Koretsky, L. A. Katz, and R. S. Balaban, *Am. J. Physiol.* **253**, H856 (1987).
[36] C. Steenbergen, G. Deleeuw, C. Barlow, B. Chance, and J. R. Williamson, *Circ. Res.* **41**, 606 (1977).
[37] E. Dora and A. G. Kovach, *J. Cereb. Blood Flow Metab.* **3**, 161 (1983).
[38] H. Franke, C. H. Barlow, and B. Chance, *Am. J. Physiol.* **231**, 1082 (1976).
[39] T. A. Fralix, F. W. Heineman, and R. S. Balaban, *FEBS Lett.* **262**, 287 (1990).

the generation of free radicals concurrent with the NADH photolysis must be considered. Despite these complications, NADH ED-FRAP is the only method available to dynamically monitor the turnover of NAD(P)H in intact systems. It has been demonstrated that this approach can work at several different spatial scales, from subcellular measures to suspensions of enzymes and cellular organelles. It will improve the understanding of NADH utilization in intact cellular systems in terms of both cellular topology and cellular metabolic regulation.

Acknowledgments

The authors thank Drs. Emily Rothstein and Paul Jobsis for many useful discussions and help in some experiments that contributed to this work. We also thank William Riemenschneider for help in figure preparation.

[16] Imaging Myocardium Enzymatic Pathways with Carbon-11 Radiotracers

By Carmen S. Dence, Pilar Herrero, Sally W. Schwarz, Robert H. Mach, Robert J. Gropler, and Michael J. Welch

Introduction

Under normal conditions, the heart utilizes a variety of metabolic pathways, such as the oxidation of carbohydrates, fatty acids, lactate, and pyruvate, to meet the high-energy demands of contraction and maintenance of cellular function. The metabolic flux through each pathway is determined by the availability of substrates for each metabolic pathway in plasma, as well as hormonal status and myocardial oxygen supply. For example, the high blood levels of fatty acids during fasting result in the oxidation of fatty acids as the principal form of energy production and account for approximately 70% of the cardiac energy requirements. Consumption of a high carbohydrate meal results in an elevation of plasma glucose levels, an increase in insulin production, and an activation of glycolysis. Exercise results in the release of lactate by skeletal muscle, which is taken up rapidly by the heart, converted to acetyl-CoA, and oxidized through the tricarboxylic acid cycle (TCA) cycle. The extraordinary ability of the heart to utilize a number of different metabolic pathways and to change its metabolic preference rapidly is necessary for the maintenance of proper mechanical function under a variety of physiological conditions. Therefore, a derangement in the balance of myocardial metabolism is expected to play a key role in a number of pathological conditions leading to abnormal cardiac function.

Much of the work that resulted in the characterization of the different enzymes involved in intermediary metabolism was carried out *in vitro*. Although this seminal research provided the foundation for the fields of biochemistry and enzymology, the techniques used in the mapping of intermediary metabolism are inadequate for studying the change in metabolic processes that underlie myocardial dysfunction in human disease. While the elucidation of the different metabolic pathways was complete by the middle of the 20th century, the study of the change in myocardial metabolism as a consequence of disease was not possible until the advent of non-invasive imaging techniques such as positron emission tomography (PET) and single photon emission computed tomography (SPECT) in the 1970s.

Positron emission tomography is an imaging technique developed for the *in vivo* study of metabolic functions in both healthy and diseased stages. The goal of the technique was best expressed in 1975 by Dr. Michel Ter-Pogossian,[1] one of the pioneers of this modality at Washington University: "Our ultimate goal is to measure *in vivo* regionally and as noninvasively as possible metabolic processes. Perhaps a term for this approach could be either *in vivo* biochemistry or functional [imaging]. The reason for seeking this goal, of course, is the application of this approach to medicine using the premise that any form of pathology either results from or is accompanied by an alteration of some metabolic pathway. Our approach to achieve the above goal consists in labeling with cyclotron-produced radionuclides, more specifically, oxygen-15, carbon-11, nitrogen-13 and fluorine-18, certain metabolic substrates, the fate of which is studied *in vivo* subsequent to their administration, by some radiation detector or imaging device, with the hope, after suitable unraveling of the metabolic model used, of measuring *in vivo* a particular pathway."

Since the early 1980s and more extensively in the mid 1990s, investigators at the Washington University School of Medicine have used PET to study the change in metabolic substrate utilization that occurs under a variety of experimental conditions, including normal aging, obesity, dilated cardiomyopathy, type 1 diabetes mellitus, and hypertension-induced left ventricular hypertrophy. This is accomplished by measuring the dynamics (i.e., uptake and washout kinetics) of radiolabeled substrates for each metabolic pathway. These studies, which utilize the radiotracers 1-[^{11}C]D-glucose, 1-[^{11}C]acetate, and 1-[^{11}C]palmitate, are collectively termed the

[1] M. Ter-Pogossian, "The Developing Role of Short-Lived Radionuclides in Nuclear Medicine," p. 9. U.S. Department of Health, Education and Welfare, 1977.

GAP studies. This Chapter provides details of the radiosynthesis, dosimetry, quality control, data acquisition, and kinetic modeling that are needed to conduct this experimental paradigm successfully. These issues are discussed to help the reader new to the field gain a broad understanding of the problems faced by the PET researchers that work with short-lived isotopes and to learn some of the approaches used to solve these problems. This Chapter does not include a discussion of the basic principles of PET. The reader interested in a general survey on the synthesis of ^{11}C-labeled compounds and of radiopharmaceuticals used for studying the heart is referred to Antoni et al.[2] and Hwang and Bergmann,[3] respectively.

Overview of the Production of Carbon-11

The physical characteristics of the radionuclides used in PET are listed in Table I, along with the most common nuclear reaction to produce them in a clinical setting. Their decay mode by positron emission allows their detection outside the body after annihilation with an electron in the body. The result is the production of two photons (0.511 MeV each) at almost 180° to each other. These two photons are detected by the imaging device, which then creates images of the tissue under study. A simplified version of these events is presented in Fig. 1.

Because GAP studies require the administration of multiple radiolabeled substrates, it is necessary to use the shorter-lived positron-emitting

TABLE I
PHYSICAL CHARACTERISTICS OF COMMONLY PRODUCED SHORT-LIVED ISOTOPES

Nuclide	Half-life	Nuclear reaction	Max energy (MeV)	Range in H_2O (mm)	Specific activity (Ci/mmol)	Decay mode
Carbon-11	20.4 m	$^{14}N(p,\alpha)^{11}C$	0.96	4.1	9.22×10^6	$\beta+$(99%)
Nitrogen-13	10.0 m	$^{16}O(p,\alpha)^{13}N$	1.19	5.42	18.9×10^6	$\beta+$(100%)
Oxygen-15	2.03 m	$^{14}N(d,n)^{15}O$	1.70	8.0	91.7×10^6	$\beta+$(100%)
Fluorine-18	109.7 m	$^{18}O(p,n)^{18}F$	0.64	2.4	1.71×10^6	$\beta+$(97%)

[2] G. Antoni, T. Kihlberg, and B. Långström, in "Handbook of Radiopharmaceuticals: Radiochemistry and Applications" (M. J. Welch and C. S. Redvanly, eds.), p. 141. Wiley, West Sussex, England, 2003.
[3] D.-H. Hwang and S. R. Bergmann, in "Handbook of Radiopharmaceuticals: Radiochemistry and Applications" (M. J. Welch and C. S. Redvanly, eds.), p. 529. Wiley, West Sussex, England, 2003.

radionuclides in order to assure that sufficient radioactive decay has occurred between imaging studies. Furthermore, because it is our goal to measure the metabolic flux through each enzymatic pathway, positron-emitting versions of the metabolic substrates are best suited for this purpose. The radionuclide chosen for these studies is carbon-11, and efficient syntheses for the preparation of 1-[^{11}C]D-glucose, 1-[^{11}C]acetate, and 1-[^{11}C]palmitate have been developed in our laboratory.

Carbon-11 is produced by the $^{14}N(p,\alpha)^{11}C$ reaction using a gas target system of 0.5% oxygen in nitrogen with typical bombardments of 20–40 min at 40 μA beam power. The product obtained from the target is [^{11}C]CO$_2$, which is then trapped under vacuum in a specially designed stainless steel coil cooled to $-196°$ with liquid nitrogen.

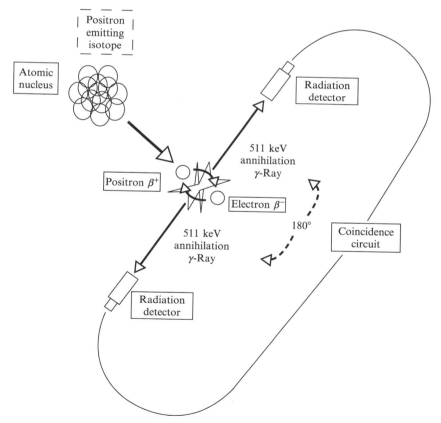

Fig. 1. Positron annihilation and coincidence detection of 0.511-MeV γ rays.

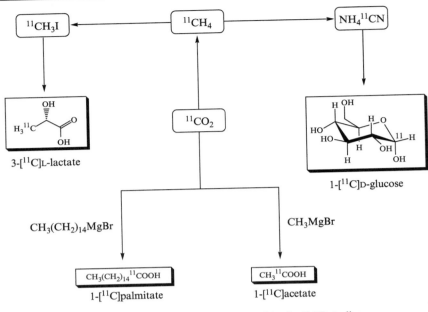

FIG. 2. Synthesis of the radiotracers used in the GAP studies.

The transformations undergone by $[^{11}C]CO_2$ to produce these cardiac tracers are depicted in Fig. 2. The primary conversions shown inside the small rectangles are fast (on average, 2–5 min are needed for the conversions) and can take place on solid support systems by catalysts, as in the case of $[^{11}C]CH_4$ and $[^{11}C]NH_4CN$, or in the gas phase, as in the case of $[^{11}C]CH_3I$. It is important to remove from the $[^{11}C]CO_2$ gas all traces of nonradioactive oxides of nitrogen (NO_x) produced as contaminants during bombardment of the target gas. These oxides, present in about 50–150 ppm after 30 min of bombardment, may contribute to catalytic poisoning of the surface of the Pt wire inside the furnace used to produce glucose. These oxides may also lead to a lowering of radiochemical yields when using organometallic reactions,[4] such as the Grignard reaction used in the synthesis of 1-$[^{11}C]$palmitate and 1-$[^{11}C]$acetate. Removal of the oxides of nitrogen has a significant advantage in the reaction of carbon-11 with Grignard reagents, as it allows the use of a smaller amount of starting material and results in a significant improvement in specific activity. Removal of the

---

[4] M. S. Kharasch and O. Reinmuth, "Grignard Reactions of Non-metallic Substances," p. 1243. Prentice Hall, New York, 1954.

oxides of nitrogen is accomplished easily by placing a commercially available NO_x scrubber inline before passing the radioactive gases through to the next conversion stage. The Kitagawa gas detector solid-phase system (Matheson Gas Products tube #175SH) has been used routinely for this purpose in all of the synthesis modules described here.[5]

Specific Activity (SA)

Specific activity is a very important parameter when working with radiopharmaceuticals. SA, defined as the amount of activity per unit mass, is expressed as the number of nuclear disintegrations per minute per mole of compound. For carbon-11, the theoretical SA of 9.2×10^6 Ci/mmol of the carrier-free radionuclide (Table I) is never achieved in practice, and dilutions to approximately 10^4 Ci/mmol are common due to the presence of traces of carbon in the gas lines, delivery systems, reagents, etc. High SA radiotracers are very important when dealing with receptor-based compounds, gene expression experiments, and other specific uptake studies where saturation with the carrier compounds may invalidate the results.

The issue of low SA for carbon-11 compounds has been documented extensively in the literature, and a number of solutions have been offered to try to improve it. Among them are the use of ultrahigh purity gases, a careful consideration of the target and window foil material used, and an adherence to rigid protocols for cleaning such targets and related glassware. The use of the highest quality material available is necessary for all ancillary equipment needed for isotope production, such as gas regulators, connectors, compressors, water lines, and vacuum pump oil, as all are potential sources of traces of carbon that will reduce the specific activity of the radiopharmaceutical. In addition, minimizing the amount of reagents used in each reaction and exploring gas-phase and solid-phase reactions have all been examined by chemists with some success. Because of the physiological presence of glucose, acetate, and palmitate in the human body, a very high SA for the GAP radiopharmaceuticals is of lesser concern.

Synthesis Modules

We have built semiautomated remote chemistry systems for the production of 1-[^{11}C]D-glucose and 1-[^{11}C]palmitate. These systems are shielded by placement within hot cells in the cyclotron area. They are

---

[5] C. S. Dence, T. J. McCarthy, and M. J. Welch, "Proceedings 6th Workshop on Targetry and Target Chemistry," p. 216. Vancouver, Canada, 1995.

relatively inexpensive and easy to construct and allow the chemist to make quick adjustments when needed during the synthesis. In contrast to the more sophisticated robotic manipulators that are also used, these remote gantries can be adapted to new, exploratory syntheses with a minimum investment of time. All the alternatives to synthesize these carbon-11 tracers (remote gantries, robotic hands, and commercially available synthesizers) share the important characteristics of reliability and very low exposure to the operator (<2 mR per synthesis). The in-house built systems are also the least expensive of the three systems.

1-[^{11}C]D-Glucose

Ever since the early 1990s, the authors have been involved in the synthesis of 1-[^{11}C]D-glucose.[6] The result has been an improved procedure for the production of the desired compound in sufficient quantities for two simultaneous human studies and an additional animal study if needed.[7] This improved synthesis involves the use of a preformed sugar–borate complex of the starting substrate, D-arabinose, to effect the condensation with [^{11}C]NH$_4$CN. The stereochemistry of this sugar–borate complex favors the formation of glucose over mannose (ratio 1.8 ± 0.6:1). The overall chemistry illustrated in Scheme 1 is performed using a semiautomated remote system illustrated in Fig. 3.

The reaction vessel A is a two-necked 10-ml conical flask (14/20 joints) filled with 0.5 ml of 0.01 N NaOH, and the pH probe is set in place in one of the side arms. Vessel B, a 10-ml conical vial that is modified to admit a side Teflon line, contains a freshly prepared Raney-nickel slurry (about 0.3 g) in 30% formic acid (2.0 ml). The purification column C (Bio-Rad column 26 cm long × 10 mm i.d.) contains 8.5–9.0 g of anion-exchange resin Bio-Rex 5, 100–200 mesh (OH form), and 3.5 g of cation-exchange resin AG50W-X8 100–200 mesh (H form). The purification column C is connected to the 50-ml vessel of a rotary evaporator by means of a three-way valve and a Teflon line. The rotary evaporator vessel is provided with 10–11 ml of acetonitrile to remove excess water azeotropically from the final mixture prior to HPLC purification.

At the end of bombardment (EOB), the [^{11}C]CO$_2$ is first converted to [^{11}C]CH$_4$ in a furnace heated with a nickel-chromium resistance wire maintained at 385°. This furnace is provided with a borosilicate glass tube (26 cm length × 9 mm i.d.) filled in the center of the tube with small pieces

---

[6] C. S. Dence, W. J. Powers, and M. J. Welch, *Appl. Radiat. Isot.* **44**, 971 (1993).

[7] C. S. Dence, W. J. Powers, R. J. Gropler, and M. J. Welch, *J. Labeled Compd. Radiopharm.* **40**, 777 (1997).

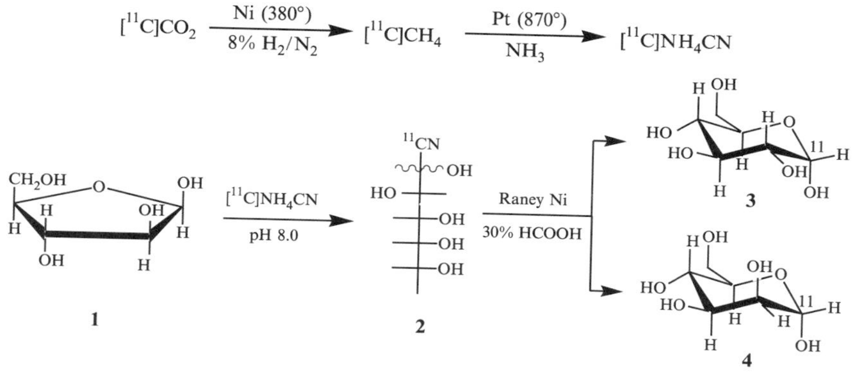

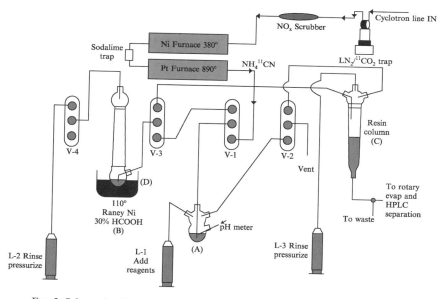

SCHEME 1. Radiosynthesis of 1-[^{11}C]D-glucose.

FIG. 3. Schematic of the gantry system used in the synthesis of 1-[^{11}C]D-glucose.

of glass wool coated with 0.5 g of nickel powder on Kieselguhr (Ventron). The nickel catalyst is held in place by putting at each end of the same another small piece of glass wool and stainless steel wool. A second furnace to convert the [^{11}C]CH$_4$ to [^{11}C]NH$_4$CN has a quartz glass column 36 cm length × 9 mm i.d., which holds 1.2 g of 0.25-mm-diameter platinum wire (Aldrich 26,717-1) wound inside the quartz tube and is kept at 870–880°. All the connections between the two furnaces and the radiator trap for the radioactive gases are done with 1/8 o.d. Teflon tubing. The connections within valves and glassware in the gantry are made with 1/16 Teflon tubing. Two small radioactive detectors located strategically near the $[^{11}C]CO_2$ trap and vessel A help follow the flow of radioactivity during the synthesis.

The $[^{11}C]CO_2$ gas is first trapped under vacuum in a radiator trap kept at −196° with liquid N_2. At the end of collection, the radiator trap is brought to atmospheric pressure, and the $[^{11}C]CO_2$ is displaced from the trap with a reducing gas mixture of 8% H_2 in nitrogen. The gas is first passed through the Kitagawa purification tube to remove nonradioactive oxides of nitrogen, as described earlier. The radioactive gas then goes through the first furnace (nickel furnace) and is converted to $[^{11}C]CH_4$ by the stream of the reducing gas at 20–30 ml/min. Any unconverted $[^{11}C]CO_2$ is removed from the gas stream by a stainless steel soda-lime tube (6–8 g) placed in between the two furnaces.

Anhydrous ammonia (2 ml/min) is then added to the $^{11}CH_4/H_2/N_2$ stream, and the mixture is passed through the second furnace (platinum furnace), where it is converted to $[^{11}C]NH_4CN$. The no-carrier-added $[^{11}C]NH_4CN$ is introduced into the gantry through valve V1 and is collected in vessel A after about 5 min from EOB. After a peak reading on the radioactive detector, the ammonia flow is stopped and the pH is adjusted to approximately 9 with about 0.18 ml of 3 M glacial acetic acid. The substrate D-arabinose (10–14 mg) in 0.45 ml of 0.033 M borate buffer (pH 8.1) is added to reaction flask A and the mixture is allowed to react for 5 min at room temperature.

At the end of this incubation time, the clear reaction mixture is transferred by air pressure into flask B, where the intermediate aldonitriles are reduced with the Raney-nickel slurry. The mixture is heated under reflux for 5 min at 110°, is cooled for approximately 1 min, and is transferred by air pressure to resin purification column C. After the first 2–3 ml of eluate are discarded, column C is flushed dry with air to eliminate as much as possible the excess solvents used: sodium hydroxide, formic acid, acetic acid, etc. Finally, the radioactive sugars are eluted from column C by the addition of about 9 ml of water. The eluate is reduced azeotropically in the rotary evaporator under vacuum/heat to less than 0.1 ml volume, diluted with deionized (DI) water to 0.8–0.9 ml, and injected onto the preparative high-performance liquid chromatography (HPLC). 1-$[^{11}C]$D-Glucose is purified on a 7.8 × 300-mm Bio-Rad HPX-87P column heated to 85° and is then eluted with water at 1.0 ml/min. The aqueous fraction containing the

1-[^{11}C]D-glucose is collected after about 8 min in a sterile 6-ml syringe. This fraction is purified further of any remaining inorganic or metal ions by filtering through an ion-exchange chromatography Chelex cartridge (Alltech 30250). The eluate is made isotonic by the addition of 3 M NaCl, diluted to 5–7 ml with DI water, and filtered through a 0.22-μm, 25-mm vented filter to produce a sterile and pyrogen-free solution ready for injection. Current yields are from 40 to 60 mCi of final product, ready for injection, following a synthesis time of about 50 min from EOB.

[1-^{11}C]Palmitate

Synthesis of [1-^{11}C]palmitate is accomplished according to the method outlined in Fig. 2 and with the remote system detailed in Fig. 4.[8,9] The system uses a single reaction vessel 13.5 cm long × 3.5 cm wide (B), equipped with two side arms. The lower section is in the shape of a 10-mm o.d. test tube marked at the 1-cc volume. The vessel is agitated using a standard laboratory mixer (C). A 1-mm i.d. Teflon tube (L1) is inserted through one side arm to add reagents from outside the hood. Another 1-mm i.d. Teflon tube (L2) is inserted through the second side arm to the bottom of the vessel (B). The other end of L2 enters a Teflon block (D) (4 × 5.5 cm, bored as indicated in Fig. 2). Block D is the common path for connecting the reaction vessel (B) to the waste receptacle (F) and the filtration gantry (E).

At the end of bombardment, the [^{11}C]CO$_2$ produced is collected by evacuating the target gas into the cooled ($-196°$) radiator trap (A). The Grignard reagent, 0.1 M 1-pentadecylmagnesium bromide in diethyl ether (3 ml), is added to the reaction vessel (B). The Dewar (I) is lowered, the radiator trap (A) is warmed, and the [^{11}C]CO$_2$/N$_2$ is bubbled through the Grignard reagent for 1–3 min. The reaction is quenched by the addition of 3 ml of 1.0 N HCl, and 3 ml of diethyl ether is added to extract the [^{11}C]palmitate. The vessel (B) is shaken in the vortex mixer (C) and the layers are allowed to separate. The lower aqueous layer is drawn off by closing valve V2 (vent), opening valve V7 briefly to slightly pressurize the vessel (B), and then opening valve V6. The aqueous layer is pushed through the Teflon line (L2) and Teflon block (D) and into the waste receptacle (F). The remaining ether layer is washed twice with 5 ml of 0.9%

[8] M. J. Welch, S. L. Wittmer, C. S. Dence, and T. J. Tewson, in "Short-Lived Radionuclides in Chemistry and Biology" (J. W. Root and K. A. Krohn, eds.), ACS Advances in Chemistry Series 197. p. 407. Washington, DC, 1981.

[9] M. J. Welch, C. S. Dence, D. R. Marshall, and M. R. Kilbourn, J. Labeled Compd. Radiopharm. 30, 1087 (1983).

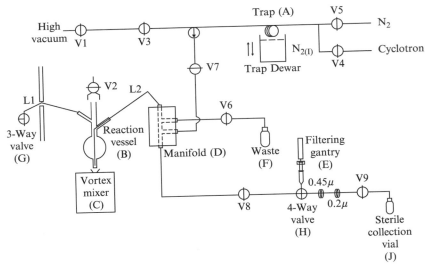

FIG. 4. Schematic of the gantry system used in the synthesis of 1-[^{11}C]palmitate.

sodium chloride (USP saline), and each time the lower aqueous layer is discarded into the waste receptacle (F). The ether solution is then diluted with 1 ml of 95% ethyl alcohol and the diethyl either is evaporated using a 100-ml/min flow of nitrogen through the solution, accomplished by opening valves V5, V7, and V2. Evaporation is continued until less than 1 ml of liquid (ethanol layer) is left, the nitrogen flow is reduced to 10 ml/min, and valve V5 is closed.

The ethyl alcohol solution is then diluted with 8.2 ml of a 3.5% solution of human serum albumin in USP saline (kept at ~50° during the synthesis) and left to stand undisturbed for 3 min to complex the fatty acid with the albumin. For the final filtration, the albumin fatty acid solution is pulled from the vessel (B) into the 12-ml syringe and is pushed through the 0.45- and 0.22-μm, 25-mm filters and into the collection vial (J). Current yields at the end of a 25- to 30-min bombardment are from 150 to 350 mCi of final product ready for injection in a synthesis time of 15–20 min from EOB.

1-[^{11}C] Acetate

We use a *robotic system*, as well as a commercially available *synthesizer*, for the production of 1-[^{11}C]acetate. The Hudson workstation (Thermo Electron Corp., Ontario, Canada) is illustrated in Fig. 5. All platforms have

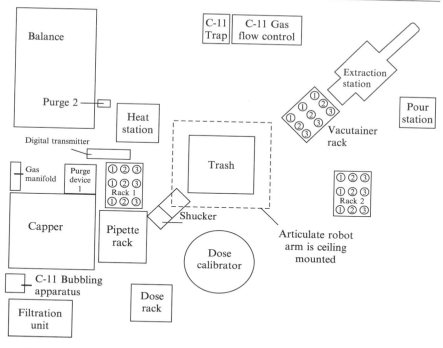

FIG. 5. Hudson robotic system for the compounding of an 1-[^{11}C]acetate injection.

been custom built and include a dedicated reagent rack constructed of Plexiglas, a capping station, a nitrogen purge device, and three 3-way valves and four 2-way valves to direct the [^{11}C]CO$_2$ target gas and nitrogen gas flow to the trapping/reaction vessel. The capping station holds a 5-ml Reactivial (Wheaton Scientific, Millville, IL) used for trapping the [^{11}C]CO$_2$. A 3.5-in. × 19-gauge needle connects the inlet line for incoming target gases and nitrogen, and a 1.5-in. × 18-gauge needle connects the outlet line for venting and trapping any unreacted [^{11}C]CO$_2$. After the [^{11}C]CO$_2$ target gas addition, the system removes the cap from the Reactivial with the top-loading capping station from Hudson. The heating station is constructed from a heating block obtained from Fisher Scientific (St. Louis, MO), and an aluminum block into which a hole is drilled to accommodate the 5-ml Reactivial. A temperature controller (Model D1311 from Omega Engineering, Stanford, CT) controls the temperature of the heating block. All liquid transfers are done with disposable 1000-μl pipette tips using a pipette tip holder attached to the gripping hand of the Hudson robot that connects to a Hamilton syringe station to control pipetting and dispensing.

After cyclotron irradiation, $[^{11}C]CO_2$ is collected in an evacuated cold trap cooled with liquid nitrogen. The trap is removed from the liquid nitrogen, warmed, and purged using a stream of nitrogen gas (10 ml/min). The stream of nitrogen gas containing the $[^{11}C]CO_2$ is bubbled through 50 μl of 3 M methylmagnesium bromide diluted with 3 ml diethyl ether in a 5-ml Reactivial. USP sterile water for injection (0.25 ml) is added, and the reaction vial is placed onto a 50° heating block to evaporate the ether using a stream of nitrogen gas. The intermediate organometallic complex is hydrolyzed by the addition of 2 ml of 0.4 N hydrochloric acid, followed by the addition of 2 ml of USP sterile water for injection. Purification of the acidic mixture by solid phase is accomplished by the robot, which pours the solution onto the PrepSep funnel, which contains 600–700 mg of C18 resin activated previously with 5 ml of ethanol and rinsed with 10 ml of sterile water and 2–3 g of AG 11A8 ion retardation resin. The final product is filtered through a 0.22-μm, 25-mm, sterile, apyrogenic filter. Batches of 220 mCi on average at the end of 20-min syntheses are obtained routinely after a 20-min bombardment.

1-$[^{11}C]$Acetate is also produced routinely in our laboratory with a commercially available *chemistry synthesizer*, the CTI acetate module (Knoxville, TN) (Fig. 6). After irradiation, the target gases are passed through a needle valve into the reagent vessel; the flow of the target gas is limited to less than 100 ml/min. The stream of nitrogen gas containing the $[^{11}C]CO_2$ is bubbled through 50 μl of 3 M methylmagnesium bromide diluted with 1 ml diethyl ether in a 10-mm o.d. × 75-mm-long test tube sealed with a red flange stopper top. Sterile water for injection, USP (0.25 ml) is added, and the reaction vial is heated to 145°. The ether is evaporated using a stream of helium gas. The intermediate is hydrolyzed by the addition of 0.5 ml of 10% phosphoric acid. The acidic mixture is then purified by heating at 145° for 420 s to distill 1-$[^{11}C]$acetate into USP saline (10 ml). The final product is passed through a 0.22-μm, 25-mm, sterile, apyrogenic filter. Average yields of 200 mCi of 1-$[^{11}C]$acetate are obtained routinely in a synthesis time of 20 min from EOB.

Overview of the Quality Assurance of C-11 Radiopharmaceuticals

Radiopharmaceuticals administered for PET procedures and which contain radionuclides of very short half-lives, such as carbon-11 (20.4 min), must be analyzed, must meet quality assurance specifications, and must be fully documented prior to administration to humans.[10,11] This requires various types of quality control (QC) determinations, such as radionuclidic identity, radionuclidic purity, radiochemical and chemical purity, sterility,

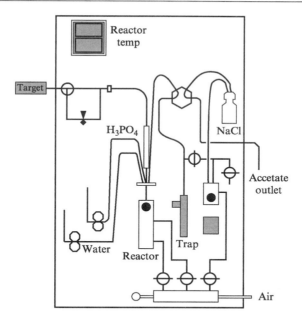

FIG. 6. CTI 1-[^{11}C]acetate synthesis module.

and pyrogen testing. It is important that each analytical test be validated and the limits for each test specified.

Radionuclidic Identity

This can be accomplished prior to release of the radiopharmaceutical by decay analysis using a dose calibrator computer program to calculate $T_{1/2}$.

Radionuclidic Purity

The fraction of total radioactivity that is present as the specified radionuclide should be determined from a γ-ray spectrum by means of a multichannel pulse height analyzer or a germanium detector to detect the presence of any γ photon energies other than 0.511 MeV and the sum peak at 1.02 MeV. The radionuclidic purity determination can be made three to four half-lives after the end of bombardment with little interference from the main 0.511-MeV photopeak of the positron emitters.

[10] M. R. Kilbourn, C. S. Dence, K. A. Lechner, and M. J. Welch, *in* "Quality Assurance of Pharmaceuticals Manufactured in the Hospital" (A. Warbick-Cerone and L. G. Johnson, eds.), p. 243. Pergamon Press, New York, 1985.

[11] H. Vera-Ruiz, C. S. Marcus, V. W. Pike *et al.*, *Nucl. Med. Biol.* **17**, 445 (1990).

Radiochemical Purity

The quality control method for determining *radiochemical purity* (the fraction of the radiopharmaceutical that is present in the desired chemical form, with the label in the specified molecular position) is usually based on a chromatographic method with simultaneous mass and radioactivity detection. Various types of chromatographic methods such as thin-layer chromatography (TLC) or HPLC can be used. HPLC allows the in-line measurement of chemical and radiochemical purity. These chromatographic methods are performed using standards for the comparison of R_f values and retention times or for generating standard calibration curves.[12]

Radioactivity Balance

For the development of new radiopharmaceuticals requiring analysis by HPLC, it is generally necessary to perform a *radioactivity balance* measurement to ensure that the total radioactivity injected onto a HPLC column is recovered at the end of the specified HPLC run time. This determination can be made by injection of an aliquot of the radiopharmaceutical (10–50 μl) onto the HPLC and collecting all the postinjection effluent solvent in a volumetric flask. The flask is then taken to the standard volume and mixed well. For example, for an HPLC run at 1.5 ml/min that lasts 10 min, a 25-ml volumetric flask is used to collect the eluate. After 10 min the flask is removed from the eluting line and a sufficient volume of water (or a solvent compatible with the HPLC solvent used) is added to bring the total volume to 25 ml. This constitutes "sample counts." The "standard counts" sample is obtained by aliquoting the original volume injected on the HPLC column (10–50 μl from the same original analyte) into another 25-ml volumetric, and adding sufficient solvent to bring the total volume to 25 ml. An aliquot from each volumetric flask is transferred into a test tube, and the samples are counted in a sodium iodide γ counter. Due to the short half-life of C-11, it is necessary to decay correct both samples to time zero and compute the percentage eluted as follows: ("counts in sample"/ "counts in standard") × 100. Acceptable values should be 85% or higher. With high specific activity radiopharmaceuticals, some radioactivity losses are unavoidable on the HPLC tubing, column, precolumn, filters, etc. Values below 85% are usually indicative of radioactivity being retained on the chromatographic column or precolumns that need to be addressed in order to avoid inaccurate radiochemical purity determinations.

[12] M. R. Kilbourn, M. J. Welch, C. S. Dence, and K. A. Lechner, *in* "Analytical and Chromatographic Techniques in Radiopharmaceutical Chemistry" (D. M. Wieland, M. C. Tobes, and T. J. Mangner, eds.), p. 251. Springer-Verlag, New York, 1986.

Chemical Purity

Analyses are required to verify the absence of any chemical impurities or solvent residues in the final preparation. Analyses also require an injection of a known standard each time a QC is performed. In this way the accuracy of the HPLC, or of any other system that is used, is determined each time. Chemical impurities separated by HPLC are generally easy to detect if they are ultraviolet (UV) absorbing. Compounds with low UV absorption characteristics can be detected by pulsed amperometric detectors (PADs) or by HPLC mass spectrometry. The *chemical purity* requirement serves to verify the chemical identity of the product and by-products, including stereoisomeric purity, if needed (e.g., 1-[^{11}C]D-glucose versus 1-[^{11}C]D-mannose). Ion chromatographic techniques are applied routinely to the detection of residual inorganic species in order to validate the absence of nickel, borate, and lead ions. Lead ions may potentially arise from the lead counterion of the Aminex resin of the HPLC column used in the preparative purification of the 1-[^{11}C]D-glucose preparation.[6]

The quantification of organic solvent residues such as ethanol, acetonitrile, and ether used in the preparation of the carbon-11 compounds is performed by gas chromatography. Separation is carried out on a Varian 3800 gas chromatograph using a capillary column DB-Wax, 30-m × 0.53-mm i.d., 1-μm film thickness (J&W Scientific). The injector and FID detector temperatures are held at 225° and 275°, respectively. Helium is used as a carrier at a flow rate of 7 ml/min. The column temperature is held at 40° for 2 min and is raised to 110° in 7 min. Analyses are performed on a 1.0-μl aliquot of the final injectable solution after sterile filtration.

Sterility, Apyrogenicity, Isotonicity, and Acidity

Tests should be performed to ensure sterility, apyrogenicity, isotonicity, and suitable acidity (pH) before administration to humans. *Sterility* should be determined postrelease on each batch of parenteral radiopharmaceutical intended for human use. The injectable drug product must be sterilized by filtration through a 0.22-μm filter as a final step in the radiosynthetic procedure. USP methods for sterility require inoculation of the radiopharmaceutical into both tryptic soy broth and fluid thioglycolate media within 24-h postend of synthesis (EOS). Because an entire lot of a PET radiopharmaceutical may be administered to one or several subjects, depending on the radioactivity remaining in the container at the time of administration, administration of the entire quantity of the lot to a single patient should be anticipated for each lot prepared. Verification of *apyrogenicity* should be made using the USP bacterial endotoxin test (BET) on each batch of every nongaseous radiopharmaceutical prepared for intravenous human administration. The USP limit for endotoxins is 175 endotoxin units (EU) per volume (V), which is the maximum volume administered in the total dose. If the $T_{1/2} \geq 20$ min, a 20-min endotoxin "limit test" must be performed prerelease (USP 46). A standard 60-min test must also be performed on each batch. The *pH* and *isotonicity* can be adjusted to physiologic values prior to the final filtration by the addition of sterile buffers or by a 0.9% sodium chloride solution.

Dosimetry Calculations

All radiopharmaceuticals that are injected into humans require that absorbed dose calculations be performed. These dose calculations are needed to predict the risk involved in the use of any ionizing radiation.[13] Medical internal radiation dose (MIRD) calculations require knowledge of certain parameters. These include the amount of *cumulative activity* in each of the organs of the body and the type of radiation administered.

Cumulative activity represents the time course of the radioactivity in the body that requires information on the rate of radiopharmaceutical uptake and removal (biological half-life), as well as the physical decay of the radionuclide injected (physical half-life).

Biological clearance data (percentage injected dose/organ over time) are often obtained by the performance of biodistribution studies, usually in rodents, over a time interval. If radiopharmaceuticals labeled with C-11 are used, a 2- to 3-h time interval is acceptable. Clearance can also be determined by the use of dynamic PET imaging. Mathematical models, such as the dynamic bladder model,[14] can also be used to calculate the clearance. Biological clearance data are used to calculate the cumulative activity in each organ. The formula for the absorbed dose is $D = \tilde{A}S$, where $\tilde{A}$ is cumulative activity and S is absorbed dose per unit cumulative activity. The absorbed doses per unit cumulative activity can be calculated for the specific isotope used ($S = \Delta\phi/m$), where

$\Delta\phi = 2.13 \Sigma n_i E_i$, n_i is the number of particles or photons per nuclear transformation, E_i is the mean energy of the radiation, and m is the mass of the organ. These S values can also be obtained from a dosimetry computer software program MIRDOSE3.[15,16]

---

[13] M. G. Stabin, M. Tagesson, S. R. Thomas, M. Ljungberd, and S. E. Strand, *Appl. Radiat. Isot.* **50**, 73 (1999).

[14] S. R. Thomas, M. G. Stabin, C. T. Chen, and R. C. Samaratunga, *J. Nucl. Med.* **40**, 102S (1999).

[15] M. G. Stabin, *J. Nucl. Med.* **37**, 538 (1996).

[16] R. N. Howell, B. W. Wessels, and R. Loevinger, *J. Nucl. Med.* **40**, 3S (1999).

Conduct of GAP Studies

Measurement of Regional Perfusion and Metabolism

Data Acquisition. Figure 7 is a representation of the imaging protocol for the GAP studies. A transmission scan is initially conducted prior to administration of the radiotracer. The transmission scan consists of a data acquisition session in which a positron-emitting point source is rotated 360° around the subject to be scanned. The function of the transmission scan is to provide an accurate measurement of photon attenuation for the attenuation correction of the ensuing emission scans (i.e., a scan acquired after the administration of the radiotracer). For the measurement of myocardial blood flow, up to 0.40 mCi/kg of ^{15}O water (prepared using the method of Welch and Kilbourn[17]) is administered as a bolus intravenously with the immediate initiation of dynamic data collection for 5 min. After allowance for decay of the ^{15}O water, up to 0.30 mCi/kg of 1-[^{11}C]acetate is injected intravenously, and a 30-min dynamic data collection is performed to measure myocardial oxygen consumption. After allowance for decay of the 1-[^{11}C]acetate, up to 0.30 mCi/kg of 1-[^{11}C]D-glucose is injected intravenously, and a 60-min dynamic data collection is performed to measure myocardial glucose metabolism. Finally, after allowance for decay of the 1-[^{11}C]D-glucose, up to 0.30 mCi/kg of 1-[^{11}C]palmitate is injected intravenously, and a 30-min dynamic data collection is performed to measure myocardial fatty acid metabolism.

During the 1-[^{11}C]acetate, 1-[^{11}C]palmitate, and 1-[^{11}C]D-glucose scans, 8–10 venous samples are obtained to measure $^{11}CO_2$ production (in the case of 1-[^{11}C]D-glucose, 1-[^{11}C]lactate is measured as well) in order to correct the arterial input function during kinetic modeling (see kinetic modeling section). Between each metabolic scan (i.e., between 1-[^{11}C]acetate and 1-[^{11}C]D-glucose), subjects are removed from the tomograph to increase their comfort during the study. Upon their return to the tomograph, another transmission scan is performed (a total of three transmission scans per study). Table II lists the total amount of radioactivity administered to the test subject in a typical GAP study, which is approximately 1475 mrem. This represents about 30% of the annual exposure limit for a radiation worker (e.g., personnel involved in the synthesis of PET radiotracers).

Data Reconstruction and Generation of Time–Activity Curves. After data reconstruction is performed, myocardial images are reformatted from the transaxial orientation to the true short-axis views on which measurements of perfusion and metabolism will be performed. Myocardial

---

[17] M. J. Welch and M. R. Kilbourn, *J. Labeled Compd. Radiopharm.* **22,** 1193 (1985).

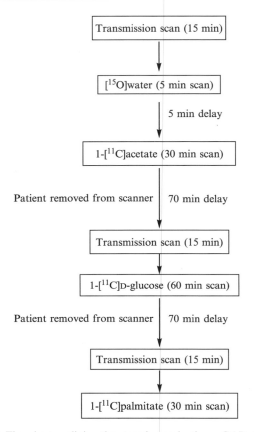

FIG. 7. Flowchart outlining the steps in conducting a GAP study.

[^{15}O]water, 1-[^{11}C]acetate, 1-[^{11}C]D-glucose, and 1-[^{11}C]palmitate images are generated and then reoriented to standard short- and long-axis views. To generate myocardial time–activity curves, regions of interest on anterior, lateral, septal, apical, and inferior myocardial walls (3–5 cm^3) are placed on three to four midventricular short and long axis slices of composite [^{15}O]water, 1-[^{11}C]acetate, 1-[^{11}C]D-glucose, and 1-[^{11}C]palmitate images. To generate blood time–activity curves for each tracer, a small region of interest (1 cm^3) is placed within the left atrial cavity on a midventricular slice in the horizontal long-axis orientation of each composite image. Within these regions of interest, myocardial and blood time–activity curves are generated for each of the sets of tracer data. Subsequently, blood and myocardial time–activity curves are used in conjunction with

TABLE II
EFFECTIVE DOSE EQUIVOLENT (EDE) FOR ALL THE RADIOPHARMACEUTICALS
ADMINISTERED IN A GAP STUDY[a]

[^{15}O]Water[b]	[^{11}C]Acetate[c]	[^{11}C]Palmitate[c]	[^{11}C]D-Glucose[c]	Total
92 mrem[d]	273 mrem[d]	320 mrem[d]	790 mrem[d]	1475 mrem

[a] A measure of the biological damage caused by "ionizing" radiation, e.g., γ rays, X rays, and β particles, is expressed by the dose equivalent, the unit of which is the rem. A special variation of the dose equivalent is the EDE, a computed uniform whole body dose that applies to nonuniform irradiation of differing organs and tissues in the body. The allowable EDE for a radiation worker in the United States is 5000 mrem. The EDE for administration of the radiopharmaceuticals listed here would be 30% of the allowable annual exposure for a radiation worker.
[b] Administered dose: 28 mCi (0.4 mCi/Kg; 70 Kg man).
[c] Administered dose: 21 mCi (0.3 mCi/Kg; 70 Kg man).
[d] EDE from unpublished dosimetry estimates.

well-established kinetic models to measure myocardial blood flow, oxygen consumption, glucose utilization, fatty acid utilization, and oxidation.

Metabolite Analysis

In order to validate the kinetic models used to study the enzymatic pathways of the heart with the carbon-11 tracers, it is necessary to account for the impact of blood acidic metabolites on the arterial input function.[18] This routinely includes the analysis of C-11 acidic metabolites present in blood, namely carbonate and lactate, following the injection of C-11 radiotracers.[19]

Determination of [^{11}C]CO$_2$. The determination is based on the loss of [^{11}C]CO$_2$ under acidic conditions. A 5-ml blood sample is withdrawn into a gray-top Vacutainer (L10330-00; Becton-Dickinson, NJ) after a specified postinjection time. Blood is spun for 5–6 min at 3500 rpm to separate the plasma. For each sample to be analyzed, a set of two test tubes are prepared, each containing 1.0 ml of 0.9 M sodium bicarbonate and 3 ml of isopropanol. To one of the test tubes, 1 ml of 0.1 N NaOH is added (labeled "basic"). To each test tube, 0.5 ml of plasma is added and the tubes are vortexed briefly and gently. The other test tube is then treated with 1 ml of 6 N HCl (labeled "acidic"). All the "acidic" test tubes are

[18] P. Herrero, T. L. Sharp, C. Dence, B. M. Haraden, and R. J. Gropler, *J. Nucl. Med.* **43,** 1530 (2002).
[19] C. S. Dence, P. Herrero, T. L. Sharp, and M. J. Welch, "12th International Symposium on Radiopharmacology," Interlaken Switzerland, Abstract Book, 2001.

placed in a custom-made manifold and purged with a stream of nitrogen for 10 min at room temperature to eliminate $[^{11}C]CO_2$. At the end of bubbling, all the tubes are counted for radioactivity (from last-to-first collected), decay corrected to time zero, and the percentage of $[^{11}C]CO_2$ is calculated (counts "acidic"/counts "basic" $\times$ 100).

Total Acidic Metabolites. Following injections of 1-$[^{11}C]$D-glucose, blood samples are analyzed for total acidic metabolites (TAM) ($[^{11}C]$carbonate and $[^{11}C]$lactate) and residual 1-$[^{11}C]$D-glucose. Analysis is based on the trapping of these acidic species on an anion-exchange column while eluting the neutral species. The authors use 1.2–1.3 g of AG1-X8, in the formate form, 100–200 mesh (Bio-Rad) placed in a 6-ml disposable syringe with a glass wool plug in the bottom. The resin is rinsed with 4–6 ml of water. These columns can be prepared 24 h ahead and kept moist at room temperature. The blood samples are collected on special Vacutainer tubes (as described earlier) containing sodium fluoride and potassium oxalate to stop further glucose metabolism. About 1.0 ml of plasma is pipetted into the resin column, and deionized water (in 2 + 3-ml fractions) is used to elute the nonmetabolized 1-$[^{11}C]$D-glucose. This eluate (5 ml total) is collected in one test tube. The acidic metabolites, $[^{11}C]$carbonate and $[^{11}C]$lactate are retained on the resin. Both eluate and resin are counted for radioactivity, and after decay correction the percentage of acidic metabolites is calculated (counts in resin/counts in eluate $\times$ 100). The percentage of $[^{11}C]$lactate is calculated by subtracting the percentage of CO_2 (determined as described earlier) from the percentage of TAM, and the percentage of residual glucose as 100 – %TAM.

Kinetic Modeling

After the blood time–activity curve has been corrected for its metabolites, the blood (input function) and tissue–time activity curves generated from PET cardiac images are used in conjunction with kinetic models to measure myocardial blood flow and metabolism. In the kinetic modeling of PET data derived from GAP studies, the method that is typically used is the *compartmental model method.* A compartmental model is generally represented by a series of compartments linked together by arrows representing transfer between the compartments. A compartment is defined as the space in which the radiotracer is distributed uniformly.[20] The number of compartments that are needed to quantify the fate of the radiotracer

[20] S.-C. Huang and M. E. Phelps, *in* "Positron Emission Tomography and Autoradiography: Principles and Applications for the Brain and Heart" (M. Phelps and J. Mazziotta, eds.), p. 287. Raven Press, New York, 1986.

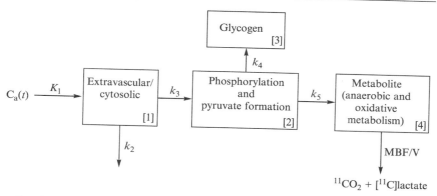

FIG. 8. Four-compartment model used for 1-[^{11}C]D-glucose. K_1 (ml/g/min), k_2, and k_3 (min^{-1}). Net 1-[^{11}C]D-glucose uptake (K, ml/g/min) = $(K_1 * k_3)/(k_2 + k_3)$.

in vivo is directly proportional to the number of steps in the metabolic process being investigated (Fig. 8). The arrows in Fig. 8 represent possible pathways the radiotracer can follow in a particular metabolic pathway. The symbol k above each arrow is a rate constant and denotes the fraction of the total radiotracer that would leave the compartment per unit time [turnover rate (min^{-1})].

The key assumptions of compartmental modeling are that the tracer is distributed homogeneously in each compartment and that the rate of tracer transfer from compartment 1 to compartment 2 is directly proportional to the tracer concentration in compartment 1. These assumptions make it possible to describe the system under study by n first-order, linear differential equations, where n is the number of compartments and each differential equation represents the rate of change of tracer concentration over time in a given compartment, defined as the rate of tracer *in* minus the rate of tracer *out* of the compartment. The solution to the system of differential equations gives the concentration of tracer in each compartment over time. The sum of the concentrations from all compartments (a function of the blood activity and the turnover rates of the tracer) represents the myocardial activity over the scanning period (theoretical myocardial time–activity curve). Using nonlinear least-square approaches, turnover rates (k_s values) are estimated by minimizing the differences between the PET derived myocardial time-activity curve and the analytical curve derived from the model. A brief description of the compartmental models used in the kinetic analysis of a GAP study is outlined next.

Myocardial Blood Flow (MBF). [^{15}O]Water is a freely diffusible radiotracer whose myocardial kinetics appears to be related solely to blood flow.

That is, the uptake of [^{15}O]water is not altered by changes in myocardial metabolism.[21–23] The kinetics of freely diffusible tracers such as [^{15}O]water can be described by a simple one compartment model. The approach for quantifying MBF is based on the model developed by Kety[24–26] for radiolabeled inert freely diffusible gases. The rate of change of myocardial tracer activity over time can be defined by a simple differential equation:

$$dC_T(t)/d(t) = (MBF)\ C_a(t) - (MBF)\ C_T(t)/\lambda \tag{1}$$

The solution to the differential equation is shown as

$$C_T(t) = MBF\ C_a(t) * \exp(MBF\ t/\lambda) \tag{2}$$

where $C_T(t)$ is tissue tracer concentration (counts/g), $C_a(t)$ is arterial tracer concentration (input function) (counts/ml), MBF is myocardial blood flow per unit of tissue volume (ml/g/min), λ is the tissue/blood partition coefficient (0.92 ml/g), and * is the convolution process.

Using least-square techniques, estimates of MBF can be obtained by fitting Eq. (2) to the myocardial tissue activity $C_T(t)$. However, because of the limited resolution of PET tomographs (8–12 mm FWHM) in relation to the thickness of the myocardium (approximately 10 mm in humans) and effects of cardiac motion, the true tissue concentration cannot be measured directly with PET. If the object under study is less than twice the resolution of the tomograph, the true activity in the region of interest is underestimated (partial volume effect) and activity from adjacent regions is detected into the region of interest (spillover effect). The relationship between the true and the observed PET activity can be defined as

$$C_{T\ PET}(t) = F_{MM} \times C_T(t) + F_{BM} \times C_a(t) \tag{3}$$

$$C_{a\ PET}(t) = F_{BB} \times C_a(t) + F_{MB} \times C_T(t) \tag{4}$$

where $C_{T\ PET}(t)$ is observed tissue activity (counts/g), $C_T(t)$ is true tissue activity (counts/g), $C_{a\ PET}(t)$ is observed blood pool activity (counts/g), $C_a(t)$ is true blood pool activity (counts/g), F_{MM} is the tissue recovery coefficient (accounts for partial-volume effects in the myocardium), F_{BM} is fraction of blood activity into tissue (accounts for spillover of blood counts into myocardium), F_{BB} is the blood recovery coefficient (accounts

[21] S. R. Bergmann, P. Herrero, J. Markham, C. J. Weinheimer, and M. N. Walsh, *J. Am. Coll. Cardiol.* **14**, 639 (1989).
[22] P. Herrero, J. Markham, and S. R. Bergmann, *J. Comp. Assist. Tomogr.* **13**, 862 (1989).
[23] P. Herrero, J. J. Hartman, M. J. Senneff, and S. R. Bergmann, *J. Nucl. Med.* **35**, 558 (1994).
[24] S. S. Kety, *Pharmacol. Rev.* **3**, 1 (1951).
[25] S. S. Kety, *Methods Med. Res.* **8**, 223 (1960).
[26] S. S. Kety, *Methods Med. Res.* **8**, 228 (1960).

for partial-volume effects in the blood), and F_{MB} is the fraction of tissue activity into blood (accounts for spillover of myocardium counts into blood).

If these correction factors are known *a priori*, the true blood and myocardial activity can be calculated from Eqs. (3) and (4). These correction factors can be calculated analytically if the resolution of the tomograph and the dimensions of the blood chambers and myocardial tissue are known and there is no motion present.[27] However, when cardiac and respiratory motion are present and/or the measurements of cardiac chambers and tissue are not accurate, then measurement of the true blood and tissue–activity curves will be inaccurate and will result in erroneous MBF estimates. This problem is circumvented by incorporating within the blood flow model [Eq. (2)] this relationship between the true tissue activity and the PET-derived tissue activity given by

$$C_{T\,PET}(t) = F_{MM} \times [MBF\ C_a(t) * \exp(-(MBF)t/\lambda)] + F_{BM} \times C_a(t) \quad (5)$$

where MBF is estimated along with F_{MM} and F_{BM} by fitting Eq. (5) to PET-derived tissue activity [$C_{T\,PET}(t)$]. This model assumes that $C_a(t)$ can be measured directly with PET [i.e., $F_{BB} = 1.0$ and $F_{MB} = 0.0$ in Eq. (4)]. This assumption has been validated in experimental human studies.[27]

Kinetic Modeling of ^{11}C-Labeled Metabolic Radiotracers

For all ^{11}C-labeled metabolic radiotracers, a correction for myocardial partial volume and spillover effects was done by estimating the corresponding model turnover rates along with F_{BM} after fixing F_{MM} to values obtained from the [^{15}O]water analysis.

Myocardial Glucose Utilization. Myocardial glucose utilization (MGU) is obtained using a four-compartment model for 1-[^{11}C]D-glucose (Fig. 8).[28] This model assumes that vascular 1-[^{11}C]D-glucose enters the interstitial and cytosolic component of the myocardium (compartment 1) at a rate of K_1 (ml/g/min). Once the tracer enters compartment 1, it either diffuses back into the vascular space at a rate of k_2 (min^{-1}) or it is phosphorylated and metabolized to [^{11}C]pyruvate (compartment 2) at a rate of k_3 (min^{-1}). Phosphorylated 1-[^{11}C]D-glucose can either form [^{11}C]glycogen at a rate of k_4 (min^{-1}) or be metabolized through anaerobic and oxidative pathways (compartment 4) at a rate of k_5 (min^{-1}). Radiotracer entering the metabolic pool is assumed to be unidirectional, and washout of [^{11}C]CO$_2$ and

[27] P. Herrero, J. Markham, D. W. Myears, C. J. Weinheimer, and S. R. Bergmann, *Math. Comp. Model.* **11**, 807 (1988).

[28] P. Herrero, C. J. Weinheimer, C. S. Dence, W. F. Oellerich, and R. J. Gropler, *J. Nucl. Cardiol.* **9**, 5 (2002).

[^{11}C]lactate to the vasculature is assumed to be proportional to blood flow. This is an important assumption, as it enables one to "lump" all the radiolabeled intermediates in the glycolytic pathway (Fig. 9) into one compartment.

Differential equations defining the model are as follows:

$$dq_1(t)/dt = K_1 C_a(t) - (k_2 + k_3)q_2(t) \tag{6}$$

$$dq_2(t)/dt = k_3 q_1(t) - (k_4 + k_5)q_2(t) \tag{7}$$

$$dq_3(t)/dt = k_4 q_1(t) \tag{8}$$

$$dq_4(t)/dt = k_5 q_2(t) - (MBF)q_4(t) \tag{9}$$

The solution of this set of differential equations results in the concentration of tracer in each compartment (q_1–q_4). The total radiotracer concentration in the myocardium as a function of time is then defined as the sum of the radiotracer concentrations in each compartment:

$$q_t(t) = q_1(t) + q_2(t) + q_3(t) + q_4(t) = f(C_a(t), K_1, k_2-k_5, MBF, V) \tag{10}$$

where $C_a(t)$ is arterial 1-[^{11}C]D-glucose concentration (i.e., the input function for 1-[^{11}C]D-glucose) K_1 and k_2–k_5 are turnover constants describing the transfer of radiotracer between compartments, q_n is the concentration of radiotracer in compartment n (counts/ml), and V is the fractional vascular volume and is assumed to be 10% of total volume (0.1 ml/g); f represents a function of the parameters in parentheses. The model transfer rate constants (K_1, k_2–k_5) and the spillover fraction are estimated using well-established least-square approaches by fitting the model Eq. (10) to PET myocardial time–activity curves. If one assumes steady-state conditions (i.e., differential equations are set to zero), then these estimated turnover rates (K_1, k_2–k_5) are used to calculate the net 1-[^{11}C]D-glucose uptake (GLU$_{uptake}$) and MGU as

$$GLU_{uptake}(ml/g/min) = (K_1 k_3)/(k_2 + k_3) \quad and \tag{11}$$

$$MGU(nmol/g/min) = (Gl_b, nmol/ml)(GLU_{uptake}, ml/g/min) \tag{12}$$

The myocardial glucose extraction fraction (EF$_{GLU}$) is calculated as

$$EF_{GLU} = GLU_{uptake}/MBF \tag{13}$$

Hence, myocardial glucose utilization can be calculated as the product of three key measurements—the plasma glucose level, myocardial blood flow, and the myocardial glucose extraction fraction:

$$MGU(nmol/g/min) = (Gl_b)(MBF)(EF_{GLU}) \tag{14}$$

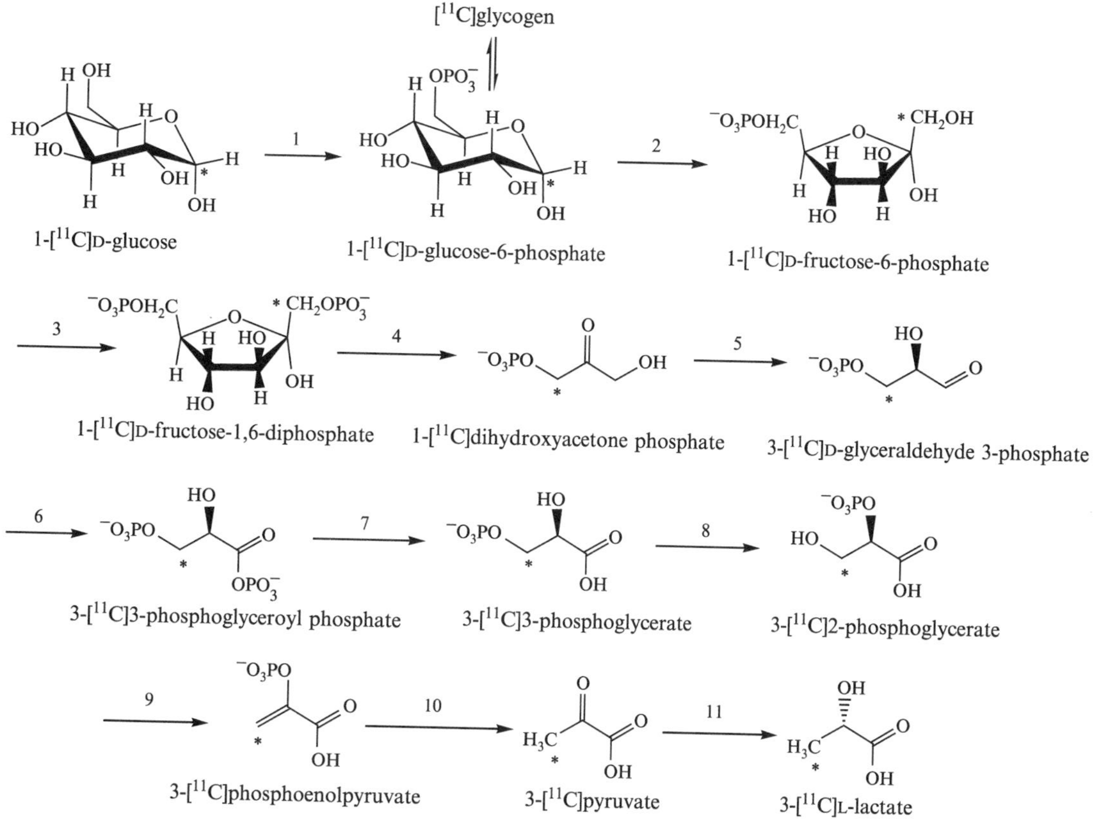

FIG. 9. Fate of the ^{11}C radiolabel in 1-[^{11}C]D-glucose in the glycolytic pathway. Enzymes: 1, hexokinase; 2, glucose phosphate isomerase; 3, 6-phosphofructokinase; 4, fructose diphosphate aldolase; 5, triosephosphate isomerase; 6, glyceraldehydephosphate dehydrogenase; 7, phosphoglycerate kinase; 8, phosphoglyceromutase; 9, enolase; 10, pyruvate kinase; 11, lactate dehydrogenase. From Nelson and Cox.^{28a}

where Gl_b is the plasma glucose level in nmol/ml measured from venous blood samples obtained during the PET study and EF_{GLU} is the myocardial glucose extraction fraction estimated from $1\text{-}[^{11}C]$D-glucose kinetics.

Myocardial Oxidative Metabolism. After correction of the PET-derived blood ^{11}C activity for the $^{11}CO_2$ contribution, the blood and myocardial time–activity curves are used in conjunction with a simple one compartment model to estimate two turnover rates: K_1 (ml/g/min) representing the net rate of $1\text{-}[^{11}C]$acetate uptake into the myocardium, and k_2 (min^{-1}) representing the rate at which $1\text{-}[^{11}C]$acetate is converted to $^{11}CO_2$. The latter has been shown to be directly proportional to myocardial oxygen consumption (MVO_2).[29,30] MVO_2 (μmol/g/min) is then calculated from an experimentally derived linear relationship between k_2 and MVO_2 in humans.[31]

Myocardial Fatty Acid Metabolism. The kinetics of $1\text{-}[^{11}C]$palmitate metabolism are described by a four-compartment model (Fig. 10). The model assumes that the radiotracer entering mitochondria is unidirectional and that metabolites are washed out of mitochondria into the vasculature at a rate proportional to blood flow. The differential equations describing the model are:

$$dq_1(t)/dt = MBF[Ca(t) - q_1/V] + k_2q_2 - k_1q_1 \qquad (15)$$

$$dq_2(t)/dt = k_1q_1 + k_4q_3 - (k_2 + k_3 + k_5)q_2 \qquad (16)$$

$$dq_3(t)/dt = k_3q_2 - k_4q_3 \qquad (17)$$

$$dq_4(t)/dt = k_5q_2 - (MBF/V)q_4 \qquad (18)$$

The solution of this set of differential equations gives the concentration of tracer in each compartment (q_1–q_4). The total radiotracer concentration in the myocardium at any given time can be defined by the sum of the radiotracer concentrations in each compartment:

$$q_t(t) = q_1(t) + q_2(t) + q_3(t) + q_4(t) = f(C_a(t), k_1 - k_5, MBF, V) \qquad (19)$$

where MBF is myocardial blood flow (ml/g/min), $C_a(t)$ is concentration of $1\text{-}[^{11}C]$palmitate in blood (input function of $1\text{-}[^{11}C]$palmitate, counts/ml), k_1–k_5 (min^{-1}) are turnover constants describing the transfer of radiotracer between compartments, V is the fractional vascular volume (assumed to be

[28a] D. L. Nelson and M. M. Cox, in "Lehninger Principles of Biochemistry" (D. L. Nelson and M. M. Cox, eds.), p. 527. Worth Publishers, New York, 2000.
[29] M. A. Brown, D. W. Myears, and S. R. Bergmann, *J. Am. Coll. Cardiol.* **12**, 1054 (1988).
[30] D. B. Buxton, C. A. Nienaben, A. Luxen *et al.*, *Circulation* **79**, 134 (1989).
[31] R. S. Beanlands, D. S. Bach, R. Raylman, W. F. Armstrong, V. Wilson, M. Montieth, C. K. Moore, E. Bates, and M. Schwaiger, *J. Am. Coll. Cardiol.* **5**, 1389 (1993).

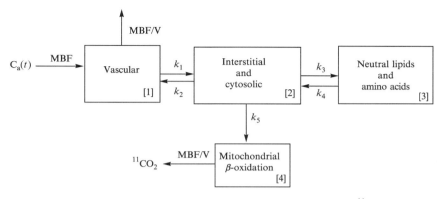

FIG. 10. The four-compartment model used in the kinetic analysis of 1-[^{11}C]palmitate.

10% of total volume) (0.1 ml/g), q_n is the concentration of radiotracer in compartment n (counts/ml), and q_t is the total concentration of radiotracer in the myocardium (counts/ml).

Because blood contains radiolabeled metabolites in addition to 1-[^{11}C]palmitate, the PET-derived ^{11}C blood activity must be corrected for the presence of ^{11}CO$_2$ as described earlier in order to derive the true input function. Myocardial blood flow is fixed to values obtained with [^{15}O]water, and the fractional vascular volume is fixed at 0.1 ml/g. Similarly to the analysis of 1-[^{11}C]D-glucose, after a correction of the blood ^{11}C activity for ^{11}CO$_2$, the turnover rates (k_1–k_5) and F_{BM} are estimated by fitting Eq. (19) to the PET myocardial time–activity curves after fixing F_{MM} to values obtained from [^{15}O]water analyses.

If one assumes steady-state conditions (i.e., each differential equation is set equal to zero), then key quantities such as fatty acid utilization, oxidation, and esterification (all in nmol/g/min) can be calculated from the estimated turnover rates. The model assumes that once the tracer is in the interstitial/cytosolic compartment, it can follow only three pathways: back diffusion into the vasculature (k_2), storage into neutral lipids and amino acids (esterification) (k_3), and β oxidation (k_5) (see Fig. 10). Because the proportion of arterial palmitate (a long chain fatty acid) to the total free fatty acid concentration in blood remains constant during a wide range of free fatty acid levels, and different chains of fatty acids have comparable myocardial extraction fractions,[32] fatty acid utilization, oxidation, and esterification can be calculated from the kinetics of 1-[^{11}C]palmitate as

$$\text{MFAU (nmol/g/min)} = (\text{FFA}_b)(\text{MBF})(\text{EF}_{11\text{C-pal}}) \qquad (20)$$

where FFA$_b$ is free fatty acid in blood (nmol/ml), MBF is myocardial blood flow (ml/g/min), $EF_{11C\text{-}pal}$ is the myocardial extraction fraction of palmitate estimated from the kinetics of $1[^{11}C]$palmitate, and

$$MFAU \ (nmol/g/min) = MFAO + MFAE \qquad (21)$$

where MFAO is myocardial fatty acid oxidation, MFAE is myocardial fatty acid that is retained by the myocardium and includes esterified fatty acids, as well as intermediary metabolites, and

$$MFAO \ (nmol/g/min) = k_5 q_2 \qquad (22)$$

$$MFAE \ (nmol/g/min) = k_3 q_2 \ (k_4 \text{ is negligible during the study period}) \qquad (23)$$

$$q_2 = \{k_1 [MBF \ FFA_b]\}/\{[(k_2 + k_5)(k_1 + MBF/V) - (k_1 k_2)]\} \qquad (24)$$

where q_2 is the concentration of unlabeled palmitate in compartment 2 (nmol/g), FFA_b is the fatty acid concentration in blood (nmol/ml), and V is the fractional vascular volume in myocardium [assumed to be 10% of total volume (ml/g)].

Finally, the fraction of extracted palmitate undergoing oxidation (F_{ox}) or esterification (F_{es}) can be calculated as

$$F_{ox} = MFAO/MFAU \qquad (25)$$

$$F_{es} = MFAE/FMAU \qquad (26)$$

Conclusion

GAP studies currently represent the most thorough application of PET in the measurement of substrate utilization and oxidative metabolism in the heart. The successful completion of this research protocol requires the detailed coordination of radiochemists, cyclotron operators, cardiologists, nurses, and technicians with mathematical modelers and data analysis personnel in order to contend with the short half-lives of the radionuclides used in the imaging studies. Analysis of PET data requires sophisticated metabolite correction, image coregistration, partial volume and spillover correction techniques in order to quantitate myocardial metabolism accurately. Nonetheless, data analyses of a typical GAP study require minimum manual data entry and, on average, are completed within 2 h from the time the myocardial PET images are reconstructed, making these types of complex studies feasible to implement in a clinical PET environment. To date, this imaging paradigm has been used to measure the changes in metabolic flux in a variety clinical research protocols, including

---

[32] S. R. Bergmann, C. J. Weinheimer, J. Markham, and P. Herrero, *J. Nucl. Med.* **37**, 1723 (1996).

normal aging,[33,34] obesity,[35] dilated cardiomyopathy,[36] type 1 diabetes mellitus, and hypertension-induced left ventricular hypertrophy.[37] Future studies will be directed toward the application of this paradigm in order to delineate the mechanisms responsible for the metabolic changes observed in these conditions and to assess the efficacy of novel therapies designed to reverse these metabolic abnormalities.

Acknowledgments

We thank Jeff Willits for his assistance with the illustrations and Dr. Joseph B. Dence for his suggestions and reading of the manuscript. This work was conducted under NIH Grants HL13851 and RO1 AG15466.

[33] A. M. Kates, P. Herrero, C. S. Dence, P. Soto, M. Srinivasan, D. G. Delano, A. Ehsani, and R. J. Gropler, *J. Am. Coll. Cardiol.* **41**, 293 (2003).
[34] P. F. Soto, P. Herrero, A. M. Kates, C. S. Dence, A. A. Ehsani, V. Davila-Roman, K. B. Schechtman, and R. J. Gropler, *Am. J. of Physiol. Heart Circ. Physiol.* **285**, H2158 (2003).
[35] L. R. Peterson, P. Herrero, K. B. Schechtman, S. B. Racette, A. D. Waggoner, Z. Kisreivaware, C. S. Dence, S. Klein, J. Marsala, T. Meyer, and R. J. Gropler, *Circulation* 2004 (in press).
[36] V. G. Davila-Roman, G. Vedula, P. Herrero, L. de las Fuentes, J. G. Rogers, D. P. Kelly, and R. J. Gropler, *J. Am. Coll. Cardiol.* **40**, 271 (2003).
[37] L. De las Fuentes, P. Herrero, L. R. Peterson, D. P. Kelly, R. J. Gropler, and V. G. Davila-Roman, *Hypertension* **41**, 83 (2003).

[17] Molecular Imaging of Enzyme Function in Lungs

By Delphine L. Chen, Jean-Christophe Richard, and
Daniel P. Schuster

Introduction

A wide variety of pulmonary transport and metabolic processes are amenable for study by positron emission tomography (PET).[1] Most PET studies of the lungs so far have focused on measuring transport functions such as pulmonary blood flow and ventilation, as the matching of these two on a regional level is the main determinant of effective gas exchange. In addition, because pulmonary edema is a common and important clinical problem

[1] D. Schuster, *in* "Positron Emission Tomography: Principles and Practice" (D. Bailey, D. V. Townsend, P. Valk, and M. Maisey, eds.), p. 465. Springer-Verlag, London, 2003.

that interferes with gas exchange, PET imaging has also been used to quantify the magnitude and extent of edema accumulation, as well as vascular "leakiness" to protein and water, a major determinant of pulmonary edema.

More recently, PET imaging has been used to study the functions of several enzymes expressed in lung tissue.[2-4] These studies not only illustrate the potential power, but also the inherent challenges and limitations of using noninvasive imaging to study *in vivo* enzyme function.

Some Issues Specific to Pulmonary PET

Several problems are intrinsic to all quantitative PET studies. These include scattered radiation, random coincidences, tissue attenuation of annihilation photons, and partial volume and "spillover" effects. Data collected by PET must be corrected for these effects. Of these, corrections for the tissue attenuation of emitted radiation can be particularly important in lung studies because intrathoracic tissue density can vary from zero, as in large air-containing structures such as bullae in severely damaged emphysematous lungs, to >1 g/ml, as in soft tissues and bone. Normal parenchymal lung density is $\sim$0.3 g/ml and increases to 0.6–0.7 g/ml during states of pulmonary edema. Ignoring such changes between PET imaging studies can lead to serious errors in the measured variables.

Several problems are particularly characteristic of lung studies with PET. For instance, as in all imaging studies, PET data are expressed in terms of tissue volume or, more specifically, per milliliter lung. However, because the lung is an air-containing structure, the value of any PET measurement may change not only because of a true change in the concentration of radioactivity in lung tissue itself, but also because of a change in regional inflation (and hence lung volume). Another problem is that nearly half of the density of the lung is blood; the result can be significant background radiation if a large fraction of the radiotracer remains in the blood at the time of scanning.

The most common "solution" for such problems is to "normalize" data with other regional data obtained by PET. For instance, dividing a regional parameter value by the density of the same lung region normalizes data for differences in regional inflation [i.e., (units/ml lung)/(g lung/ml lung) = units/g lung]. By measuring both regional tissue density and blood

[2] F. Qing, T. McCarthy, J. Markham, and D. Schuster, *Am. J. Respir. Crit. Care Med.* **161,** 2019 (2000).

[3] J. C. Richard, Z. Zhou, D. E. Ponde, C. S. Dence, P. Factor, P. N. Reynolds, G. D. Luker, V. Sharma, T. Ferkol, D. Piwnica-Worms, and D. P. Schuster, *Am. J. Respir. Crit. Care Med.* **167,** 1257 (2003).

[4] D. Schuster, J. Kozlowski, L. Hogue, and T. Ferkol, *Exp. Lung Res.* **29,** 45 (2003).

volume, it is also possible to express data in terms of *extravascular* tissue weight, eliminating both blood volume and inflation effects. Other strategies include expressing data *within* a single scan as a ratio of regional values or as a fraction of the mean or total value for the slice. Such values emphasize differences in regional physiology or metabolism but fail to take into account absolute differences between scans or subjects. Of course, data can also be expressed as the change between temporally separated scans.

Respiratory motion is another problem peculiar to pulmonary studies (similar to artifacts created by cardiac motion). As with the partial-volume effect, the movement of structures with heterogeneous activity in and out of a region of interest degrades spatial resolution. This problem is greatest for scans of long duration, taken during ventilation with large tidal volumes, in areas where tissue activity is most heterogeneous. While respiratory gating is the obvious solution to this problem, dosimetry considerations may not always allow effective compensation for the proportionate decrease in counting statistics. Thus far, respiratory gating has not been employed routinely in PET imaging studies of the lung.

General Approaches to Evaluating Enzyme Function with PET

There are two general strategies to measuring enzyme function with PET imaging. One strategy involves the use of a specific radiolabeled inhibitor of the enzyme. The other involves the use of a radiolabeled substrate, which is trapped intracellularly when metabolized by the enzyme of interest. Both strategies result in the accumulation of radioactive tracer in the target tissue, which can be measured directly with PET.

The simple measurement of tissue radioactivity can give misleading information about enzyme function because it fails to take into account such factors as limitations on tracer delivery (perfusion) to the target tissue, the relative contribution of blood activity to the total tissue activity measurement, inactive radiolabeled metabolites in the imaging field of view, protein and other nonspecific binding of the tracer, or the presence of endogenous (nonradiolabeled) inhibitors, among other issues. To address such problems, alternative imaging protocols are employed, usually involving the dynamic acquisition of data over multiple time points. Simultaneously, time–activity data are also acquired in an appropriate blood pool to estimate an "input function" into the target tissue. When necessary, separate measurements of, or assumed corrections for, inactive radiolabeled metabolites are also included.

When a radiolabeled inhibitor of an enzyme is used to study enzyme function, receptor models can be used to analyze the acquired time–activity data. The estimated rate for receptor binding in many PET studies is the so-called "combined forward rate constant" (CFRC), which is actually the product of the association constant k_a and the number (or concentration) of receptor sites available for binding (B_{max}). Often, one would like to estimate B_{max} itself as a measure of enzyme protein expression, but doing so with PET is generally possible only if two conditions are met: (1) the number of occupied receptors is small relative to B_{max} during a portion of the study and (2) the number of unoccupied receptors is made to change during another portion of the study.

To meet these criteria, it is usually necessary to perform a study that involves multiple injections of both radiolabeled and unlabeled ligand (or, in the case of enzyme studies, an inhibitor). A common protocol involves an initial injection of a trace amount of the radiolabeled ligand followed by a second injection of a large or saturating amount of unlabeled ligand. Time–activity data can then be modeled with the assumption that the second injection changes only the number of available binding sites and not the relevant association and dissociation constants. Of course, a potential problem with such a protocol is that the second injection can have pharmacologic effects that alter organ or system physiology.

An alternative strategy to evaluating enzyme function with PET is to use a radiolabeled substrate that is trapped in the target tissue if metabolized by the enzyme under study. The assumption is that increases in enzyme expression will result in measurable increases in trapping of the radiolabeled substrate. The prototypical example of such an approach is the use of F-18-labeled fluorodeoxyglucose ([^{18}F]FDG). In this particular case, [^{18}F]FDG moves into cells via the same GLUT family of transporter proteins as glucose (primarily through GLUT-1). There, in the presence of hexokinase, it is phosphorylated to [^{18}F]FDG-6-monophosphate ([^{18}F] FDG-6-P). However, unlike glucose, which continues to be metabolized via oxidative metabolism, [^{18}F]FDG-6-P cannot be metabolized by the next enzyme in the respiratory chain because [^{18}F]FDG lacks a hydroxyl group in the second carbon position. Therefore, in tissues that lack dephosphorylases (most tissues other than the liver), [^{18}F]FDG is trapped. A PET signal is generated as [^{18}F]FDG accumulates in the tissue over time.

Pulmonary Angiotensin-Converting Enzyme (ACE) Expression

This enzyme, which is mainly situated on the luminal side of the pulmonary endothelium, is present in high concentrations in the lung. It catalyzes the conversion of angiotensin I to the vasoactive compound angiotensin II. Tissue ACE plays an important role in vascular remodeling, including that associated with pulmonary vascular disease. An *in vivo* method of measuring

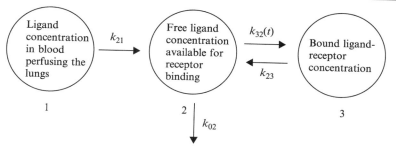

FIG. 1. Three-compartment model used to analyze angiotensin-converting enzyme (ACE) kinetics using PET and F-18-labeled captopril. Compartment 1, blood containing tracer unavailable for ACE binding. Compartment 2, blood containing tracer available for ACE binding. Compartment 3, tissue with tracer reversibly bound to ACE.

changes in enzyme concentration in the lungs could be relevant to any of a group of diseases characterized by pulmonary hypertension.

To assay the pulmonary ACE system with PET, a derivative of the ACE inhibitor captopril (2-D-methyl-3-mercapto-propanoyl-L-proline) has been labeled with fluorine-18 ([18]FCAP).[5] The tracer has high binding specificity for ACE (Fig. 1). Less ideal aspects, however, include the fact that a relatively high proportion of the tracer dimerizes to an inactive form within minutes of administration and that the tracer itself is administered as a mixture of *cis* and *trans* conformers, each with a different binding affinity for ACE.

Despite these handicaps, an imaging protocol has been developed and implemented, and estimates of ACE density (B_{max}) have been calculated from time–activity data obtained with this protocol that are consistent with *in vitro* estimates.[6,7] As with other receptor-based imaging protocols, time–activity data are obtained after two injections of ligand, in this case the first involving [18]FCAP and the second a displacing dose of unlabeled captopril.

Using these methods, it has been shown that *pulmonary* ACE per se can be blocked with relatively low doses of ACE inhibitors (Figs. 2 and 3). This study represents a novel way to determine the optimum dose of drug needed to specifically block activity of a key enzyme in a targeted tissue such as the lungs. The only alternative in the past to this approach to dose determination for treatment of conditions such as primary pulmonary hypertension

[5] D. R. Hwang, W. C. Eckelman, C. J. Mathias, E. W. Pertrillo, J. Lloyd, and M. J. Welch, *J. Nucl. Med.* **32,** 1730 (1991).

[6] J. Markham, T. J. McCarthy, M. J. Welch, and D. P. Schuster, *J. Appl. Physiol.* **78,** 1158 (1995).

[7] D. P. Schuster, T. J. McCarthy, M. J. Welch, S. Holmberg, P. Sandiford, and J. Markham, *J. Appl. Physiol.* **78,** 1169 (1995).

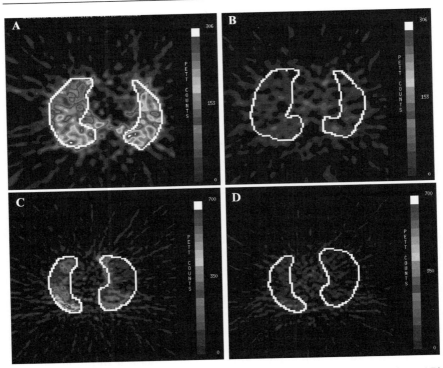

FIG. 2. Lung PET images (transverse orientation) obtained in a normal subject (A and B) and in a patient with pulmonary hypertension (C and D). A and C were obtained approximately 1 h after iv administration of F-18-labeled captopril. B and D were obtained similarly, approximately 1 week after oral ingestion of daily enalapril (5 mg). Note the reduction in tracer uptake after oral enalapril, indicating pulmonary specific blockade of ACE. (See color insert.)

has been to target changes in pulmonary artery pressure. Often, such changes require substantially more drug, and in any case, changes in pulmonary artery pressure per se may be irrelevant to long-term benefits of vascular remodeling.

Inflammation Imaging of Lungs

In vivo glucose metabolism can be measured using [^{18}F]FDG and PET imaging. Because [^{18}F]FDG follows the same metabolic pathways as glucose (as described earlier), an increase in [^{18}F]FDG uptake presumably measures the activity of hexokinase in the cell and therefore can be used to approximate glucose metabolic rates. Approximation of glucose metabolic rates requires use of a "lumped constant" (LC), which accounts for

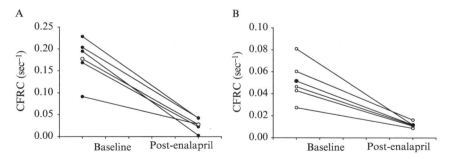

FIG. 3. Values for the "combined forward rate constant" (CFRC) of [18F]FCAP with lung ACE in five patients with pulmonary hypertension studied before (A: mean [●] and individual [O]) and after (B: mean [O] and individual [●]) 1 week of treatment with oral enalapril, as measured with PET imaging. The mean values are significantly different. (Reproduced with permission from Qing et al.[2])

differences in the rate of transport and phosphorylation of [18F]FDG when compared with glucose itself.

[18F]FDG-PET imaging may be useful clinically to detect or characterize various inflammatory lesions.[8–11] The normal reported value for [18F]FDG uptake in the lungs in humans is about 0.6–1.9 μmol h^{-1} ml^{-1}.[12] In contrast, [18F]FDG uptake is elevated in two chronic inflammatory conditions: sarcoidosis[13] and idiopathic pulmonary fibrosis (IPF).[14] This increased uptake can fall during treatment with corticosteroids. During inflammatory processes, cells of various types are activated and recruited to the site of inflammation. As a result, these inflammatory cells increase their uptake of [18F]FDG.[15–18] Despite this potential nonspecificity, evidence to date

[8] N. Manthey, P. Reinhard, F. Moog, P. Knesewitsch, K. Hahn, and K. Tatsch, *Nucl. Med. Commun.* **23**, 643 (2002).

[9] J. Meller, F. Strutz, U. Siefker, A. Scheel, C. Sahlmann, K. Lehmann, M. Conrad, and R. Vosshenrich, *Eur. J. Nucl. Med. Mol. Imaging* **30**, 730 (2003).

[10] K. Stumpe, H. Dazzi, A. Schaffner, and G. von Schulthess, *Eur. J. Nucl. Med.* **27**, 822 (2000).

[11] Y. Sugawara, D. Braun, P. Kison, J. Russo, K. Zasadny, and R. Wahl, *Eur. J. Nucl. Med.* **25**, 1238 (1998).

[12] C. G. Rhodes and J. M. B. Hughes, *Eur. Respir. J.* **8**, 1001 (1995).

[13] L. H. Brudin, S. O. Valind, C. G. Rhodes *et al.*, *Eur. J. Nucl. Med.* **21**, 297 (1994).

[14] C. F. Pantin, S. O. Valind, M. Sweatman *et al.*, *Am. Rev. Respir. Dis.* **138**, 1234 (1988).

[15] R. Chakrabarti, C. Jung, T. Lee, H. Liu, and B. Mookerjee, *J. Immunol.* **152**, 2660 (1994).

[16] R. Gamelli, H. Liu, L. He, and C. Hofmann, *J. Leukocyte Biol.* **59**, 639 (1996).

[17] T. Mochizuki, E. Tsukamoto, Y. Kuge, K. Kanegae, S. Zhao, K. Hikosaka, M. Hosokawa, M. Kohanawa, and N. Tamaki, *J. Nucl. Med.* **42**, 1551 (2001).

[18] L. Sorbara, F. Maldarelli, G. Chamoun, B. Schilling, S. Chokekijcahi, L. Staudt, H. Mitsuya, I. Simpson, and S. Zeichner, *J. Virol.* **70**, 7275 (1996).

suggests that the neutrophil is the primary cell type responsible for the increase in [^{18}F]FDG uptake in tissues during states of *acute* inflammation, such as that occurring in pneumonia.[19–21] The contribution to FDG uptake by other resident lung cells in response to the inflammatory stimulus is still unclear.

In vitro studies have shown that in the presence of inflammatory mediators such as tumor necrosis factor-α (TNF-α), neutrophils increase their uptake of extracellular glucose.[22] However, the mechanisms underlying *increased* glucose (and therefore [^{18}F]FDG) uptake and its regulation in neutrophils or other cells in response to inflammatory stimuli are not yet well defined.

To make it possible to study these problems in an intact, whole animal model of lung inflammation, we have used [^{18}F]FDG and microPET imaging in a murine model of pulmonary infection and inflammation after the airway instillation of *Pseudomonas aeruginosa*,[4,23] in which neutrophilic inflammation in the lungs of these mice is a prominent histopathologic feature.[24] For these studies, *P. aeruginosa* organisms (strain M57-15) were embedded in agarose beads and instilled into the right lung in different groups of mice in varying doses, from 1.5×10^4 to 10^6 CFU. The animals were allowed to recover and then, 72 h later, [^{18}F]FDG imaging was performed in a Concorde model microPET scanner. Images were obtained 1 h after tail vein administration of ~100 μCi [^{18}F]FDG over a 15-min period.

The amount of [^{18}F]FDG taken up by the lungs was calculated as simply the amount of radioactivity in the tissue at the end of the scan period, normalized for the injected dose of radioactivity. This measurement is equivalent to the standardized uptake value (SUV), which allows differentiation of areas of abnormally high uptake from normal surrounding tissue. Comparison of this value in postmortem tissue from both infected and uninfected portions of lung shows a good correlation with estimates of neutrophil infiltration obtained with an assay for the neutrophil-specific enzyme myeloperoxidase (MPO) (Fig. 4).

[19] H. Jones, J. Schofield, T. Krausz, A. Boobis, and C. Haslett, *Am. J. Respir. Crit. Care Med.* **158,** 620 (1998).

[20] H. Jones, S. Sriskandan, A. Peters, N. Pride, T. Krausz, A. Boobis, and C. Haslett, *Eur. Respir. J.* **10,** 795 (1997).

[21] H. A. Jones, R. J. Clark, C. G. Rhodes, J. B. Schofield, T. Krausz, and C. Haslett, *Am. J. Respir. Crit. Care Med.* **149,** 1635 (1994).

[22] H. Jones, K. Cadwallader, J. White, M. Uddin, A. Peters, and E. Chilvers, *J. Nucl. Med.* **43,** 652 (2002).

[23] J. Starke, M. Edwards, C. Langston, and C. Baker, *Pediatr. Res.* **22,** 698 (1987).

[24] A. van Heeckeren, T. Ferkol, and M. Tosi, *Gene Ther.* **5,** 345 (1998).

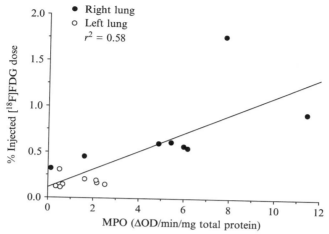

FIG. 4. Correlation of percentage injected [18F]FDG dose found in infected (right) lung vs normal (left) lung in a mouse model of *Pseudomonas aeruginosa* infection with estimates of neutrophil activity using the myeloperoxidase (MPO) assay.

Quantitation Issues

More detailed information on the kinetics of glucose metabolism can be determined by using so-called "dynamic" PET imaging. By acquiring PET data as a series of images over varying lengths of time throughout the entire scan period, time–radioactivity relationships can be determined for [18F]FDG in either blood or lung tissue. To acquire such data, PET imaging is initiated immediately after the intravenous administration of [18F]FDG and is continued for ~75 min. The radioactivity measurements from these sequential scans can then be used to estimate individual rate constants by mathematical modeling of the movement of [18F]FDG between different tissue compartments.[25] As illustrated in Fig. 5, the compartments defined in [18F]FDG transport are the extracellular space, intracellular space, and the "space" in which [18F]FDG is phosphorylated, and hence trapped, in the cell. In some tissues, where dephosphorylation of [18F]FDG is possible due to the presence of the enzyme glucose-6-phosphatase, a fourth rate constant, k_4, is also included in the analysis because dephosphorylation can lead to a loss of [18F]FDG from tissue during the scan period. In the lung, this process is assumed to be negligible, as both inflammatory cells and lung cells do not contain significant amounts of glucose-6-phosphatase.

[25] R. Carson, *in* "Positron Emission Tomography: Basic Science and Clinical Practice" (P. Valk, D. Bailey, D. W. Townsend, and M. Maisey, eds.), p. 147. Springer, London, 2003.

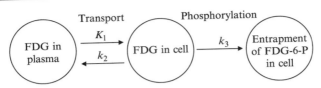

FIG. 5. Diagram of the three-compartment model used to interpret [^{18}F]FDG tracer kinetics in tissue. K_1 and k_2 are the rates of free [^{18}F]FDG transport into and out of cells, and k_3 is the rate of phosphorylation, and hence entrapment, of [^{18}F]FDG in the cell. In the lungs, the rate of dephosphorylation of phosphorylated FDG is assumed to be negligible (and hence no k_4 is included in the model).

A model-independent approach to estimating the overall rate of FDG influx into tissue is to use a multiple time point graphical analysis, first described by Patlak and colleagues.[26,27] Assuming that at some point during the scan period that the concentration of [^{18}F]FDG in the tissue and blood pools reaches a steady state, the equations used in the three-compartment model can be simplified to the following equation relating the concentration of [^{18}F]FDG in the plasma and tissue[27,28]:

$$\frac{A_m(t)}{C_p(t)} = K_i \frac{\int_0^t C_p(t)dt}{C_p(t)} + f V_e + V_p \tag{1}$$

This equation has a linear form, making a purely graphical analysis of data possible. Thus, the slope of this equation, K_i, represents the net influx of [^{18}F]FDG into the tissue, and the intercept, $(f V_e + V_p)$, corresponds to the apparent volume of distribution of FDG in both the tissue and the blood pools. $A_m(t)$ is the amount of activity in the region of interest (ROI) at time t, and $C_p(t)$ is the concentration of activity in the plasma at time t. Therefore, when tissue activity divided by plasma activity is plotted against cumulative plasma activity divided by plasma activity at time t, the slope and intercept of the linear portion of the plot will give an estimate of the influx rate of [^{18}F]FDG uptake into the tissue, as well as its volume of distribution.

Because the Patlak equation can be derived from equations that define the compartmental model, K_i can also be expressed as a function of the individual rate constants:

[26] C. Patlak, R. Blasberg, and J. Fenstermacher, *J. Cereb. Blood Flow Metab.* **3**, 1 (1983).
[27] C. S. Patlak and R. G. Blasberg, *J. Cereb. Blood Flow Metab.* **5**, 584 (1985).
[28] K. Mori, K. Schmidt, T. Jay, E. Palombo, T. Nelson, G. Lucignani, C. Pettigrew, C. Kennedy, and L. Sokoloff, *J. Neurochem.* **54**, 307 (1990).

$$K_i = \frac{K_1 * k_3}{k_2 + k_3} \qquad (2)$$

For both compartmental model or graphical analyses, an accurate "input function," representing the availability of [^{18}F]FDG to the tissues, must be determined, either by drawing blood samples and measuring the radioactivity in them or by using image-derived functions from a ROI located within the cardiac blood pool (right atrium or right ventricle) or over a suitable major vessel, such as the main pulmonary artery. Because of limitations in spatial resolution, especially during microPET imaging, significant degrees of inaccuracy can be anticipated for image-derived estimates of the input function. Thus, for the time being, input functions to the lung must still be obtained via some form of direct blood sampling.

Figure 6 shows an example of dynamic PET images collected using the same murine model of pneumonia described earlier. The graphs show time–activity curves for the infected and normal lung, the input function from blood samples, and results of the Patlak graphical analysis. From the Patlak analysis, it is clear that the rate of FDG uptake (as indicated by the slope of the regression line) in the infected lung is nearly threefold higher than that in the normal lung. The compartmental model analysis for this same mouse showed that K_i was 23×10^{-3}/ml /blood/ml lung/min in infected lung vs 6.7×10^{-3}/ml /blood/ml lung/min in uninfected lung. The calculated K_i from the compartmental model is essentially the same as that calculated by the Patlak analysis, thus mathematically validating the results.

Despite a good correlation between the SUV and tissue assays of MPO, quantitative estimates of [^{18}F]FDG kinetics may still be useful. For instance, the SUV does not take into account the contribution of residual blood radioactivity as a source of background radiation to the PET imaging signal. This source of error may be important where rates of [^{18}F]FDG uptake are low. Thus, estimates of K_i might improve measurement sensitivity compared to simple SUV calculations. In addition, estimating the individual rate constants of the compartmental model should allow one to distinguish whether the imaging signal is dependent primarily on transport or trapping (hexokinase) steps. While phosphorylation by hexokinase is the determining step in [^{18}F]FDG uptake in tissues such as the heart and brain, it remains to be determined whether this is also true for other tissues, such as inflammatory cells.

Validation Issues

Nevertheless, [^{18}F]FDG-PET imaging appears to be a promising new tool to quantify inflammation by using [^{18}F]FDG metabolism as a proxy for the presence of inflammation. However, additional validation studies

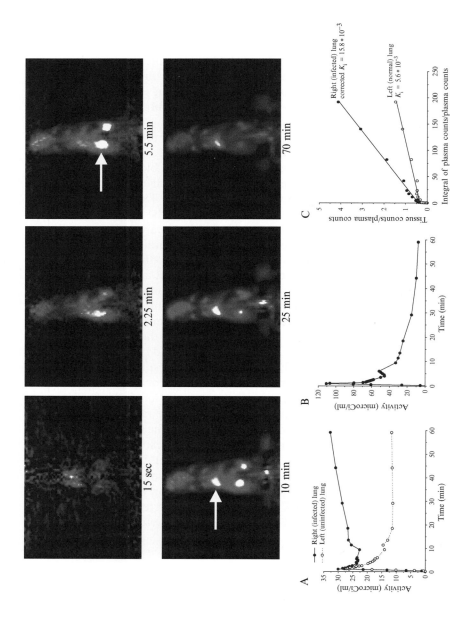

are still necessary, and correlation with standard markers of inflammation must be approached with caution. For example, when neutrophils are simply sequestered in the lungs (e.g., in response to exposure to bacterial lipopolysaccharide), they are not necessarily accessible by bronchoalveolar lavage unless they also penetrate into the airspaces. Therefore, correlations with BAL neutrophil counts may be poor.

Finally, changes in the volume of distribution of [^{18}F]FDG may occur as a result of the disease process being studied. The lung is particularly vulnerable to these changes, as the normal density of the lungs is approximately 0.35 g/ml, but can range from 0 (e.g., in areas of bullous emphysema) to 1 g/ml when completely consolidated. It is not clear how such changes (whether due to atelectasis or pathologic alveolar filling processes such as inflammatory exudate or edema) will affect the calculation of K_i, either by Patlak or by compartmental modeling analyses.[20] Jones et al.[20,29] have approached this issue by dividing K_i by the intercept value from the Patlak graphical analysis to normalize it. Despite its theoretical appeal, no validation of this approach is yet available. Thus, the optimal form of quantitation during [^{18}F]FDG-PET imaging still remains to be determined.

Gene Expression Imaging

Noninvasive monitoring of gene expression *in vivo* is one of several challenges still to be overcome before gene therapy can be used routinely in the clinical setting. Using a reporter gene strategy in which the expression level of a "reporter" gene is used to infer the level of expression of a linked "therapeutic gene," several imaging techniques have emerged as powerful new tools to detect the magnitude, spatial distribution, and timing of transgene expression *in vivo* in experimental studies.[30] Most of these methods, however, are not particularly suitable for pulmonary imaging, either because, as is the case for magnetic resonance imaging (MRI), proton density is too low in the lungs or, as in optical imaging, signal attenuation is too significant in deep tissues like the lungs. PET imaging is much more

[29] H. A. Jones, P. S. Marino, B. H. Shakur, and N. W. Morrell, *Eur. Respir. J.* **21**, 567 (2003).
[30] T. F. Massoud and S. S. Gambhir, *Genes Dev.* **17**, 545 (2003).

FIG. 6. Dynamically acquired PET images (coronal orientation) of a supine mouse with experimental right lung infection. The arrow in the 5.5-min frame shows high concentrations of activity in the kidneys, whereas the arrow in the following frame shows increasingly high activity in the right (infected) lung. (A) Time–activity curves for regions placed over the right and left lungs. (B) Time–activity curve from blood samples drawn throughout the scanning period. (C) Patlak plot illustrating a threefold increased uptake in the infected right lung compared with the normal left lung. (See color insert.)

attractive as a method for pulmonary gene expression imaging, as signal generation is nearly isotropic and therefore not dependent on the location of the radiation source within the organ. Furthermore, any signal attenuation due to variations in tissue density, as is common within the thorax, can be quantified and corrected.

The first evidence that PET imaging could be used to monitor tissue gene expression was reported by Tjuvajev et al.[31] Since then, several different PET reporter systems using a specific reporter gene with a corresponding reporter probe have been described.[30,32–34] These so-called "PET reporter genes" include exogenic enzymes,[35,36] membrane-bound receptors,[37] or cell-membrane transporters.[38] Of these, enzyme-based strategies have the theoretical advantage of signal amplification in which each reporter protein metabolizes several molecules of radioactive probe. Receptor-based systems, however, do not require possibly rate-limiting intracellular transport of the radioactive probe. As these techniques are developed, which PET reporter gene and probe constitute the optimal combination will continue to be a matter of lively debate.[39,40] Ultimately, the proper choice may depend on the gene transfer method[39] or the PET reporter probe pharmacokinetics in the targeted organ. Nevertheless, the use of PET imaging to monitor exogenous gene transfer in pulmonary tissue by measuring the expression of a suitable enzyme is already proving to be a fruitful experimental strategy.

[31] J. G. Tjuvajev, G. Stockhammer, R. Desai, H. Uehara, K. Watanabe, B. Gansbacher, and R. G. Blasberg, Cancer Res. 55, 6126 (1995).

[32] R. G. Blasberg and J. G. Tjuvajev, J. Clin. Invest. 111, 1620 (2003).

[33] S. S. Gambhir, E. Bauer, M. E. Black, Q. Liang, M. S. Kokoris, J. R. Barrio, M. Iyer, M. Namavari, M. E. Phelps, and H. R. Herschman, Proc. Natl. Acad. Sci. USA 97, 2785 (2000).

[34] D. C. MacLaren, T. Toyokuni, S. R. Cherry, J. R. Barrio, M. E. Phelps, H. R. Herschman, and S. S. Gambhir, Biol. Psychiat. 48, 337 (2000).

[35] S. S. Gambhir, J. R. Barrio, L. Wu, M. Iyer, M. Namavari, N. Satyamurthy, E. Bauer, C. Parrish, D. C. MacLaren, A. R. Borghei, L. A. Green, S. Sharfstein, A. J. Berk, S. R. Cherry, M. E. Phelps, and H. R. Herschman, J. Nucl. Med. 39, 2003 (1998).

[36] U. Haberkorn, F. Oberdorfer, J. Gebert, I. Morr, K. Haack, K. Weber, M. Lindauer, G. van Kaick, and H. K. Schackert, J. Nucl. Med. 37, 87 (1996).

[37] D. C. MacLaren, S. S. Gambhir, N. Satyamurthy, J. R. Barrio, S. Sharfstein, T. Toyokuni, L. Wu, A. J. Berk, S. R. Cherry, M. E. Phelps, and H. R. Herschman, Gene Ther. 6, 785 (1999).

[38] T. Groot-Wassink, E. O. Aboagye, M. Glaser, N. R. Lemoine, and G. Vassaux, Hum. Gene Ther. 13, 1723 (2002).

[39] J. J. Min, M. Iyer, and S. S. Gambhir, Eur. J. Nucl. Med. Mol. Imaging 30, 1547 (2003).

[40] J. G. Tjuvajev, M. Doubrovin, T. Akhurst, S. Cai, J. Balatoni, M. M. Alauddin, R. Finn, W. Bornmann, H. Thaler, P. S. Conti, and R. G. Blasberg, J. Nucl. Med. 43, 1072 (2002).

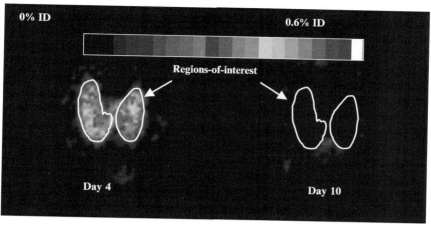

FIG. 7. PET images of lungs in a rat obtained 4 and 10 days after adenovirus-mediated gene transfer of a mutant herpes simplex virus type 1 thymidine kinase (mHSV1-*tk*). Images are transverse slices obtained at the midchest level in the same animal, scanned 1 h after injection of [^{18}F]FHBG. Despite significant tracer uptake 4 days after gene transfer, the lung PET signal was not different from background after 10 days, suggesting a reduction in TK expression by this latter time. *In vitro* assays obtained at the same time points in different animals confirmed the kinetics of transgene expression assessed by PET imaging. [^{18}F]FHBG, 9-(4-[^{18}F]-fluoro-3-hydroxymethylbutyl)guanine; ID, injected dose. (See color insert.)

To date, mutant variants of the herpes simplex virus type 1 thymidine kinase (mHSV1-*tk*) have been the only reporter genes detected successfully in the lungs with PET imaging[3,41] using 9-(4-[^{18}F]fluoro-3-hydroxymethyl-butyl)guanine ([^{18}F]FHBG), an imaging substrate for the mutant thymidine kinase,[42,43] as the PET reporter probe. Using this reporter gene/probe combination, very low levels of viral thymidine kinase (TK) expression can be detected, making this detection system even more sensitive than *in vitro* assays such as TK enzyme activity assays. Repeated measurements of reporter gene expression with PET using this combination correlate well with tissue-based assays of enzyme activity (Fig. 7) over time. This system was also used successfully in another study to define the intrapulmonary spatial distribution of transgene expression[44] and to compare different vehicles for pulmonary gene transfer (Fig. 8).

[41] M. Iyer, M. Berenji, N. S. Templeton, and S. S. Gambhir, *Mol. Ther.* **6**, 555 (2002).
[42] M. M. Alauddin and P. S. Conti, *Nucl. Med. Biol.* **25**, 175 (1998).
[43] D. E. Ponde, C. S. Dence, D. P. Schuster, and M. J. Welch, *Nucl. Med. Biol.* **31**, 133 (2004).
[44] J. C. Richard, P. Factor, L. C. Welch, and D. P. Schuster, *Gene Ther.* **10**, 2074 (2003).

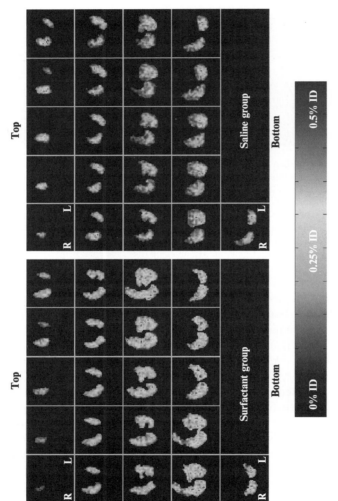

FIG. 8. Transverse PET images of lungs obtained in two representative rats 3 days after intratracheal pulmonary gene transfer using surfactant (left) or saline (right) as a vehicle for adenovectors. The transferred gene encoded for a mutant herpes simplex virus type 1 thymidine kinase (mHSV1-*tk*) and [¹⁸F]FHBG was used as a PET reporter probe. Note higher mHSV1-*tk* expression in the lungs, as assessed by [¹⁸F]FHBG uptake, in the surfactant group. The spatial distribution of transgene expression is more homogeneous and more peripheral when gene transfer was performed with surfactant. ID, injected dose; R, right; L, left; [¹⁸F]FHBG, 9-(4-[¹⁸F]-fluoro-3-hydroxymethylbutyl]guanine. (See color insert.)

Pulmonary Gene Transfer

In gene expression imaging protocols, gene transfer is a critical step, and transfer efficiency is one of the main determinants of reporter protein levels in the target tissue. Because of its high transfer efficiency, adenovirus has been the viral vector of choice to date. The most common route of administration used in studies so far has been intravenous injection. However, because hepatocytes have high concentrations of the coxsackie-adenovirus receptor (CAR), which is required for adenoviral entry into cells, adenoviral delivery to the lungs is limited by significant first-pass filtration in the liver. As a consequence, intravenous administration of adenoviral vectors results in very low levels of gene transfer to the lungs. Transductional and transcriptional retargeting have been proposed as a means of reducing this first-pass filtration and redirecting adenoviral-mediated gene transfer toward the lungs[45,46] but have not yet been applied in PET imaging studies. Iyer *et al.*[41] have also used naked DNA encapsulated in cationic lipids as a vehicle for pulmonary gene transfer in mice in an attempt to address this issue.

An alternative method, administering adenovector via intratracheal instillation, promises to be an efficient means of selectively achieving high levels of transgene expression in alveolar epithelial cells. Using this route, the viral vector is diluted in an appropriate volume of vehicle, such as normal saline, to facilitate virus dispersion toward distal alveolar spaces.[47] However, the vehicle used may affect not only dispersion but also the efficiency of gene transfer. We have shown with microPET imaging that adenovector administration in a surfactant vehicle improved gene transfer by 40% over saline.[44]

PET Imaging Protocol

Ideally, for any of the PET reporter systems already mentioned, the radiotracer will have been cleared from the blood by the time of PET imaging, eliminating background radiation as one potential source of noise. This factor is especially important for lung imaging, as pulmonary blood volume may comprise more than 20% of a lung ROI, with air accounting for 60–70% of the region in the normal lung. However, the longer the time period

[45] P. N. Reynolds, S. A. Nicklin, L. Kaliberova, B. G. Boatman, W. E. Grizzle, I. V. Balyasnikova, A. H. Baker, S. M. Danilov, and D. T. Curiel, *Nat. Biotechnol.* **19**, 838 (2001).

[46] P. N. Reynolds, K. R. Zinn, V. D. Gavrilyuk, I. V. Balyasnikova, B. E. Rogers, D. J. Buchsbaum, M. H. Wang, D. J. Miletich, W. E. Grizzle, J. T. Douglas, S. M. Danilov, and D. T. Curiel, *Mol. Ther.* **2**, 562 (2000).

[47] D. J. Weiss, G. M. Mutlu, L. Bonneau, M. Mendez, Y. Wang, V. Dumasius, and P. Factor, *Mol. Ther.* **6**, 43 (2002).

between tracer injection and the start of imaging, the lower the number of available counts are as a result of radioactive decay, introducing additional variability in the PET data collected. Thus, the pharmacokinetic behavior of the PET reporter probe will be a major determinant in choosing an appropriate time to start imaging after tracer injection. However, studies of the currently available probes for gene expression imaging have given this area little attention.

Image Analysis

Because PET imaging provides functional but not anatomical information, a complete description of the intraorgan spatial distribution of gene expression can be challenging. This issue may be especially relevant to lung studies, where disease heterogeneity makes the ability to control the distribution of gene transfer important. With traditional PET imaging, reconstruction of the transmission scan, initially performed to evaluate photon attenuation in biological tissues, can provide a parametric image of thoracic density (similar to the one provided with traditional computed tomography scans). ROIs may then be drawn on the images to include only lung tissue and later superimposed on the emission scan to measure lung radioactivity specifically.[48] Unfortunately, current microPET devices cannot yet generate transmission images of sufficient quality to provide useful anatomical information. As a consequence, ROIs must be drawn over regions where radioactivity is visually greater than the background. Thus, areas of low gene transfer may go undetected. Even so, useful spatial information on pulmonary reporter gene expression has been reported.[44]

TABLE I

FACTORS AFFECTING THE QUANTITATION OF MUTANT HERPES SIMPLEX VIRUS TYPE 1
THYMIDINE KINASE (mHSV1-*tk*) EXPRESSION BY PET IMAGING

Limitation of PET reporter probe (e.g., [^{18}F]FHBG) to gain access to mHSV1-*tk*
  Rapid clearance of probe from blood entering the lungs
    Limited access to tissues expressing the viral kinase across tissue or cell barriers not
    expressing the viral kinase
Membrane nucleoside transporter kinetics or affinity
Affinity of viral kinase for radioactive probe
Competition between intracellular thymidine and probe as a substrate for mHSV1-*tk*[39]

---

[48] D. P. Schuster, *Am. Rev. Respir. Dis.* **139,** 818 (1989).

Quantitation Issues

Accurate quantitation of transgene expression with PET imaging relies primarily on the ability of PET imaging to quantify pulmonary lung radioactivity accurately *in vivo*. Despite the potential limitations (including spatial resolution, attenuation artifacts, and partial-volume averaging artifacts, among others), we have shown a strong linear correlation between microPET estimates of pulmonary radioactivity and corresponding measurements obtained *ex vivo* in a gamma counter.[3] Theoretically, PET [^{18}F]FHBG uptake should be dependent on the level of expression of mHSV1-*tk* in tissue. As a member of the acycloguanosine family, this tracer enters cells via nucleoside transporters[49] and is trapped intracellularly after selective phosphorylation by cells expressing the viral kinase. The correlation between PET estimates of the absolute amount of gene enzyme expression, as indicated by the radioactivity signal, and that estimated by tissue-based assays has been variable.[3,50,51] Several factors may explain this variability (Table I), some of which are currently under investigation.

Summary

Although PET imaging has been used for decades to evaluate enzyme activity in various tissues, its use for either inflammation or gene expression imaging is very new. Despite current limitations, PET imaging is likely to be invaluable as a tool to detect and quantify both inflammation and transgene expression noninvasively in the lungs and other organs.

Acknowledgments

This work was supported in part by NIH Grants HL32815 and T32 GM08795.

[49] C. E. Cass, J. D. Young, S. A. Baldwin, M. A. Cabrita, K. A. Graham, M. Griffiths, L. L. Jennings, J. R. Mackey, A. M. Ng, M. W. Ritzel, M. F. Vickers, and S. Y. Yao, *Pharm. Biotechnol.* **12,** 313 (1999).

[50] M. Inubushi, J. C. Wu, S. S. Gambhir, G. Sundaresan, N. Satyamurthy, M. Namavari, S. Yee, J. R. Barrio, D. Stout, A. F. Chatziioannou, L. Wu, and H. R. Schelbert, *Circulation* **107,** 326 (2003).

[51] Q. Liang, J. Gotts, N. Satyamurthy, J. Barrio, M. E. Phelps, S. S. Gambhir, and H. R. Herschman, *Mol. Ther.* **6,** 73 (2002).

[18] Brain Uptake and Biodistribution of [11C]Toluene in Nonhuman Primates and Mice

By M. R. GERASIMOV

Introduction

Inhalants are commonly abused by adolescents due to easy access to an array of products containing these volatile substances.[1,2] Despite the reported increasing prevalence of inhalant abuse in the United States and the medical consequences associated with it, there is surprisingly little research on the acute effects of inhalants and the mechanisms underlying their abuse, particularly when compared to other addictive drugs. As the mammalian brain exhibits significant anatomical and functional heterogeneity, an *in vivo* examination of the distribution and pharmacokinetics for distinct brain areas may reveal regional differences in the uptake of a drug with far-reaching potential benefits.

Inhalants, which partition preferentially into lipid-rich areas of the brain and body due to high lipophilicity, were previously thought to exert their central nervous system effects through nonspecific interactions with cell membranes. Under the hypothesis that the abuse liability of toluene can be related to its pharmacokinetic properties and the pattern of regional brain uptake, we developed a methodology for radiolabeling and purifying [11C]toluene for use in positron emission tomography (PET) studies. This technology has been applied successfully to study therapeutic and toxic properties of various substances of abuse. It allows for quantitative noninvasive measurements of the distribution and kinetics of positron emitter-labeled compounds in the brain and whole body of a living organism. Positron emission tomography is an effective tool for studying drug pharmacokinetics, as C-11 labeling does not alter the properties of the parent compound. Therefore, measurements of the regional distribution and kinetics of a radiotracer are similar to that of the unlabeled compound.

This Chapter reports on pharmacokinetic studies using [11C]toluene in nonhuman primates with the goal of further elucidating the potential neuronal mechanisms underlying the reported behavioral, addictive, and toxic effects of this inhalant. Several methods for purification and formulation of this volatile tracer are also reported. These studies are supported and extended by experiments in mice characterizing the regional distribution

[1] K. E. Espeland, *Pediat Nurs.* **23,** 82 (1997).
[2] R. J. Flanagan and R. J. Ives, *Bull. Narc.* **46,** 49 (1994).

and kinetics of [^{11}C]toluene in various regions of the brain and the whole body.

Synthesis, Purification, and Formulation of [^{11}C]Toluene

No-Carrier-Added Synthesis of [^{11}C]Toluene

We have used the rapid coupling of methyl iodide with tributylphenylstannane mediated by the palladium(0) complex[3] to synthesize no-carrier-added [^{11}C]toluene starting with ^{11}CH$_3$I. The time for synthesis, purification, and formulation was 40 min. [^{11}C]Toluene was produced in a radiochemical yield of 40–45% (decay corrected to end of bombardment), with a radiochemical purity of >99% and a specific radioactivity of >1 Ci/μmol.

Briefly, the starting carbon-11 as [^{11}C]CO$_2$ is prepared using a 41-in. medical cyclotron (Japan Steel Works). The [^{11}C]CO$_2$ is unloaded and converted automatically to [^{11}C]CH$_3$I using the GE Medical Systems PET trace MeI synthesis module. The [^{11}C]CH$_3$I is trapped in 0.5 ml of DMA solvent at 0° containing 13 μl of tributylphenylstannane, mixed with 0.9 mg of tris(dibenzylideneacetone)-dipalladium(0), and 1.2 mg of tri-o-toylphosphine. All chemicals are from Aldrich Chemical Co. and used without further purification.

Once trapped, the solution is heated in a sealed vessel at 100° for 6 min while stirring. At the completion of this step, the contents are cooled to ambient temperature and 0.5 ml of distilled water is added. The contents are then passed through a vented 2-μm filter (Millipore Corp.) to yield roughly 1 ml of a clear liquid that could be injected either onto a HPLC column for further purification and formulation suitable for intravenous injection.

[^{11}C]Toluene-specific activity is measured by counting an aliquot of the formulated tracer dose using a Capintec dose-monitoring chamber. Known volume aliquots are then analyzed for toluene mass using capillary gas chromatography on a Hewlett Packard 5890A gas chromatograph equipped with a flame ionization detector. Toluene is eluted off a 60-m × 0.25-mm id SE-30 fused silica capillary column (Alltech, Inc.). The detector response

[3] M. Suzuki, H. Doi, M. Bjorkman, Y. Andersson, B. Langstrom, Y. Watanabe, and R. Noyori, *Chem. Eur. J.* **3,** 2039 (1997).

to mass is referenced against a standard calibration curve that is generated on each day of use.

Typically a 15-min target irradiation to generate the carbon-11 produces [^{11}C]toluene with a specific activity of 1.2 Ci/μmol decay corrected to the end of bombardment.

Tracer Purification Using a Conventional HPLC System

This method involves HPLC with a C18 250 × 4.6-mm id Novopak column and a mobile phase containing 50% dimethylacetamide (DMA) in water. At a flow rate of 1 ml/min, the retention of [^{11}C]toluene is 10 min. The final formulation consists of 5% DMA solution.

The potential interference of DMA with the brain delivery and uptake of toluene prompted us to develop an alternative method of tracer purification and formulation. Thus, [^{11}C]toluene is separated from the starting materials using a conventional HPLC column and supercritical CO_2 fluid as the mobile phase and consequently trapped in a 1.5% cyclodextrin (a cyclic oligosaccharide) solution suitable for intravenous injection.

Radiochromatography System for Tracer Purification Using Supercritical CO_2 Fluid as the Mobile Phase

Supercritical-grade carbon dioxide is from Scott Specialty Gas, Inc. as a pressurized liquid. We use a high-volume stainless-steel syringe pump (Isco, Inc.; Model 500 D) equipped with a flow meter and a pressure-monitoring head. The system is integrated with an isotemp refrigerated circulator (Fisher Scientific, Inc.; Model 9100) that circulates chilled water through a jacket located around the pump head. Samples up to 1 ml in volume can be injected through a standard six-port HPLC valve (Rheodyne; Model 7110). The valve is integrated into an Alltech column heater module (Model 530) that also houses the chromatography column and a 2-ml volume preheated column that is simply a long coil of 1/16″ stainless-steel tubing. This preheat column is essential to system performance and serves to equilibrate the fluid to the desired critical temperature before it is introduced into the sample injection valve. The outlet from the separation module feeds into a second heated detector module that consists of a radiation detector (Carroll Ramsey Associates; Model 101-H PIN diode detector), a mass detector (Knauer UV K-2501 spectrophotometer connected via fiber optics to a Knauer high-pressure absorption cell; 3 μl volume, 2 mm path length, 4500 psi pressure limit), and two flow-diverting valves (High Pressure Equipment Co.) that are opened and closed through motor drives. The module is heated resistively and is controlled using thermocouple feedback (Omega PN-9000 controller). Valve control is accessed directly using the pump control panel that operates either as a stand-alone system or via RS-232 through a PC. In addition, 1-V analog signals from both UV and radiation detectors are acquired using an SRI (Model 202) peak simple chromatography data system.

The design of the product collection module took special consideration due to the fact that it was essential to render the [^{11}C]toluene free of any organic solvents in the final stage of formulation. This requirement precluded the use of organic solvents for trapping agents. In addition, [^{11}C]toluene is not sufficiently soluble in pure water to allow for its efficient trapping upon exiting the chromatography system. A solid support trap was adapted that would allow quantitative trapping of the [^{11}C]toluene as it elutes the system while allowing the CO_2 gas to pass through unhindered. We use Tenax GR material (Supelco, Inc.) in the trap, which is constructed of a 76 × 6.25-mm o.d. Pyrex glass tube packed with 1 g of molecular sieve 4 Å (pellet form), followed in series with 5 g of Tenax GR (60–80 mesh) and held by a Mel-Temp melting point apparatus (Laboratory Devices, Inc.). This apparatus has a 300-W cartridge heater that allows for rapid heating during tracer desorption. Helium gas serves both to not sweep the trap during bake out and to sweep out the radiotracer during thermal desorption. The eluting [^{11}C]toluene is dissolved in 10 ml of the 1.5% cyclodextrin solution and rendered sterile and pyrogen-free by passing through a sterile 2-μm filter (Gelman, Inc.) into a multi-injection vial that contains 0.65 ml of sterile saline.

PET Studies of [^{11}C]Toluene in Baboons

Adult female baboons (*Papio anubis*, 10–15 kg weight) are anesthesized and prepared for PET studies as described previously.[4] Briefly, animals are initially sedated with an intramuscular injection of ketamine hydrochloride (10 mg/kg). Catheters are placed in an antecubital vein and a femoral artery. The arterial line is maintained with heparinized saline. The venous line is used for radioisotope injections only. [^{11}C]Toluene (1.3–1.0 mCi, specific activity >1 Ci/μmol) is injected intravenously, with each baboon receiving one injection. In our studies, we use the Siemen's HR + high-resolution, whole body PET scanner (4.5 × 4.5 × 4.8 mm at center of the field of view) in three-dimensional acquisition mode. A transmission scan is performed using a ^{68}Ge rotating rod source before each emission scan to correct for attenuation before radiotracer injection.

During the study, arterial blood is drawn continuously during the first 2 min of the scan and then at selected times throughout the scanning period using an automated blood sampling machine (Ole Dich). Blood

[4] S. L. Dewey, R. R. MacGregor, J. D. Brodie, B. Bendriem, P. T. King, N. D. Volkow, D. J. Schlyer, J. S. Fowler, A. P. Wolf, S. J. Gatley *et al.*, *Synapse* **5,** 213 (1990).

samples are placed into preheparinized vials for subsequent radioassay and metabolite analysis. During the entire scanning procedure, vital signs are monitored and recorded automatically.

Animals are maintained on gaseous anesthesia [isoflurane (Forane 1.0–4.0%)], oxygen (1500 ml/min), and nitrous oxide (800 ml/min). In a separate study, the animal is kept under Saffan intravenous anesthesia without ventilation, and dimethylacetamide/saline serves as a vehicle for the radiotracer. In addition, we carry out a series of studies where [^{11}C]toluene is administered to the same baboon eight times: dimethylacetamide/saline serves as a vehicle four times and β-cyclodextrin/saline solution is used four times. The PET scanning protocols are identical in all eight experiments except for the composition of the vehicle.

This research is conducted under NIH Assurance Number A3106-01 and is fully accredited by the American Association for Accreditation of Laboratory Animal Care (AAALAC approval date 06/16/99–05/31/04). All baboon studies are reviewed and approved by the Brookhaven National Laboratory (BNL) Institutional Animal Care and Use Committee.

Assay of [^{11}C]Toluene and Its Metabolites in Baboon Plasma

To determine the role of peripheral metabolism in interpreting the observed brain kinetics after the tracer injection and to correct the image analysis with respect to unchanged tracer present in plasma, the chemical identity of the radioactivity in plasma is needed. Two methods for directly assaying unmetabolized [^{11}C]toluene in plasma were developed. These include a manual HPLC method (A) and an automated solid-phase extraction (SPE) method (B) that rely on a Zymark robot.[5] In a typical study, 0.5 ml of baboon plasma is obtained at intervals of 1, 5, 10, and 30 min after the injection of the tracer. Plasma samples for the HPLC method A are spiked with toluene, mixed with 0.5 ml of acetonitrile, ultrasonicated, and centrifuged. An aliquot of the supernatant is counted in a Packard Auto-Gamma counter. A second aliquot is then injected onto a Phenomenex Prodigy Phenyl-3 (250 × 10 cm; 5 μm) column and eluted with 50% acetonitrile/water (flow of 4 ml/min). The HPLC effluent is assayed for radioactivity by collecting fractions every 3 min until UV absorbence indicates the elution of toluene. At this point the entire toluene peak is collected and its radioactivity is assayed using a Packard AutoGamma counter.

Plasma samples processed according to method B are first counted in a Picker well counter to assess the total ^{11}C radioactivity and are then extracted over 500 mg of C8 in a 10-ml cartridge. Extraction of dissolved [^{11}C]toluene with C8 is >98% for all plasma volumes (0.05–1.0 ml).

[5] D. Alexoff, C. Shea, R. A. Ferrieri, M. R. Gerasimov, V. Garza, and P. King, *J. Nucl. Med.* **42**, 267P (2001).

Cartridges are then rinsed with 5 ml of water followed by 5 ml of 50% methanol/water, and the radioactivity of each rinse and the cartridge is measured in a Picker well counter.

An HPLC method C was developed to provide direct measurement of the individual labeled metabolites in plasma. This method relies on a 250 × 4.6-mm id Phenomenex Hypersil 5 μm C18 (ODS) column with a mobile phase composed of 69% water:28% methanol:3% acetic acid and provides baseline resolution of hippuric acid, benzyl alcohol, benzaldehyde, and benzoic acid (Fig. 1 and Table II). [¹¹C]Toluene does not elute from the column under these conditions. According to this method, plasma samples are processed by counting an aliquot in the Packard AutoGamma counter and injecting an aliquot spiked with the carrier amounts of the metabolites into the HPLC system. The column effluent is collected at 2-min intervals and assayed for radioactivity until the IV absorbance indicates the elution of a specific metabolite. At this point the entire peak is collected and assayed for radioactivity. The summed ¹¹C radioactivity of these compounds is then subtracted from the volume-corrected total ¹¹C radioactivity injected to yield a value that closely corresponds to the unmetabolized [¹¹C]toluene values obtained from methods A and B (Table I). Therefore, it is safe to assume that all of the metabolite products are accounted for in this analytical method.

Finally, the only route that provides *unlabeled* metabolites would be the cleavage of the C–C bond connecting the ¹¹C methyl group to the aromatic ring. This reaction yields $^{11}CO_2$. We ruled out this possibility by assaying the blood for $^{11}CO_2$. Specifically, the contribution of $^{11}CO_2$ to total plasma radioactivity is assessed by taking two separate 1-ml aliquots of plasma and

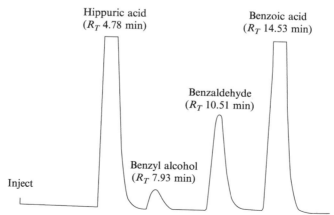

Hippuric acid
(R_T 4.78 min)

Benzoic acid
(R_T 14.53 min)

Benzaldehyde
(R_T 10.51 min)

Benzyl alcohol
(R_T 7.93 min)

Inject

FIG. 1. UV response to metabolite plasma sample spiked with authentic compounds and injected on a 250 × 4.6-mm id Phenomenex Hypersil 5-μm C18 (ODS) column using a mobile phase composed of 69% water:28% methanol:3% acetic acid and a flow rate of 1.5 ml/min.

TABLE I

PERCENTAGE UNCHANGED [¹¹C]TOLUENE MEASURED BY DIFFERENT HPLC METHODS

	% Unchanged [¹¹C]toluene	
Minutes after injection	Method A	Method B
1	92 ± 8	85 ± 8
5	58 ± 6	57 ± 4
10	44 ± 10	42 ± 8
30	32 ± 14	38 ± 8

mixing each with 3 ml of isopropyl alcohol and 1 ml of 0.9 M sodium bicarbonate. One milliliter of 6 N HCl is added to one sample in the pair, and 1 ml of 0.1 N NaOH is added to the other. Both samples are then heated at 85° in an ultrasonic bath for 10 min. The radioactivity in each sample is then measured using a gamma well counter with correction for carbon-11 decay. If $^{11}CO_2$ were present in the plasma sample, it would be noted by the release of radioactivity in the acidified sample.

Precautions are taken to avoid the loss of the tracer from the analyte during the manipulations, i.e., samples are kept at low temperature (ice bath). The demonstrated close correlation between numbers for the unchanged tracer obtained from different analytical methods suggests that the solvent loss during the chromatographic analysis manipulations was negligible.

Tracer Kinetic Modeling and PET Data Analysis

The plasma input function that drives the uptake of tracer into tissue is the product of the measured radioactivity and the fraction of that activity due to unchanged [¹¹C]toluene. The integrated uptake (IU) at 45 min expressed as percentage dose ($IU = \int_0^{45} A(t)dt/dose \times 100$) is used as a measure of drug exposure. The 45-min time point is chosen for calculating IU so as to minimize differences in uptake due to the blood flow that affects IU at early time points. The distribution volume (DV), which is a measure of the capacity of the tissue to bind the tracer, is calculated using the model-independent graphical method. Differences in IU and DV values between cyclodextrin and DMA studies for the corresponding brain regions are tested by comparing the average values using independent student's t tests.

Image ROI Analysis

We obtained 63 planes from 20 time frames for each PET study. Circular or elliptical ROIs were chosen with a size appropriate for the scanner resolution. Selected regions were projected onto each dynamic frame to obtain time–activity curves. We averaged right and left regions of interest.

TABLE II
ANALYSIS OF ^{11}C RADIOACTIVITY IN BABOON ($n = 3$) PLASMA SAMPLES

Minutes after injection	Percentage of unchanged [11C]toluene	% [11C]Benzoic acid	% [11C]Benzaldehyde	% [11C]Benzyl alcohol	% [11C]Hippuric acid	% [11C]Benzoyl glucuronide
1	98 ± 8	Not detectable levels				
5	55 ± 6	8 ± 3	2 ± 2	2 ± 1	11 ± 2	21 ± 4
10	49 ± 4	5 ± 2	2 ± 2	2 ± 1	12 ± 3	30 ± 5
30	34 ± 4	3 ± 2	1 ± 1	1 ± 1	18 ± 3	40 ± 6

Regions were taken on at least two contiguous planes. Regional tissue carbon-11 concentrations was normalized to the total injected dose to obtain percentage injected dose/cc.

Tissue Distribution Experiments in Mice

[^{11}C]Toluene is administered by tail vein injection in 0.2 ml of saline. Mice are sacrificed by cervical dislocation, followed by decapitation at various time points after the radiotracer administration. The brain is removed quickly on a chilled block to minimize the potential loss of radioactivity due to evaporation. Sacrifice times for brain distribution experiments are 0.25, 0.5, 1, 2, 5, 10, 30, and 60 min. There are five to six mice for each time point. Radioactivity measurements are performed with a Packard Auto Gamma 5500 instrument. In a separate group of experiments, animals are injected with 800 mg/kg ip of nonlabeled toluene 5 min prior to administration of [^{11}C]toluene. These animals ($n = 5$) are sacrificed 1 min following the injection of radioactive compound, and brains are processed in the same manner as described earlier for the brain distribution experiments. Extracerebral tissues and fluids taken at the time of sacrifice (30 min postinjection) include blood, urine, body fat, lungs, heart, spleen, liver, kidneys, and gonads. This time point is chosen deliberately in order to assess the tissues that can retain the compound over a prolonged period of time.

Solvent Extraction Experiments in Mice

Mice (two per time point) are injected and sacrificed at three time points (1, 10, and 30 min) as described earlier. Brains are homogenized in 3 ml of acetonitrile/methanol (4:1, v/v) with a Polytron. Homogenates are spun at 14,000g for 2 min. Pellet and supernatant are separated. An aliquot (0.2 ml) of supernatant is vortex mixed with 0.5 ml of toluene plus 0.5 ml of borate buffer, pH 10. The mixture is spun as before, and aliquots of organic and aqueous phases are assayed for carbon-11 using a gamma counter for the pellet and supernatant and either a gamma counter or a liquid scintillation counter for the toluene and buffer phases. Results are expressed as percentage of total counts in supernatant and percentage of total counts in the toluene layer. Extraction efficiency for [^{11}C]toluene is >90%.

Results

Baboon Studies

Analysis of the chemical form of ^{11}C in plasma revealed fast metabolism of the tracer (see Tables I and II). Regional kinetic data from three baboon PET studies after [^{11}C]toluene injection are given in Table III.

TABLE III
KINETIC DATA FROM BABOON [^{11}C]TOLUENE STUDIES[a]

ROI	IU (% dose/cc/min)	IU/PI	K_1	DV (ml/ml)	$T_{1/2}$	Peak uptake (% dose/cc)
		Study 1. Plasma integral 0.18% dose/cc/min				
STR	0.550	3.07	0.29	3.34	18	0.025
THL	0.527	2.90	0.25	3.29	18	0.021
FC	0.458	2.51	0.18	3.18	20	0.017
CB	0.404	2.21	0.23	2.31	20	0.019
WM	0.522	2.90	0.18	4.00	30	0.017
		Study 2. Plasma integral 0.20% dose/cc/min				
STR	0.634	3.17	0.34	3.65	18	0.026
THL	0.626	3.16	0.27	3.63	19	0.025
FC	0.506	2.54	0.19	3.13	20	0.017
CB	0.348	1.75	0.33	1.87	20	0.021
WM	0.635	3.19	0.18	4.75	37	0.018
		Study 3. Plasma integral 0.15% dose/cc/min				
STR	0.535	3.61	1.00	3.58	10	0.056
THL	0.607	4.07	1.3	4.05	10	0.059
FC	0.482	3.25	0.81	2.85	12	0.047
CB	0.436	2.94	0.70	2.35	12	0.036
WM	0.553	3.68	0.50	3.65	13	0.031

[a] STR, striatum; FC, frontal cortex; CB, cerebellum; WM, white matter; ROI, region of interest; IU, integrated uptake; PI, plasma integral (at $T = 45$ min); K_1 tissue influx constant; DV, distribution volume; $T_{1/2}$, half-time of clearance from the peak uptake).

Animals in studies 1 and 2 were maintained on gaseous anesthesia, and in study 3 the animal was kept under intravenous anesthesia without ventilation as described in the previous section. Uptake curves illustrating the time course of [^{11}C]toluene in five brain regions are shown in Figs. 2 and 3. In study 1, toluene exhibits rapid uptake into all five regions studied, reaching a maximum value of 0.017 to 0.026% of the injected dose 1 to 2 min postinjection (Figs. 2 and 3 and Table III). In studies 1 and 2 the maximum uptake in STR is 0.025% of the injected dose occurring at 3 to 4 min. Clearance from the brain is also rapid. The half-time ($T_{1/2}$) (measured from the peak uptake) is on the order of 20 min for studies 1 and 2 but 10 min for study 3. In column 1 the dose-corrected integrated radioactivity at $T = 45$ min is given as integrated uptake (IU). Although similar values were found for the DV and IU (see Table III), there were significant differences in $T_{1/2}$, maximum uptake, and K_1 for study 3 versus studies 1 and 2.

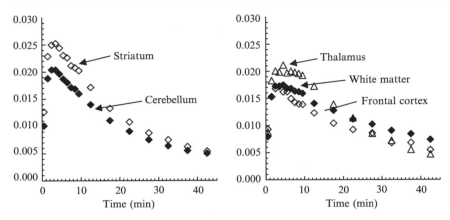

FIG. 2. Uptake curves of [^{11}C]toluene in a baboon (study 1) expressed as percentage injected dose per cc over time. *Left:* Striatum and cerebellum uptake. *Right:* Thalamus, frontal cortex, and white matter.

Additionally, we compared the general pharmacokinetic parameters of [^{11}C]toluene when administered in two different vehicles (dimethylacetamide/saline or β-cyclodextrin/saline solution). Integrated uptake was higher in all studies for all five regions reported ($p < 0.05$) with cyclodextrin as a vehicle compared to the corresponding ones where DMA was used (Fig. 3A). Distribution volumes were also significantly higher for all cyclodextrin studies in all studied brain regions (Fig. 3B).

Average plasma integral values of total radioactivity at 45 min were 0.34 ± 0.04 and 0.36 ± 0.06 (% dose/cc/min) for DMA and cyclodextrin groups of studies, respectively. Correcting for the presence of metabolites, the integrated radioactivity of [^{11}C]toluene was 0.20 ± 0.025 (DMA) and 0.18 ± 0.05 (CD).

Mouse Dissection Experiments

Results from the brain dissection study demonstrated similar kinetics as the baboon studies, with rapid uptake and clearance of radioactivity from the brain (Table IV).

Roughly half of the ^{11}C was found in the urine 30 min after tracer injection (i.e., even though collection of the urine cannot be quantitative, the radioactivity found in the urine accounted for $46 \pm 3.1\%$ of injected activity as opposed to $0.32 \pm 0.09\%$ found in the brain at this time point). Data on extracerebral tissues are presented as percentage of injected radioactivity per gram of tissue in Table V. Kidneys have the highest concentration (6.8% IA/g), consistent with the excretion of polar metabolites through

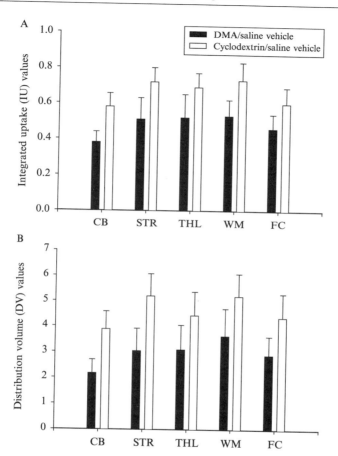

FIG. 3. Integrated uptake (IU) (A) and distribution volumes (DV) (B) for the five brain regions from baboon [^{11}C]toluene studies performed using as a vehicle either DMA/saline or cyclodextrin/saline. CB, cerebellum; STR, striatum; THL, thalamus; WM, white matter; FC, frontal cortex. The student's t test performed on average values demonstrated significant differences between the corresponding brain regions with the CD studies being significantly higher than DMA ones ($p < 0.05$). Plasma integral $= 0.20 \pm 0.025$ (% dose/cc/min).

kidneys and urine. A relatively high liver concentration (2.56%) is consistent with hepatobiliary excretion and production of toluene metabolites. We confirmed the existing reports of nonpermeability of the blood–brain barrier to polar toluene metabolites by using a selective solvent extraction experiment (see Methods section). The decay-corrected radioactivity was fully accounted for by the toluene (nonpolar) extract layer at all time points (1, 10, and 30 min).

TABLE IV

TIME COURSE OF WHOLE BRAIN [^{11}C]TOLUENE UPTAKE IN MICE EXPRESSED AS
PERCENTAGE OF INJECTED ACTIVITY (IA) PER GRAM OF TISSUE $\pm$ SE[a]

Time of sacrifice (min)	% IA/g	SE
0.25	32.19	5.28
0.5	20.78	3.20
1	13.53	1.34
1.5	6.02	0.39
2	5.76	0.77
30	0.32	0.09
60	0.20	0.08

[a] $n = 5$–7 mice for every time point.

TABLE V

WHOLE BODY [^{11}C]TOLUENE UPTAKE IN MICE EXPRESSED AS PERCENTAGE OF
INJECTED ACTIVITY (IA) PER GRAM OF TISSUE

Tissue[a]	% IA/g	SE
Blood	1.59	0.36
Heart	0.98	0.24
Liver	2.56	0.54
Spleen	0.75	0.59
Kidney	6.80	2.67
Lung	3.41	0.60
Testes	1.40	0.71
Fat	4.31	1.23
Brain	0.32	0.09

[a] Animals were sacrificed 30 min postinjection. $n = 6$ mice.

In the experiments where the animals were pretreated with the nonradiolabeled compound prior to administration of the radioactivity, we did
not observe any difference in brain radioactivity uptake (13.2% injected
activity/g of tissue $\pm$ 0.92 vs 13.5 $\pm$ 1.9% IA/g, $p = 0.85$).

Discussion

[^{11}C]Toluene was prepared in high radiochemical yield, purity, and specific activity and was formulated to allow intravenous injection in baboons.
It should be noted that even though the synthesis of [^{11}C]toluene has been

reported,[3] this procedure was not adapted for synthesis and formulation suitable for a PET study. To our knowledge, this is the first reported study of the *in vivo* brain pharmacokinetics of labeled toluene in nonhuman primates. Previous reports detailing toluene brain levels during or after inhalation were performed in rodents, where animals were exposed to either low/moderate concentrations of toluene and were subsequently sacrificed at specific time intervals following exposure. However, this approach limits the interpretation of these studies with respect to toluene abuse liability.

Although toluene is self-administered solely through inhalation by drug abusers, we used an intravenous route in our experiments. This was due to technical considerations and, more importantly, because this route allowed the reliable and accurate measurement of the amount and rate of delivery of the tracer to the brain. The finding that regional uptake and clearance kinetics of radioactive gases in humans do not differ significantly between intravenous bolus and single-breath administration[6] supports the validity of the conclusions drawn from the present study.

Generally, rapid absorption, rapid entry into the brain, high bioavailability, short half-life, and high rate of drug metabolism and clearance are pharmacokinetic characteristics that predict a high potential for abuse, as these factors increase positive reinforcement. Our PET studies revealed for the first time that toluene possesses all the aforementioned properties. That is, maximum uptake into brain regions in both primates and rodents is reached at 1–3 min, with half-time from the maximum on the order of 20 min (see Tables III–V). The rapid delivery of toluene to the brain is partially related to the high lipophilicity and solubility of this agent in the blood with the blood/air partition coefficient being on the order of 120.[7] Additionally, fast drug metabolism with rapid elimination is known to contribute to the abuse liability, and our data indicate fast conversion of toluene into polar and nonpolar (benzaldehyde) metabolites (see Table II). In fact, [^{11}C]toluene entered the kidneys rapidly in our baboons studies (data not shown) with accumulation of radioactivity in the urine ducts by 20 min postinjection, consistent with the known formation of polar metabolites, including benzoic and hippuric acids that are excreted in urine. It should be noted that we observed one nonpolar metabolite, benzaldehyde; nevertheless, the absolute contribution of [^{11}C]benzaldehyde to the total brain radioactivity is essentially negligible at later time points, as the overall amount of radioactivity is also lower.

[6] I. Prohovnik, C. D. Metz, and H. L. Atkins, *J. Nucl. Med.* **36**, 1458 (1995).
[7] V. A. Benignus, K. E. Muller, J. A. Graham, and C. N. Barton, *Environ. Res.* **33**, 39 (1984).

The present study observed a rapid significant uptake and clearance in the striatal region, the brain area directly associated with the reinforcing properties of psychostimulants and other abused drugs. A similar time course was observed for the frontal cortex region, also implicated in reward-related behaviors.

Interestingly, we observed significant differences between studies 1 and 2 and study 3 (see Table III). Changes in the pharmacokinetic parameters can be attributed to higher blood flow, resulting in the higher peak uptake and more rapid washout observed in study 3, which was performed under Saffan anesthesia without mechanical ventilation. This is consistent with diminished oxygen demand and thus diminished blood flow to the brain produced by ventilator-induced saturation of the blood with oxygen.

Finally, we observed a high uptake and slower clearance in white matter compared to other brain regions (see Table III), indicating a higher exposure of white matter to toluene, consistent with a higher lipid content and neurotoxicological evidence, indicating restricted and diffuse white matter changes in toluene abusers.

In those experiments where we compared whole brain uptake of [^{11}C]toluene in a group of mice treated with the tracer alone or preinjected with a cold compound, we did not observe any difference in radioactivity uptake. This suggests that brain [^{11}C]toluene uptake is insensitive to pretreatment with a pharmacologically relevant dose of cold compound, and thus radioactivity uptake observed in PET studies using radiotracers with high lipophilicity such as toluene may reflect the combination of specific and nonspecific binding. Therefore, one cannot use [^{11}C]toluene to image and quantify putative binding sites because the affinities of toluene for these sites are too low (micromolar rather than nanomolar).

Finally, we demonstrated that administration of [^{11}C]toluene in β-cyclodextrin formulation increases its brain uptake as compared to the tracer uptake following administration of a DMA/saline formulation. The similarity of the integrated plasma values indicates that the increase in integrated brain uptake observed in the studies with cyclodextrin is not due to an increased plasma input. This is also consistent with greater bioavailability of the [^{11}C]toluene–cyclodextrin complex. The vehicle did not affect the trend for rapid metabolism significantly ($T_{1/2}$ approximately 15–20 min for both vehicles) of the tracer.

The trend for higher uptake and slower clearance of [^{11}C]toluene in white matter was observed when either DMA or β-cyclodextrin was used as a vehicle. It appears then that the choice of a vehicle affected predominantly the degree of brain uptake rather than the regional and temporal aspect of this drug pharmacokinetics (Fig. 3).

In summary, the use of PET and a novel radiotracer, [^{11}C]toluene, allowed us to demonstrate, for the first time, rapid uptake and clearance of radioactivity into specific brain regions in nonhuman primates with kinetic patterns that parallel the reported intoxicating effects of toluene.

Acknowledgments

This research was carried out under contract with the U.S. Department of Energy Office of Biological and Environmental Research (USDOE/OBER DE-AC02-98CH10886) and National Institute on Drug Abuse (NIDA) Grant DA-03112.

[19] Optimizing Luciferase Protein Fragment Complementation for Bioluminescent Imaging of Protein–Protein Interactions in Live Cells and Animals

By Kathryn E. Luker and David Piwnica-Worms

Introduction

Protein–protein interactions regulate a variety of cellular functions, including cell cycle progression, signal transduction, and metabolic pathways. On a whole organism scale, protein–protein interactions regulate signals that affect overall homeostasis, patterns of development, normal physiology, and disease in living animals.[1–3] In addition, protein–protein interactions have considerable potential as therapeutic targets.[4,5] Evidence is accumulating that pathways of protein interactions in specific tissues produce regional effects that cannot be investigated fully with *in vitro* systems, and thus there is considerable interest in evaluating protein interactions in living animals.

Fundamentally, the detection of physical interaction among two or more proteins can be assisted if association between the interactive partners leads to the production of a readily observed biological or physical readout.[6]

[1] H. Zhang, G. Hu, H. Wang, P. Sciavolino, N. Iler, M. Shen, and C. Abate-Shen, *Mol. Cell. Biol.* **17,** 2920 (1997).

[2] G. Stark, I. Kerr, B. Williams, R. Silverman, and R. Schreiber, *Annu. Rev. Biochem.* **67,** 227 (1998).

[3] H. Ogawa, S. Ishiguro, S. Gaubatz, D. Livingston, and Y. Nakatani, *Science* **296,** 1132 (2002).

[4] C. Heldin, *Stem Cells* **19,** 295 (2001).

[5] J. E. Darnell, Jr., *Nat. Rev. Cancer* **2,** 740 (2002).

Compared with studies of protein interactions in cultured cells, strategies to interrogate protein–protein interactions in living organisms impose even further constraints on reporter systems and mechanisms of detection. This Chapter summarizes briefly various strategies for detecting protein-binding partners using conventional cell biology assays with the intent of identifying properties that might be exploited for imaging applications.

Detecting Protein Interactions in Intact Cells

Most strategies for detecting protein–protein interactions in intact cells are based on fusion of the pair of interacting molecules to defined protein elements to reconstitute a biological or biochemical function. Examples of reconstituted activities include activation of transcription, repression of transcription, activation of signal transduction pathways, or reconstitution of a disrupted enzymatic activity.[6] A variety of these techniques have been developed to investigate protein–protein interactions in cultured cells. The two-hybrid system is the most widely applied method used to identify and characterize protein interactions. However, several features of protein fragment complementation make it attractive as an approach for *in vivo* imaging of protein interactions in cells, particularly in live animals. Major features of these two methods are described, and their potential utility for *in vivo* imaging in relation to other strategies is compared based on our experience and published work.

Two-Hybrid Systems

Two-hybrid systems exploit the modular nature of transcription factors, many of which can be separated into discrete DNA-binding and activation domains.[7] Proteins of interest are expressed as fusions with either a DNA-binding domain (BD) or activation domain (AD), creating hybrid proteins. If the hybrid proteins bind to each other as a result of interaction between the proteins of interest, then the separate BD and AD of the transcription factor are brought together within the cell nucleus to drive expression of a reporter gene. In the absence of specific interaction between the hybrid proteins, the reporter gene is not expressed because the BD and AD do not associate independently. Two-hybrid assays can detect transient and/or unstable interactions between proteins, and the technique is reported to be independent of expression of endogenous proteins.[8] Although the

---

[6] G. Toby and E. Golemis, *Methods* **24,** 201 (2001).

[7] S. Fields and O. Song, *Nature* **340,** 245 (1989).

[8] C. von Mering, R. Krause, B. Snel, M. Cornell, S. Oliver, S. Fields, and P. Bork, *Nature* **471,** 399 (2002).

two-hybrid assay originally was developed in yeast, commercial systems (BD Biosciences Clontech) are now available for studies in bacteria and mammalian cells. We and other investigators have shown that two-hybrid systems can be used to image protein interactions in living mice with positron emission tomography (PET)[9–12] or bioluminescence imaging.[13] However, the two-hybrid method has some limitations. Some types of proteins do not lend themselves to study by the two-hybrid method. For example, because the production of signal in the two-hybrid method requires nuclear localization of the hybrid proteins, membrane proteins cannot be studied in their intact state. Also, the time delay associated with both transcriptional activation of the reporter gene and degradation of the reporter protein and mRNA limits kinetic analysis of protein interactions.[14]

Protein–Fragment Complementation

Protein–fragment complementation (PFC) assays depend on division of a monomeric reporter enzyme into two separate inactive components that can reconstitute function upon association. When these reporter fragments are fused to interacting proteins, the reporter is reactivated upon association of the interacting proteins. PFC strategies based on several enzymes, including β-galactosidase, dihydrofolate reductase (DHFR), β-lactamase, and luciferase, have been used to monitor protein–protein interactions in mammalian cells.[15–20] A fundamental advantage of PFC is that the hybrid proteins directly reconstitute enzymatic activity of the reporter. In principle, therefore, protein interactions may be detected in any subcellular compartment, and assembly of protein complexes may be monitored in real

[9] G. Luker, V. Sharma, C. Pica, J. Dahlheimer, W. Li, J. Ochesky, C. Ryan, H. Piwnica-Worms, and D. Piwnica-Worms, *Proc. Natl. Acad. Sci. USA* **99**, 6961 (2002).
[10] G. Luker, V. Sharma, C. Pica, J. Prior, W. Li, and D. Piwnica-Worms, *Cancer Res.* **63**, 1780 (2003).
[11] G. Luker, V. Sharma, and D. Piwnica-Worms, *in* "Handbook of Proteomic Methods" (P. M. Conn, ed.), p. 283. Humana Press, Totowa, NJ, 2003.
[12] G. Luker, V. Sharma, and D. Piwnica-Worms, *Methods* **29**, 110 (2003).
[13] P. Ray, H. Pimenta, R. Paulmurugan, F. Berger, M. Phelps, M. Iyer, and S. Gambhir, *Proc. Natl. Acad. Sci. USA* **99**, 2105 (2002).
[14] F. Rossi, B. Blakely, and H. Blau, *Trends Cell Biol.* **10**, 119 (2000).
[15] F. Rossi, C. Charlton, and H. Blau, *Proc. Natl. Acad. Sci. USA* **94**, 8405 (1997).
[16] I. Remy and S. Michnick, *Proc. Natl. Acad. Sci. USA* **96**, 5394 (1999).
[17] I. Remy, I. Wilson, and S. Michnick, *Science* **283**, 990 (1999).
[18] T. Ozawa, A. Kaihara, M. Sato, K. Tachihara, and Umezawa, *Anal. Chem.* **73**, 2516 (2001).
[19] A. Galarneau, M. Primeau, L.-E. Trudeau, and S. Michnick, *Nat. Biotechnol.* **20**, 619 (2002).
[20] T. Wehrman, B. Kleaveland, J. H. Her, R. F. Balint, and H. M. Blau, *Proc. Natl. Acad. Sci. USA* **99**, 3469 (2002).

time. A disadvantage of complementation approaches is that reassembly of an enzyme may be susceptible to steric constraints imposed by the interacting proteins. Another potential limitation of PFC for application in living animals is that transient interactions between proteins may produce insufficient amounts of active enzyme to allow noninvasive detection. Nonetheless, because most PFC strategies are based on reconstituting active enzymes, these systems offer the potential benefits of signal amplification to enhance sensitivity for detecting interacting proteins in living animals.

Other Strategies

The split-ubiquitin system enables signal amplification from a transcription factor-mediated reporter readout.[21,22] In one application, the interaction of two membrane proteins forces the reconstitution of two halves of ubiquitin, leading to a cleavage event mediated by ubiquitin-specific proteases that release an artificial transcription factor to activate a reporter gene. As mentioned earlier, indirect readout of the reporter limits kinetic analysis, and the released transcription factor must translocate to the nucleus.

Several variations of recruitment systems have been developed for use in whole cells, including the Ras recruitment system[23,24] and interaction traps.[25] Cells that coexpress a test protein fused to a membrane localization signal, such as a myristoylation sequence, and a protein-binding partner fused to cytoplasmic protein, such as activated Ras devoid of its membrane targeting signal, will localize mammalian Ras to the membrane only in the presence of interacting proteins. However, Ras recruitment systems, as configured originally, cannot be applied to mammalian cells, and while readout is not directly dependent on transcriptional activation, indirect readout by colony growth nonetheless limits kinetic analysis and interrogation of subcellular compartmentation of the interactions.

An interesting variation of the recruitment approach applicable to mammalian cells is the cytokine receptor-based interaction trap. Here, a signaling-deficient receptor provides a scaffold for the recruitment of interacting fusion proteins that phosphorylate endogenous STAT3. Activated STAT complexes then drive a nuclear reporter.[25] This system permits

[21] N. Johnsson and A. Varshavsky, *Proc. Natl. Acad. Sci. USA* **91,** 10340 (1994).
[22] I. Stagljar, C. Korostensky, N. Johnsson, and S. te Heesen, *Proc. Natl. Acad. Sci. USA* **95,** 5187 (1998).
[23] A. Aronheim, E. Zandi, H. Hennemann, S. Elledge, and M. Karin, *Mol. Cell. Biol.* **17,** 3094 (1997).
[24] Y. Broder, S. Katz, and A. Aronheim, *Curr. Biol.* **8,** 1121 (1998).
[25] S. Eyckerman, A. Verhee, J. Van der Heyden, I. Lemmens, X. Van Ostade, J. Vandekerckhove, and J. Tavernier, *Nat. Cell Biol.* **3,** 1114 (2001).

detection of both modification-independent and phosphorylation-dependent interactions in intact mammalian cells, but the transcriptional readout again limits kinetic analysis.

Other approaches to detecting protein–protein interactions in live mammalian cells include fluorescence resonance energy transfer (FRET) and bioluminescence resonance energy transfer (BRET).[26,27] For FRET, fluorescently labeled proteins, one coupled to a donor fluorophore and the other coupled to an acceptor fluorophore, produce a characteristic shift in the emission spectrum when the protein-binding partners interact. Limitations of FRET are inter- and intramolecular spatial constraints and sensitivity of detection, as there is no amplification of the signal. For BRET, the donor molecule is firefly luciferase or a related bioluminescent protein, whereas the acceptor is green fluorescent protein or a color variant. While intermolecular spatial constraints are thought to be less restrictive with BRET, similar limitations related to sensitivity may apply. However, because the photon donor in BRET is produced by an enzymatic activity (luciferase), the potential for substrate-dependent signal amplification exists. Nonetheless, both suffer from issues related to spectral overlap that can render quantitative analysis of the two interacting fragments difficult in whole cells when expression levels are not exactly matched.

Detecting Protein–Protein Interactions in Living Animals

Of the available strategies, complementation of firefly and *Renilla* luciferases is readily amenable to near real-time applications in living animals,[18,28] but the available fragments suffer from considerable constitutive activity of the N terminus fragments, thereby precluding general use. Thus, no enzyme fragment pair has yet been found that satisfies all criteria for noninvasive analysis of protein–protein interactions and enables interrogation in cell lysates, intact cells, and living animals.

To develop an optimized protein fragment complementation imaging system for broad use in living cells and animals, we screened a combinatorial incremental truncation library for reconstitution of the enzymatic activity of a heterodimeric firefly (*Photinus pyralis*) luciferase (Fig. 1). This Chapter describes the rationale and methods for the application of luciferase complementation imaging (LCI) to protein pairs of interest. We give special attention to considerations specific to LCI in the context of luciferase

[26] I. Gautier, M. Tranier, C. Durieux, J. Coppey, R. Pansu, J. Nicolas, K. Kernnitz, and M. Coppey-Moisan, *Biophys. J.* **80,** 3000 (2001).

[27] N. Boute, R. Jockers, and T. Issad, *Trends Pharmacol. Sci.* **23,** 351 (2002).

[28] R. Paulmurugan and S. Gambhir, *Anal. Chem.* **75,** 1584 (2003).

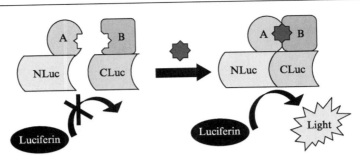

Fig. 1. Schematic of luciferase complementation imaging. Spontaneous association or drug-induced association of proteins A and B brings inactive fragments of luciferase into close proximity to reconstitute bioluminescence activity.

bioluminescence imaging in whole cells and living animals. In addition to technical aspects of LCI, we address issues related to the selection of protein pairs to be evaluated, as well as cellular and physiologic contexts for the study of protein–protein interactions *in vivo*. These studies demonstrate that the noninvasive molecular imaging of protein–protein interactions may enable investigators to determine how intrinsic binding specificities of proteins are regulated in a wide variety of normal and pathophysiologic conditions.

Methods

Optimizing Luciferase Fragments for Complementation

To identify an optimal pair of firefly luciferase fragments that reconstitutes an active (bioluminescent) heterodimer only upon association, one can construct and screen a comprehensive combinatorial incremental truncation library as shown in Fig. 2.[29] This library employs a well-characterized protein interaction system: rapamycin-mediated association of the FRB domain of human mTOR (residues 2024–2113) with FKBP-12.[16,19,30] Initial fusions of FRB and FKBP with N- and C-terminal fragments of luciferase, respectively, are designed such that the enzymatic activity of the individual overlapping fragments is weak or absent. The fragments of *P. pyralis* luciferase (derived from pGL3; Promega) are fused to FRB and FKBP by a linker containing a flexible Gly/Ser region. A multiple cloning site

[29] M. Ostermeier, A. Nixon, J. Shim, and S. Benkovic, *Proc. Natl. Acad. Sci. USA* **96,** 3562 (1999).

[30] J. Chen, X. F. Zheng, E. J. Brown, and S. L. Schreiber, *Proc. Natl. Acad. Sci. USA* **92,** 4947–4951 (1995).

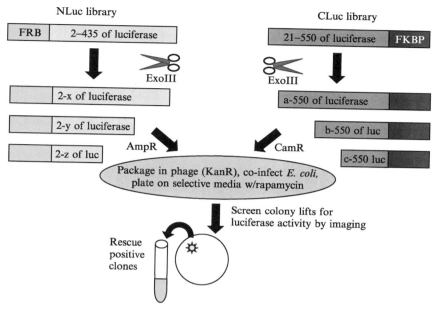

FIG. 2. Schematic of library construction and screening for optimizing LCI.

can also be inserted between each pair of fused proteins. From these constructs, N- and C-terminal incremental truncation libraries are generated by undirectional exonuclease digestion and are validated essentially as described.[29] The libraries are coexpressed in *Escherichia coli* and screened in the presence of rapamycin for bioluminescence. From this screen, one can identify an optimal pair of overlapping amino acid sequences for the NLuc fragment and for the CLuc fragment. The optimized combination of fragments should produce no signal in the absence of rapamycin and strong bioluminescence in the presence of the dimerizing agent rapamycin.

Mammalian Expression Constructs

For studies in mammalian cells, the two optimized LCI fusion constructs, termed FRB-NLuc and CLuc-FKBP, are expressed in separate mammalian expression plasmids, pcDNA3.1 TOPO (FRB-NLuc) and pEF6-TOPO (CLuc-FKBP) (Invitrogen). To generate control constructs, site-directed mutagenesis (QuikChange; Stratagene) is used to create point mutations in FRB (S2035I) and to introduce a stop codon in CLuc-FKBP at the 3′ end of the linker for the expression of unfused CLuc.

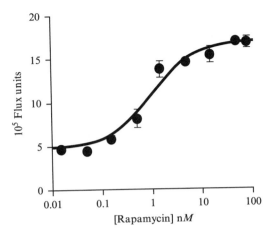

FIG. 3. Determination of the apparent K_d of rapamycin-induced FRB/FKBP association in cells. Cells transfected with FRB-NLuc/CLuc-FKBP were pretreated 22 h prior to LCI with the indicated concentrations of rapamycin. Data expressed as mean photon flux $\pm$ SEM of quadruplicate wells. Curve fit shows an apparent K_d of 1.5 nM.

Determining the Fidelity of Optimized LCI for Reporting Protein Association

For initial tests of responsiveness to rapamycin, the dynamic range and time dependence of the bioluminescence signal for the FRB-NLuc/CLuc-FKBP pair expressed in live HEK-293 cells and in lysates are measured. In cell lysates, the maximal dynamic range of the bioluminescence signal is approximately four orders of magnitude for optimal pairs. The LCI signal is generated rapidly in cell lysates upon the addition of rapamycin, with a half-time of approximately 2 min. In live cells, maximal induction of bioluminescence by rapamycin is achieved at 6 h, a delay attributable to the slow accumulation of rapamycin in cells.

In live cells and in cell lysates, our optimized LCI system successfully reproduced published apparent K_d values for rapamycin (Fig. 3).[16,19,30] To test the specificity of the LCI system, we mutated the FRB fragment in the FRB-NLuc construct to a form FRB(S2035I),[30] which should be insensitive to rapamycin. We showed that this mutation, even in the presence of rapamycin, produced a low bioluminescence signal similar to the optimal pair in the absence of rapamycin. An unfused CLuc fragment coexpressed with FRB-NLuc produced a threefold lower signal than coexpressed FRB-NLuc/CLuc-FKBP in the absence of rapamycin, consistent with the weak, rapamycin-independent association of FRB and FKBP,[16,30] whereas the expression of single constructs produced no detectable signal relative to untransfected cells. Thus, our optimized LCI pair eliminated the substantial bioluminescence activity of the N-terminal luciferase fragment that was problematic with previous split-luciferase systems based on the simple bisection of luciferase.[31]

Technical Considerations for LCI in Cells and Lysates. While transfection of cells and expression of the fusion constructs used in LCI are routine, a few special considerations for assaying protein association by bioluminescence imaging should be noted. Individual fusion proteins should be tested for background bioluminescence activity. To assess the specificity of protein interactions, an irrelevant or mutant protein association partner is helpful. Measurement of protein association by LCI can also be confounded by changes in protein levels or cell number rather than the association state. To control for events not related directly to protein association, it may be useful to incorporate an independent marker into LCI reporter cells or to include in the study a parallel set of cells or animals expressing intact luciferase. Assays should be planned to allow for the kinetics of light production by luciferase upon the addition of luciferin in the given assay format. Unlike reporter enzymes, which produce a colorimetric output that accumulates steadily with time, luciferase produces a transient signal. By integrating the signal over an appropriate time period with relatively stable light production, good reproducibility can be attained. Commonly, 1-min acquisition times taken 10 min after the addition of substrate result in reproducible signals. Overall, the LCI signal is generally lower than that for intact luciferase, although maximal activity is likely to depend on the protein pair under investigation. Therefore, it is important to maximize the signal by ensuring that reporter cells express both LCI fusion proteins. It should also be noted that luciferase activity is dependent on magnesium ATP, and thus variations in cellular ATP affect luminescence output. Therefore, the addition of magnesium ATP to the buffer is necessary for the measurement of luciferase activity in cell lysates or permeabilized cells. Commercially available luciferase substrate buffers (Promega) are appropriate for assaying luciferase activity under these conditions, as these buffers contain appropriate amounts of ATP in addition to luciferin.

Technical Considerations for LCI in Mouse Models. Bioluminescence imaging of animals using CCD cameras such as the IVIS (Xenogen) has been adopted rapidly as a broadly applicable and facile means to quantify the relative expression of luciferase reporter activity (Fig. 4). Because firefly luciferase bioluminescence imaging has very high sensitivity and a broad

[31] R. Paulmurugan, Y. Umezawa, and S. S. Gambhir, *Proc. Natl. Acad. Sci. USA* **99**, 15608 (2002).

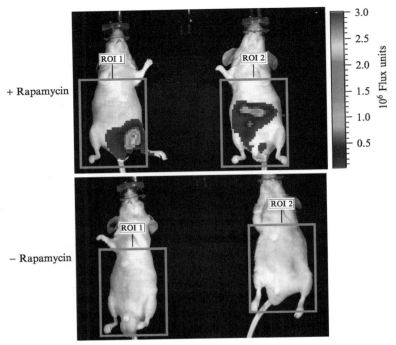

FIG. 4. LCI of representative pairs of *nu/nu* mice implanted intraperitoneally with HEK-293 cells expressing FRB-NLuc/CLuc-FKBP. LCI images were taken 10 h after receiving a single dose of rapamycin (4.5 mg/kg, intraperitoneal) (top) or vehicle control (bottom). Mice were anesthetized with isoflurane, injected ip with D-luciferin (150 μg/g in phosphate-buffered saline), and then imaged 10 min later with an IVIS CCD camera (1-min exposure, binning 8, f-stop 1, FOV 15 cm). Regions of interest (ROI) for analysis are shown.

dynamic range relative to other techniques for imaging live animals, such as fluorescence or microPET, it is perhaps the optimal means for detecting protein interactions in live animals. However, spatial and temporal characteristics of bioluminescence imaging must be considered in planning an LCI experiment. Typical luminescence images obtained by CCD cameras have resolution on the order of 2 mm, suitable for many, but not all, anatomic imaging purposes. Temporal considerations for imaging include the time needed for image collection and the practical frequency for reimaging the same animal. Depending on the intensity of the luminescence signal obtained, the time needed to acquire an image may be as little as 1 s or as long as 10 to 20 min. In live mice, bioluminescence signals from firefly luciferase reach a transient plateau phase approximately 10 min after the ip injection of luciferin. For a given mouse, firefly luciferase bioluminescence may be detectable for as much as 4 to 6 h after the administration of luciferin.

Considerations for Application of Luciferase Complementation Imaging

Selection of Protein Pairs for LCI

LCI can theoretically be used to study any process that alters the association state of a pair of proteins, including conformational changes, compartmentation changes, posttranslational modifications such as phosphorylation, or protein association mediated by small molecules. However, the practical utility of LCI to study a particular pair of interacting proteins should be considered carefully. While any reporter strategy may perturb the system under investigation, the expression of certain types of protein pairs is more likely to influence cellular function. For example, cellular homeostasis is likely to be disrupted by the overexpression of proteins that regulate cellular processes through the titration of binding partners. Proteins expressed as LCI fusions may also act as competitors for endogenous proteins. In addition to the effects of enforced protein expression on a system, the nature of the luciferase reporter must be considered. For example, protein interactions that occur in the extracellular space, where ATP levels are very low, are not likely to be detectable by LCI, as luciferase activity is dependent on ATP. Finally, to reduce the likelihood that fusion to the luciferase fragments will impact the function of the protein pair negatively, existing structural information should be considered. Introduction of steric bulk, perturbation of folding patterns, or masking of functional domains and intracellular localization sequences are among the potential causes of poor performance of LCI reporters. Thus, performance of previous fusions (GFP, CFP, etc.) to the proteins of interest should be examined, in addition to any crystallographic information that might be available. Previous success with a protein pair in two-hybrid assays may not be predictive for success with LCI, as reporter gene expression in two-hybrid assays is not strongly dependent on the exact orientation of the DNA-binding domain and the transactivation domain, whereas luciferase complementation requires correct folding and direct apposition of the luciferase fragments.

Selection of Applications and Model Systems for LCI

In general, LCI in live animals permits the study of protein interactions in a specific cell population in response to a physiologic event. For example, LCI could be applied to measure pharmacokinetics and pharmacodynamics of inhibition of protein association in tumor xenografts in response to a therapeutic agent. Considerations that affect the design of LCI experiments in live animals include the means of incorporation of LCI protein pairs into the animal, expected dynamic range of the LCI signal obtained, and time dependence of the events to be imaged.

Introduction of LCI Pairs into Animals

There are three main approaches to incorporating LCI protein pairs into live animals. The most facile of these methods is transient or stable introduction of the LCI constructs into cells *ex vivo*, followed by implantation by a variety of commonly used routes, including subcutaneous, intravenous, or intraperitoneal. LCI constructs may also be introduced directly into live animals, for example, by gene therapy strategies based on infectious agents (viral) or transduction reagents (liposomes, peptide permeation motifs) or by the hydrodynamic injection of pure plasmid into tail veins to incorporate the LCI constructs into liver cells. In addition, transgenic mouse models may be useful in certain cases. *Ex vivo* preparation of cells with LCI constructs permits the investigator to maximize the signal by selecting or sorting cells to ensure the incorporation of both LCI constructs into the same cell (if the constructs are not incorporated into single DNA plasmids). *Ex vivo* preparation of cells also permits introduction of the LCI pair into the widest variety of tissue types and genetic backgrounds.

Conclusions

The detection of protein interactions in living animals can provide useful information about the molecular basis of physiologic and pathophysiologic events, as well as the molecular response to therapeutic agents. Luciferase complementation imaging of protein interactions in cells and small animal models has been developed to permit the rapid and repetitive measurement of protein pairs of interest. Optimized luciferase complementation, quantified by imaging with a CCD camera equipped with appropriate software, can provide an accurate measure of relative protein association in animals and quantitative measurement of protein association in live cells.

Acknowledgments

We thank colleagues of the molecular imaging center for insightful discussions and excellent technical assistance. Work reviewed herein was supported by a grant from the National Institutes of Health (P50 CA94056).

[20] Mitochondrial NADH Redox State, Monitoring Discovery and Deployment in Tissue

By BRITTON CHANCE

Introduction: The Respiratory Chain

Imaging of mitochondrial NADH and flavoprotein florescence signals characteristic of mitochondrial function was one of the focal activities of the Johnson Foundation at the University of Pennsylvania. In his historic study from 1935 to the present of the cytochromes of cells, Keilin failed to interpret that the intense background fluorescence of cells (blue "auto-fluorescence") when excited by ultraviolet light, nor another green fluorescence when excited by blue light due to fluorescent FAD. Since then, these two fluorescent components have provided an unusual amount of data for one- and two-photon studies of cell and tissue function. These two components also provided the basis for the high-resolution imaging of metabolic function in heart, liver, and, more recently, tumors. A number of others have taken up this study, particularly Webb, Piston, Balaban, and Elle Kohen with striking results, especially Kohen, who studied the function of perinuclear mitochondria in single cells. The purpose of this Chapter is to indicate the origins of the study of NADH and fluoroprotein to show how they provide a window into the cell function that could not be provided readily by studies of cytochrome action.

The NADH Compound of Mitochondria

The invention of the dual-wavelength spectrophotometer[1] permitted delineation of the cytochromes of the highly scattering suspension of intact mitochondria as never before. The salient features of the oxidized minus reduced difference spectra[2] corroborated the visions of the spectroscope as seen by Keilin's keen eyes except for the flavin trough and the NADH peak, which were quite novel. The functionality of both of these could be studied, as indeed Keilin had already verified the function of the cytochromes.[3] He vividly portrayed the disappearance and reappearance of cytochrome bands in yeast cell suspensions as observed with the microspectroscope as evidence of their functionality. However, surprisingly, Keilin,

---

[1] B. Chance, *Rev. Sci. Inst.* **22**, 634 (1951).
[2] B. Chance and G. R. Williams, *J. Biol. Chem.* **217**(1), 395 (1955).
[3] M. Dixon, R. Hill, and D. Keilin, *Proc. R. Soc. Lond. Ser. Biol. Sci.* **109**, 29 (1931).

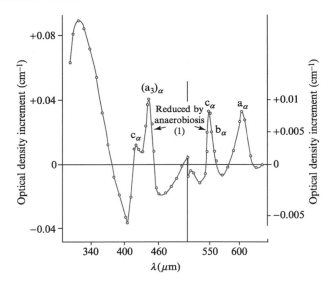

Fig. 1. The first spectroscopic recordings of the oxidized-reduced components of the respiratory chain, which revealed not only cytochromes in the alphaband region, but also the Soretbands and, unknown to Keilin, the NADH peak and flavin trough.

the cell biologist, had overlooked the fact that the biologist's "cell autofluorescence" also changed in the yeast, as did the cytochrome absorption bands. The dual-wavelength spectrometer linked NADH and flavin absorption and fluorescence to the respiratory chain and were the subject of intense experimentation in the Chance laboratory from 1955 onward[4,5] (Fig. 1).

Transference of the optical techniques from isolated mitochondrial preparations to tissues[5] was the focus of activity of the Chance laboratory for the next several decades. The first and foremost method was "activational function," where the absorption and fluorescence changes of NADH were observed with stimulation. The activation of muscle function caused a rise of ADP and P_i (inorganic phosphate) concentrations and oxidation of NADH simultaneously with changes in other carriers of the respiratory chain known as "crossovers."[6] The spectroscopic study of Lardy and Wellman's tightly coupled preparations of guinea pig mitochondrial liver showed the five stages of metabolic activity[7] [endogenous (i), starved (ii), fully

[4] B. Chance, in "Flavins and Flavoproteins" (E. C. Slater, ed.), p. 496. Elsevier, New York, 1996.

[5] B. Chance, Annu. Rev. Biophys. Biophys. Chem. 20, 1 (1991).

[6] B. Chance and G. R. Williams, J. Biol. Chem. 217, 409 (1955).

[7] H. A. Lardy and H. Wellman, J. Biol. Chem. 195, 215 (1952).

active (iii), resting (iv), and hypoxic (v)]. These "states" formed the basis for many studies of the functional activity of isolated mitochondria and mitochondria in many intact tissues: liver, brain, kidney, adrenal, skeletal muscle, and so on[8–10] and served as markers of metabolic intensity and patency of energy coupling.

Mitochondria Function *In Vivo*

Ramirez, Jobsis, Weber, and others demonstrated the function of cytochromes in muscle strips and showed elegantly how they responded to energy demand, thereby extending the studies of mitochondrial cyto-chromes, and NADH fluorescence to heart, liver, and tumors. This has provided a novel window to the function of mitochondria in normal and cancer tissues in highly oxidative cardiac tissue and in minimally oxidized electroplaque. This contribution is divided into two sections: (1) *in vivo* studies of isolated and intact organs at low resolution and (2) high-resolution cryoimaging of serial sections of cells and tissues, especially cancers. Webb, Piston, and Balaban have taken up these discoveries with avidity.

Functional Activation

The first experiment to demonstrate the functionality of the novel "pyridine nucleotide" absorption band was performed by Clancy Connelly on strips of the sartorius muscle of the frog bathed in Ringer's solution in a special holder and electrically stimulated and illuminated by dual-wavelength light beams.[11] The observation of deceased absorption by several percentile resulting from NADH oxidation in muscle mitochondria upon stimulating the frog's sartorius muscle (Fig. 2) was, to Chance and Connelly, just as exciting as Keilin's microspectroscopic observations of the disappearance of the absorption bands of cytochromes on the oxygenation of yeast. The mitochondrial redox states were coupled to muscle contractions!

The ADP activation of muscle mitochondria was the focus of muscle biochemistry due to the work of Davis, Mommaerts, Fleckenstein, and Kushmeric,[12,13] tests unavailable to David Keilin. The uncoupling of mitochondria with dinitrophenol afforded a typical control of muscle

---

[8] B. Chance, F. Jobsis, B. Schoener, and P. Cohen, *Science* **137**, 499 (1962).
[9] B. Chance, B. Schoener, and V. Legallais, *Nature* **195**, 1073 (1962).
[10] B. Chance and B. Schoener, *Biochem. Zeitsc.* **341**, 340 (1965).
[11] B. Chance and C. M. Connelly, *Nature* **179**, 1235 (1957).
[12] W. F. H. M. Mommaerts, *Annu. Rev. Biochem.* **23**, 381 (1954).
[13] A. Fleckenstein, J. Janke, R. E. Davies *et al.*, *Nature* **174**, 1081 (1954).

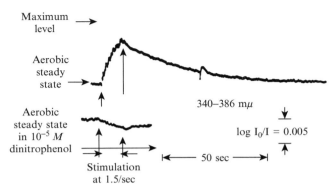

FIG. 2. Functional activation of mitochondrial NADH oxidation in the stimulation of sartorius muscle and of glycolytic NADH reduction in uncoupled mitochondria.

studies and revealed that the oxidative response of NADH to stimulation was not only ablated by increased uncoupling agents, but that a small reduction of NAD occurred (i.e., just the opposite happened). This was attributed to increased glycolytic activity and was modeled *in vivo* in studies of the mainly glycolytic tissue electroplax of electrophorus electricus.[14]

Chance and Baltscheffsky[15] followed up Warburg's observation that the pyridine nucleotides were fluorescent and found that, indeed, mitochondria not only showed a strong fluorescence emission band at 450 nm, but this fluorescence emission band was also linked to changes of metabolic activity observed in the cytochrome chain with the dual-wavelength spectrophotometer, connecting the NADH pool (the major absorption band of mitochondria) closely to cytochrome function (Fig. 3).

Because the mitochondrial function in tissue was labeled by highly sensitive fluorescence, nearly all available tissues were studied, even the salt gland of the herring gull, considered an example of a high NADH, high mitochondrial tissue, which contrasted with the electric organ of electrophorus, which was characterized by a "low content." The fluorometric method turned out to be extremely convenient and sensitive because tissue transmission was not required and measurements could be made by reflectance or "remission."

[14] X. Aubert, B. Chance, and R. D. Keynes, *Proc. R. Soc. Lond. Ser. Biol. Sci.* **160**, 211 (1964).
[15] B. Chance and Baltscheffsky, *J. Biol. Chem.* **233**, 736 (1958).

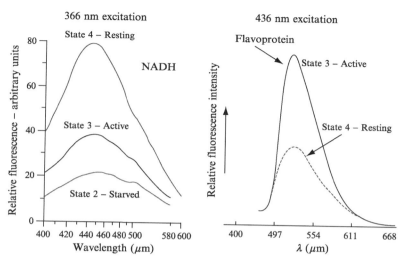

FIG. 3. Fluorescence emission spectra of NADH and flavoprotein using 366- and 436-nm excitation wavelengths, respectively, with responses of these components of the respiratory chain to metabolic activation where NADH is oxidized and flavoprotein is reduced, permitting calculation of the redox ratio.

NADH and Flavoprotein Fluorescence

Following the discovery of Chance and Balscheffsky on the mitochondrial fluorescence signal, Chance and Schoener devised a number of tissue fluorometers with which they investigated the relationship of the mitochondrial NADH signal to the function of tissues, particularly its sensitivity to oxygen. In addition, Connelley originated activation spectra for the function of skeletal and cardiac muscle and thus afforded a veritable "activity window" into the metabolic activity of cells and tissues, particularly the brain. The traces of Fig. 2 are duplicated with a high fidelity using NADH fluorescence[16] (Fig. 4).

Bleaching of Fluorescence

In microscopy of single cells with cardio condenser illumination, no difficulty was experienced, even though the high-pressure mercury arc 366-nm line was used for illumination. However, with Abbé condenser illumination with the suspension of yeast cells, photobleaching was observed readily as indicated in the enclosed record from the 1960s. It is first seen that the

[16] B. Chance, *Texas Rep. Biol. Med.* **22,** 836 (1964).

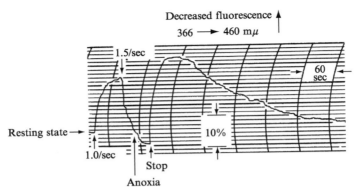

Fig. 4. The use of NADH fluorescence to quantify functional activity in the sartorius muscle of the frog showing oxidation on functional activation as the fluorescent signal of NADH decreases.

bleaching is not complete but reaches a steady state and (not shown) recovered with a time constant related to the rate of reduction of NADH as has been rediscovered independently by Balaban and colleagues. This discovery has led Balaban and others to take advantage of photobleaching and recovery kinetics as a measure of the nucleotide redox metabolism.

In Vivo Location of the Fluorescent Signal

Perhaps the most vexing question in *in vivo* studies was to verify that the fluorescent signal originated from mitochondria and not from the cytosol. Development of fluorescent microscopy, which included the concept of a vibrating reed in the image plane, compared a cytosolic or extracellular signal. This caused Chance and Schoener to resort to a meiotic telephase of the arachnoid sperm in which a mitochondrial aggregate or "nebenkern" was formed in meiotic telephase. They showed that not only that 80% of the signal originated from the mitochondrial aggregate (Fig. 5), but that the mitochondrial aggregate was also responsive to hypoxia (Fig. 6), whereas measurements of the cytoplasm with reference to the extracellular space showed little or no change. Interestingly enough, Piston's two-photon method applied to the pancreatic islet failed to distinguish mitochondria and cytosolic signals, as did the nebenkern study and results of the biochemical analysis of NADH (DPNH) and NADPH (TPNH) (Figs. 7 and 8).

The exceptionally good signal-to-noise ratio of the NADH fluorescent signal, together with its relatively robust resistance to overillumination, led

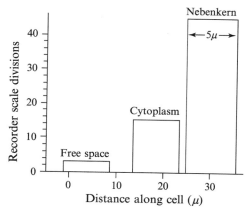

FIG. 5. A scan of a grasshopper spermatid by means of the flexible quartz plate. In the three positions of the scan indicated, the reference aperture is in the free space. In the first recording, the measuring aperture is in the free space; in the second it is in the cytoplasm; and in the third it is in the nebenkern. The scale of distances along the cell and the span of the measuring aperture are indicated. The cell wall was located ~12 μm from the origin of the measurements.

Chance and Barlow, together with Williamson, to exploit flash photography of the myocardium of the perfused heart and to demonstrate not only patchy ischemia in an underperfused heart, but also the spatiotemporal pattern of the distribution of patchy ischemia (Fig. 9).

High-Resolution Cryoimaging

Whereas normal tissues (liver, heart, skeletal, and muscle) provide repeat patterns of anatomy and histopathology, cancer presents a heterogeneous display of necrotic active and inactively growing tissues in which three-dimensional imaging to identify regions of functional interest seems essential. Furthermore, the fact that such regions are identified best by their differing metabolic rather than anatomic differences makes the measurement of NADH and flavoprotein fluorescences more valuable. Being dissatisfied with the low resolution of the fluorescence images of cardiac and other organs, Chance and Quistorff developed a cryoimager that has remained the paragon of three-dimensional high-resolution metabolic imaging ever since. Taking the queue that immobilization was necessary to avoid motion artifacts and at the same time recognizing that Chance and Spencer had shown that the steady state of electron transfer in the cytochromes was trapped by rapid freezing, they exploited the methods

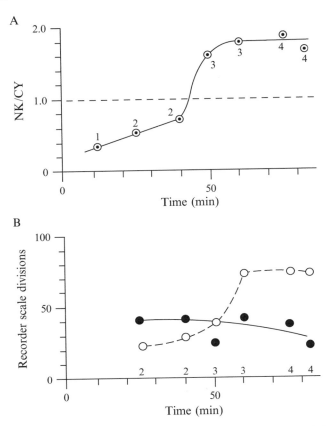

Fig. 6. Time course of the fluorescence changes of the nebenkern and of the cytoplasm in the aerobic–anaerobic transition. (A) The ratio of nebenkern to cytoplasmic fluorescence. Numbers refer to the cell studied. The abrupt upward discontinuity of the record at approximately 45 min occurs when anaerobiosis is expected. (B) Individual measurements of cytoplasmic (●) and nebenkern (○) fluorescence. The number of the cell used for measurement is also indicated along the scale of the abscissa.

of Von Herrevald and Wollenberger for fast freezing and built an apparatus that imaged the redox states and the serial sections of cells and tissues with micrometer resolution. The method took advantage of the fact that particular freezing mixtures gave multiple small refractory crystals, which prevented penetration of the light into the sample and enabled serial sections at 50–100 μm to be cut with a mill wheel at liquid nitrogen temperatures in such a way that the reflected light originated in the top layer of the frozen tissue. This scanner, using single photon technology, gives a voxel of

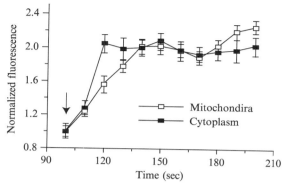

FIG. 7. Temporal separation of the NAD(P)H response. The mitochondrial and cytoplasmic NAD(P)H signals were measured in flattened pancreatic islets at 10-s intervals for 100 s at 1 mM glucose before the addition of a bolus of 20 mM glucose (denoted by the arrow). The islets were monitored for an additional 500 s. Only the 100 to 200-s time segment is shown here. Data are represented by the mean $\pm$ SE ($n = 40$ regions of interest from the four islets).

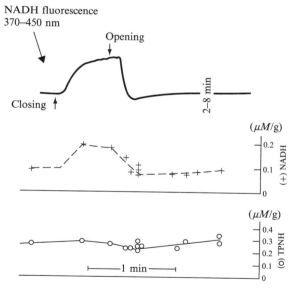

FIG. 8. Validation of *in vivo* NADH fluorescence by serial biopsy and analytical biochemistry in an anesthetized rat. Fluorometric trace of an ischemic cycle together with analytical values of AMP (bars), NADH (crosses), and NADPH (circles) of several analogous experiments.

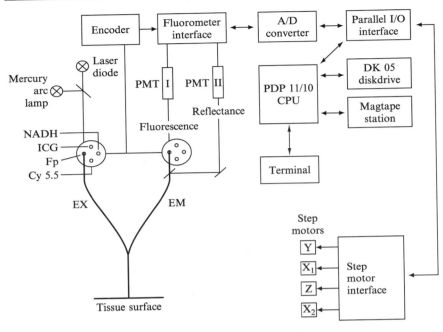

FIG. 9. Diagram of the low temperature fluorometer.

$50 \times 30 \times 10$ μm and permits the construction of three-dimensional metabolic images that illustrate the heterogeneity of metabolic states of small animal cancer cells.

The very rapid development of tissue fluorometry, particularly NADH and flavoprotein, has sparked the development of exogenous fluorochromes termed "molecular beacons," which seek out particular genetic expressions or label intrinsic substances present in tumors. These beacons provide a novel approach to the detection, tracking, and therapy of cancer using either one- or two-photon detection methods.

The Future—High Resolution Metabolic Histopathology

The increment of resolution obtainable with multiphotons and imaging (to 1-2μ) is uniquely suited to the cryo-imager where not only is biological noise immobilized by cryo-sectioning, but 2-D and 3-D sliced images of a heterogeneous material such as cancer affords exploitation of the x, y, z resolution of the multiphoton method, and cryo-sectioning affords successive images.

Author Index

Numbers in parentheses are footnote reference numbers and indicate that an author's work is referred to although the name is not cited in the text.

A

Aasar, O. S., 8
Abate-Shen, C., 349
Abdel-Nabi, H., 9
Aboagye, E. O., 242, 328
Absil, P., 41, 45(7)
Achlama, A. M., 33
Ackerman, J. J., 37, 41
Ackerman, J. L., 77
Adams, D. F., 30
Adams, G. R., 29
Adcock, J., 63
Adriany, G., 57, 63
Agdeppa, E. D., 13
Agranoff, B. W., 150, 156(15)
Ahier, R. G., 228, 229(7)
Ahrens, E. T., 241, 256(5)
Aine, C. J., 103
Akaike, H., 208
Akeroyd, M. A., 141
Akhurst, T., 328
Alauddin, M. M., 328, 329
Albert, M. S., 149, 155, 156
Albrecht, D. G., 103
Albright, T., 84, 86(7), 104, 119(25)
Alderman, D. W., 52, 54(36), 57(36)
Alexoff, D. L., 174, 178(16), 213, 219, 220, 223(27), 238, 338
Alkire, M. T., 120, 121(70)
Allegrini, P. R., 240, 254
Allen, P. J., 64, 64(16), 65, 82
Allman, J. M., 131
Allott, P. R., 162
Allred, E., 56
Alpert, N. M., 186, 190, 193(9), 211
Altes, T. A., 150, 152(7)
Altschuld, R. A., 261
Anami, K., 103

Andersen, A. H., 104, 128, 151
Andersen, M. B., 104, 109(23), 114(23), 117(23), 118(23)
Andersen, P., 63
Andersen, R. A., 84, 86(6), 104, 109(23), 111, 111(32), 113(22; 32), 114(23), 115(32), 117(23), 118(22; 23), 119(22), 120(22), 122(22), 125(22)
Anderson, A. W., 38
Anderson, C. T., 120, 121(70)
Anderson, J., 144
Anderson, M. E., 165
Andersson, J. L., 97
Andersson, Y., 335, 347(3)
Andrew, E. R., 41
Andriambeloson, E., 256
Angulo, M. C., 103
Anllo-Vento, L., 103
Anno, I., 33
Ansari, M. S., 236
Antoni, G., 288
Applegate, B., 29
Araujo, D. M., 214, 228(12), 229
Archer, B. T., 36
Armony, J. L., 64
Armstrong, W. F., 312
Arnese, M., 14
Arnold, H., 256
Arnold, K., 212
Aronen, H. J., 254
Aronheim, A., 352
Artemov, D., 242
Arthurs, O. J., 103
Asai, Y., 228, 229(4)
Ashburner, J., 93
Ashe, J., 64
Ashruf, J. F., 284(34), 285
Ashworth, S., 220, 221(30), 228, 228(9; 10), 229, 229(7; 8)

371

S

Subject Index

A

ACE, *see* Angiotensin-converting enzyme

[^{11}C]Acetate

dosimetry calculations, 302

myocardium enzyme pathway imaging with GAP positron emission tomography studies

energy substrate utilization, 286–287

kinetic modeling

calculations, 310

compartmental model, 306–307

myocardial blood flow, 307–309

metabolite analysis, 305–306

prospects, 314–315

regional perfusion and metabolism measurement

data acquisition, 303

data reconstruction, 303

time-activity curve generation, 303–305

quality assurance

acidity, 302

apyrogenicity, 301

chemical purity, 301

isotonicity, 302

radioactivity balance, 300

radiochemical purity, 300

radionuclidic identity, 299

radionuclidic purity, 299

sterility, 301–302

specific activity determination, 291

synthesis

carbon-11 production, 289

modules, 291–292

overview, 289–291

reaction conditions, 296–298

yield, 298

AD, *see* Alzheimer's disease

ADC, *see* Apparent diffusion coefficient

Alzheimer's disease

magnetic resonance imaging of biomarkers, 253

positron emission tomography, 13, 253

Angiogenesis

magnetic resonance imaging in drug evaluation, 253–254

positron emission tomography quantification, 17

Angiotensin-converting enzyme

expression imaging in lungs with positron emission tomography

captopril tracer, 319

dose determination, 319–320

time-activity curve generation, 319

functions, 318

Apoptosis, positron emission tomography assay, 18

Apparent diffusion coefficient

magnetic resonance imaging, 27–28

prediction of tumor therapy response, 254

B

Bioluminescence, *see* Protein–fragment complementation assay

Blood oxygenation level-dependent effect, *see also* Functional magnetic resonance imaging

correlation with neural activity, 103

drug evaluation, 248–249

hemodynamic parameters, 107, 110

principles, 26–27, 46, 102–103

small animal studies of exercising muscle, 37–38

Bold effect, *see* Blood oxygenation level-dependent effect

Brain activation, *see* Functional magnetic resonance imaging

Brain atlases

humans, 91

nonhuman primates

baboon b2k atlas development

image registration, 93

magnetic resonance imaging, 92–93

subjects, 92

N

O

P

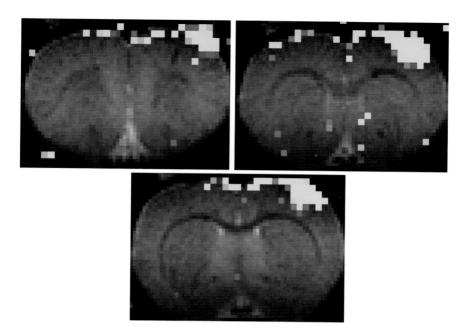

HEERSCHAP *ET AL.*, CHAPTER 3, FIG. 4. Functional activation of rat brain after left forepaw stimulation. Gradient echo NMR images are obtained before and during electrical stimulation. The images were subtracted; the subtraction image was color coded and overlaid on a control image. Areas of brain activation are visible in the right cortex.

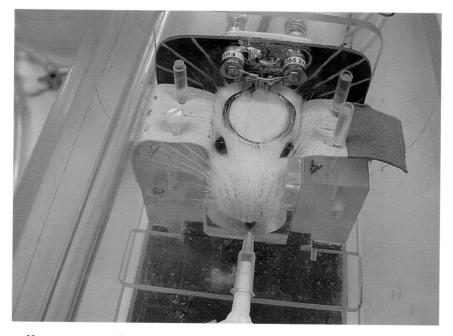

HEERSCHAP *ET AL.*, CHAPTER 3, FIG. 13. Setup for NMR of rat brain. The ^{1}H coil for the brain is clearly seen, as well as the oral tube for ventilation of the animal.

Visual cortex activation – Macaque

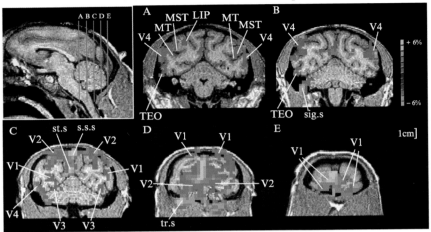

Visual cortex activation – Human

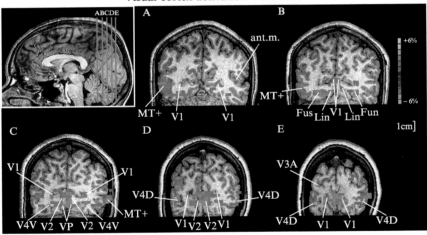

DUBOWITZ, CHAPTER 7, FIG. 3. Upper panel shows functional activation in macaque brain at 1.5 T. Visual paradigm is 25-s alternating blocks of video cartoon and darkness. Activated pixels are 2 × 2 mm superimposed on T$_1$-weighted MPRAGE images of 0.7 × 0.7 mm resolution (coronal images A–E are at 5-mm spacing ranging from 25 to 5 mm anterior to the occipital pole). Labeled areas are visual cortex (V1, V2, V3, V4), medial temporal area (MT), medial superior temporal area (MST), lateral intraparietal area (LIP), superior sagittal sinus (s.s.s), straight sinus (st.s), transverse sinus (tr.s), and sigmoid sinus (sig.s). Lower panel shows functional activation in a human visual cortex during the same stimulus. Activated pixels are 3.5 × 4 mm superimposed on T$_1$-weighted MPRAGE images of 1-mm isotropic resolution. Coronal images A–E are at 7-mm spacing ranging from 44 to 16 mm anterior to the occipital pole. Green bars on the sagittal image indicate the position of coronal images A–E. Labeled areas are visual cortex (V1, V2, VP, V3A, V4-; dorsal and ventral), medial temporal complex (MT+), fusiform gyrus (Fus), lingual gyrus (Lin), and anterior motion area (ant.m). See Table I for imaging parameters (from Dubowitz et al. with permission).

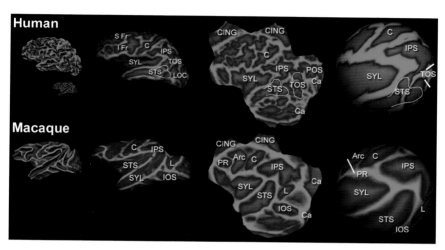

DUBOWITZ, CHAPTER 7, FIG. 9. Representations of the cortical sheet in macaques and humans. (Left to right) Gray matter/white matter boundary showing the sulcal and gyral pattern, inflated cortex, "cut" cortex mapped onto a 2D sheet, and cortex mapped onto the surface of a sphere. The small inset brain is a macaque brain at the same scale as the human brain for comparison. The sulcal pattern is shown in red; the gyral architecture is in green. Following inflation, the entire cortex is clearly seen and no areas are obscured inside the sulci. Each inflated brain is unique and is not bounded to a particular geometric shape. To compare across subjects, or across species, the flat and spherical representation provides a common reference. The 2D image is generated by "cutting" the cortical sheet and presenting it as a single sheet. The human brain is mapped onto an averaged brain in the spherical representation. The macaque brain is mapped onto the surface of a sphere and is rotated so that the view matches the human view, but is not constrained to any averaged sulcal representation. Note the close relationship of sulcal morphology between the macaque and the human brain. Sulcal areas labeled in the human brain are sylvian (SYL), superior temporal (STS), transverse occipital (TOS, human "lunate" sulcus), intraparietal (IPS), central (C), cingulate (CING), calcarine (Ca), parietooccipital (POS), lateral occipital (LOC), inferior frontal (I fr), and superior frontal (S Fr). Additional sulcal areas labeled on the macaque brain are lunate (L), arcuate (Arc), principal (PR), and inferior occipital (IOS). Macaque MRI raw data kindly provided by Nikos Logothetis and Margaret Sereno; image reconstruction and postprocessing courtesy of Martin Sereno.

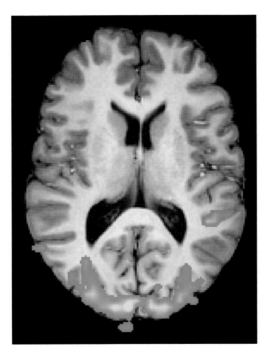

CLARE, CHAPTER 8, FIG. 1. Example of an fMRI result showing activation in the visual areas resulting from the subject looking at a contrast reversing checkerboard stimulus.

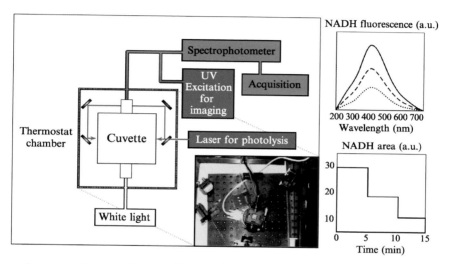

COMBS AND BALABAN, CHAPTER 15, FIG. 2. Schematic diagram of hardware for ED-FRAP experiments involving enzyme solutions or organelle suspensions in a cuvette as described in the text. *Left:* Schematic diagram of system. Photolysis light is from a laser that is split and impinged on a temperature-controlled cuvette. The fluorescence or absorbance within the cuvette can then be followed using the remaining two ports on the holder. (Inset) An actual picture of the cuvette-holding system. *Top right:* Fluorescence spectra of NADH during control and after two photolysis pulses from the laser. *Bottom right:* Time course of NADH fluorescence after two sequential photolysis pulses. Note that the fluorescence signal is constant after a photolysis reaction, attesting to the homogeneity of the irradiation field within the entire cuvette.

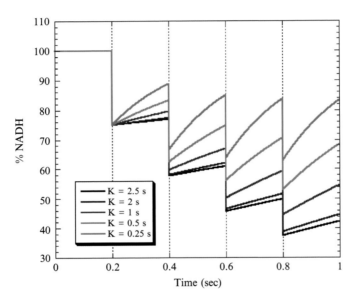

Combs and Balaban, Chapter 15, Fig. 11. Simulation of NADH recovery during a confocal microscopy imaging experiment. The plot is NADH concentration as a function of time. Calculations were based on Eq. (8), which assumes that the rate of recovery of NADH is constantly occurring over the imaging process. The rate constant for the metabolic recovery of NADH was varied as shown in the figure key. The framing rate was 200 ms, whereas each photolysis reaction consumed 25% of the NADH. Simulation was run numerically using IDL (RSI) running on a laboratory PC. This model likely underestimates the effect of metabolic recovery on NADH, as the recovery rate is assumed constant and it is known that the recovery rate increases with decreasing NADH (see Fig. 9).

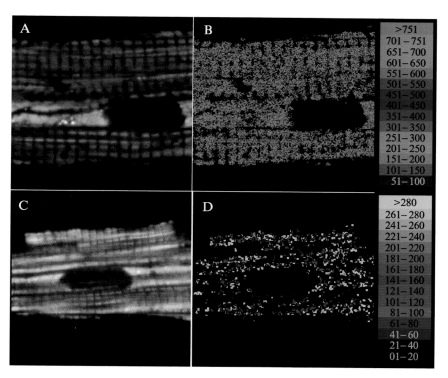

COMBS AND BALABAN, CHAPTER 15, FIG. 14. Pix maps of NADH recovery rate show local NADH flux rates within isolated cardiac myocytes. (A and C) NADH fluorescence intensity images (1.5 μm slice thickness). (B) A false color pix map of exponential recovery rates. (D) A false color pix map of the difference in exponential recovery rate between superfusion with nutrient medium and 5 mM DCA of the cell depicted in C. From Combs and Balaban.

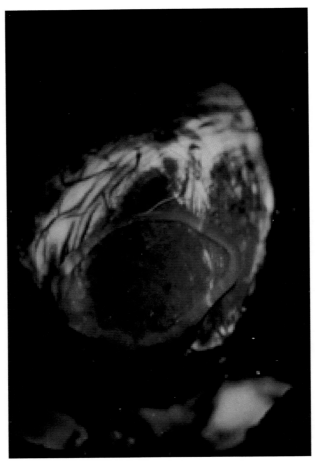

COMBS AND BALABAN, CHAPTER 15, FIG. 15. Whole-heart NADH fluorescence image of a beating canine heart *in vivo*, as described in the text. Basically, this image was collected by gating a 360-nm filtered flash xenon light source to the EKG. The image was collected by simply leaving the shutter of a camera open for the four heartbeats used to create this image. The image was collected with 100 ASA color film. The yellow fluorescence is from epicardial fat, especially around the vessels. The arterial and venous blood vessels can be observed directly, with arterial vessels being brighter than venous vessels, presumably due to the hemoglobin oxygenation state. The transparent ring on the surface of the heart is made of a silicon-based material to make a well on the surface of the heart to hold drugs and extrinsic probes on the surface of the heart.

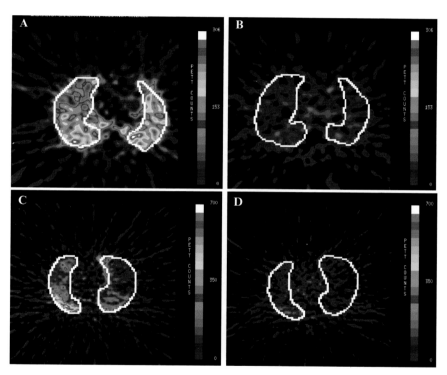

CHEN *ET AL.*, CHAPTER 17, FIG. 2. Lung PET images (transverse orientation) obtained in a normal subject (A and B) and in a patient with pulmonary hypertension (C and D). A and C were obtained approximately 1 h after iv administration of F-18-labeled captopril. B and D were obtained similarly, approximately 1 week after oral ingestion of daily enalapril (5 mg). Note the reduction in tracer uptake after oral enalapril, indicating pulmonary specific blockade of ACE.

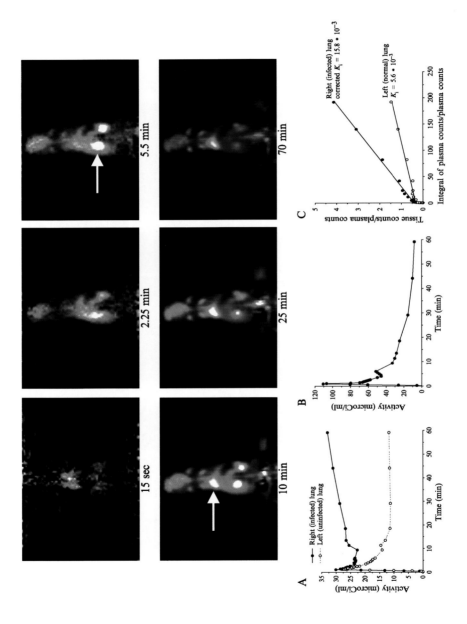

5.5 min

2.25 min

15 sec

70 min

25 min

10 min

A

Activity (microCi/ml)

35
30
25
20
15
10
5

Right (infected) lung
Left (uninfected) lung

0   10   20   30   40   50   60
Time (min)

B

Activity (microCi/ml)

120
100
80
60
40
20
0

0   10   20   30   40   50   60
Time (min)

C

Tissue counts/plasma counts

5
4
3
2
1
0

Right (infected) lung
corrected $K_i = 15.8 * 10^{-3}$

Left (normal) lung
$K_i = 5.6 * 10^{-3}$

0   50   100   150   200   250
Integral of plasma counts/plasma counts

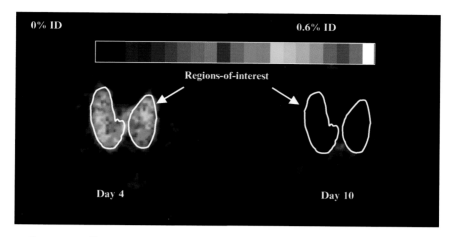

CHEN *ET AL.*, CHAPTER 17, FIG. 7. PET images of lungs in a rat obtained 4 and 10 days after adenovirus-mediated gene transfer of a mutant herpes simplex virus type 1 thymidine kinase (mHSV1-*tk*). Images are transverse slices obtained at the midchest level in the same animal, scanned 1 h after injection of [^{18}F]FHBG. Despite significant tracer uptake 4 days after gene transfer, the lung PET signal was not different from background after 10 days, suggesting a reduction in TK expression by this latter time. *In vitro* assays obtained at the same time points in different animals confirmed the kinetics of transgene expression assessed by PET imaging. [^{18}F]FHBG, 9-(4-[^{18}F]-fluoro-3-hydroxymethylbutyl)guanine; ID, injected dose.

CHEN *ET AL.*, CHAPTER 17, FIG. 6. Dynamically acquired PET images (coronal orientation) of a supine mouse with experimental right lung infection. The arrow in the 5.5-min frame shows high concentrations of activity in the kidneys, whereas the arrow in the following frame shows increasingly high activity in the right (infected) lung. (A) Time–activity curves for regions placed over the right and left lungs. (B) Time–activity curve from blood samples drawn throughout the scanning period. (C) Patlak plot illustrating a threefold increased uptake in the infected right lung compared with the normal left lung.

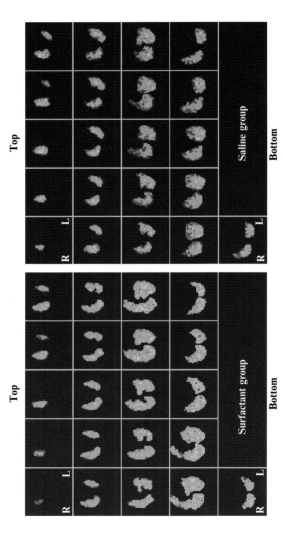

CHEN *ET AL.*, CHAPTER 17, FIG. 8. Transverse PET images of lungs obtained in two representative rats 3 days after intratracheal pulmonary gene transfer using surfactant (left) or saline (right) as a vehicle for adenovectors. The transferred gene encoded for a mutant herpes simplex virus type 1 thymidine kinase (mHSV1-*tk*) and [^{18}F]FHBG was used as a PET reporter probe. Note higher mHSV1-*tk* expression in the lungs, as assessed by [^{18}F]FHBG uptake, in the surfactant group. The spatial distribution of transgene expression is more homogeneous and more peripheral when gene transfer was performed with surfactant. ID, injected dose; R, right; L, left; [^{18}F]FHBG, 9-(4-[^{18}F]-fluoro-3-hydroxymethylbutyl)guanine.